Modern Control
Systems Engineering

**Prentice Hall International
Series in Systems and Control Engineering**

M. J. Grimble, Series Editor

BAGCHI, A., *Optimal Control of Stochastic Systems*
BENNETT, S., *Real-time Computer Control: An introduction*, second edition
BITMEAD, R. R., GEVERS, M. and WERTZ, V., *Adaptive Optimal Control*
BROWN, M. and HARRIS, C., *Neurofuzzy Adaptive Modelling and Control*
COOK, P. A., *Nonlinear Dynamical Systems*, second edition
GRIMBLE, M. J., *Robust Industrial Control*
ISERMANN, R., LACHMANN, K. H. and MATKO, D., *Adaptive Control Systems*
KUCERA, V., *Analysis and Design of Discrete Linear Control Systems*
MARTÍN SÁNCHEZ, J. M. and RODELLAR, J., *Adaptive Predictive Control:
From the concepts of plant optimization*
MARTINS DE CARVALHO, J. L., *Dynamical Systems and Automatic Control*
MATKO, D., ZUPANČIČ, B. and KARBA, R., *Simulation and Modelling of
Continuous Systems: A case study approach*
OLSSON, G. and PIANI, G., *Computer Systems for Automation and Control*
ÖZGÜLER, A. B., *Linear Multichannel Control*
PARKS, P. C. and HAHN, V., *Stability Theory*
PETKOV, P. H., CHRISTOV, N. D. and KONSTANTINOV, M. M.,
Computational Methods for Linear Control Systems
SÖDERSTRÖM, T. D., *Discrete-time Stochastic Systems*
SÖDERSTRÖM, T. D. and STOICA, P., *System Identification*
SOETERBOEK, A. R. M., *Predictive Control: A unified approach*
VANĚČEK, A. and ČELIKOVSKÝ, S., *Control Systems: From linear analysis to
synthesis of chaos*
WATANABE, K., *Adaptive Estimation and Control*
WILLIAMSON, D., *Digital Control and Instrumentation*

Modern Control Systems Engineering

Z. Gajić and M. Lelić

PRENTICE HALL

*London   New York   Toronto   Sydney   Tokyo   Singapore
Madrid   Mexico City   Munich*

First published 1996 by
Prentice Hall Europe
Campus 400, Maylands Avenue
Hemel Hempstead
Hertfordshire, HP2 7EZ
A division of
Simon & Schuster International Group

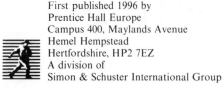

Printed and bound in Great Britain by
T. J. Press (Padstow) Ltd.

Library of Congress Cataloging-in-Publication Data

Gajić, Zoranc
    Modern control system engineering: with MATLAB Laboratory
experiments / Z. Gajić, M. Lelić.
       p.   cm. — (Prentice Hall International series in systems and
control engineering)
    Includes bibliographical references and index.
    ISBN 0–13–134116–2
    1. Automatic control.   2. Systems engineering.   3, MATLAB.
I. Title. II. Series.
TJ213.G26 1996
629.8–dc20   95–51206
                                         CIP

British Library Cataloguing in Publication Data

A catalogue record for this book is available from the British Library

ISBN 0-13-134116-2

Dedicated to my parents Olga and Radivoj Gajić
Z. Gajić

To my parents
M. Lelić

Contents

Preface

This book represents a modern treatment of classical control theory and its applications. Theoretically, *it is based on the state space approach*, where the main control theory concepts are derived by using only elementary knowledge from sophomore/junior level courses in differential equations and linear algebra. Practically, it is based on the MATLAB®1 package for computer-aided control system design, so that presentation of design techniques is simplified. The inclusion of MATLAB allows deeper insight into dynamical behavior of real physical control systems, which are quite often of high dimensions. The book treats simultaneously continuous- and discrete-time control systems with emphasis on continuous-time systems.

Our motivation for writing this book is twofold: (1) Undergraduate students usually take only one control systems course, studying first of all design techniques (mostly in the frequency domain). Students are very often left without full understanding of the standard control theory concepts such as state space, controllability, observability, and system stability. Most of nowadays control system textbooks originated in the sixties or kept the structure of the books written in the sixties. Namely, they emphasize the frequency domain approach and the corresponding design methods. With inclusion of MATLAB, which simplifies design techniques, we are able to spend more time explaining theoretical concepts. (2) Undergraduate laboratories are increasingly software oriented. This book provides at the same time the *laboratory manual*. After each chapter, the corresponding MATLAB laboratory experiments are formulated. Since there is no control systems laboratory manual in the literature, course instructors are faced with the problem of designing laboratory experiments. The appearance of MATLAB, a

---

1   MATLAB is a registered trademark of The MathWorks, Inc.

broadly accepted software package at almost all universities around the world, allows development of a unified control systems laboratory manual. The MATLAB laboratory experiments developed present clearly considered theoretical issues in view of numerous control system applications.

The book contains a lot of examples, case studies, and problems. Most of these are of analytical nature. Some of them, especially those referring to high-order systems, are done (or ought to be performed) by the MATLAB package. Real world examples and problems are given in state space forms (system, input, and output matrices). In several cases we explain also the physics of the control systems under consideration and outline the corresponding mathematical modeling. Since this book is intended for students in all engineering fields (electrical, mechanical, aerospace, systems, chemical, industrial, and general engineering), we have presented real world examples for most of these areas. Many of the real-world control problems presented in the book are also perfectly suited for fast and easy analysis by another modern computer package for simulation of dynamic systems known as SIMULINK®[2].

Outstanding features that distinguish this book from other undergraduate control system books treating the same subjects are: (a) Chapter 5 on controllability and observability, where these fundamental control theory concepts are completely derived and explained by using only elementary knowledge about systems of linear algebraic equations; (b) inclusion of MATLAB laboratory experiments, designed after each chapter, to be used either in an associated control system laboratory or as supplements for instructions; (c) MATLAB case studies, examples, and problems are given in each chapter; (d) an extensive chapter on the state space approach based only on sophomore/junior level courses in differential equations and linear algebra; (e) the stability concept is thoroughly explained through the notion of the system minimal polynomial so that the unstable nature of the multiple poles (system eigenvalues) on the imaginary axis is completely clarified; (f) complete proofs of all rules for the root locus technique (Chapter 7) using only elementary mathematics; (g) complete set of controller design techniques based solely on the root locus method, which have much simpler forms than those based on the Bode diagrams; and (h) Chapter 10, entitled "Control System Theory Overview," which gives to students better insight into this extremely broad and multi-disciplinary engineering area.

---

[2]   SIMULINK is a registered trademark of The MathWorks, Inc.

This book is intended for senior students in engineering. *Prerequisites for this course are either undergraduate sophomore/junior level courses in differential equations and linear algebra or linear systems course* (taught to juniors majoring in electrical engineering). In some schools, without undergraduate control courses, this book can be used as the first year graduate control theory text. In addition, the theoretical concepts presented in Chapters 1–5 and 10 are very helpful for graduate students interested in control system theory since they represent the required background for other graduate control theory courses and can be used for preparing master comprehensive and Ph.D. qualifying examinations. Chapters 6–9 are useful for practicing engineers who are applying control system design techniques to real physical systems.

The material presented in this book has been class-tested during several semesters at Rutgers University, Department of Electrical and Computer Engineering in the required senior level course on Automatic Control. The book includes a teacher's solution manual for problems and laboratory experiments and a computer disk with all MATLAB programs, laboratory experiments, and numerical data for vectors and matrices necessary to run MATLAB examples, problems, and laboratory experiments. The computer disk is available at no cost through The MathWorks Inc. The interested reader should fill in the card at the back of this book and send it to the address given. The software may also be retrieved from the MathWorks anonymous FTP server at `ftp.mathworks.com` in `pub/books/gajic/`. In addition, the MATLAB programs and numerical data used in this book may be obtained via anonymous FTP from the Internet site `ece.rutgers.edu` in the directory `/pub/gajic` or by pointing a Web browser to the book's WWW homepage on `http://www.ece.rutgers.edu/~gajic/control.html`.

Finally, we would like to clarify why the book is titled *Modern Control Systems Engineering*. This book is structured to cover thoroughly the fundamental control theory concepts (state space, controllability, observability, stability) deeper than any other undergraduate textbook through the use of only elementary mathematics. These concepts are used nowadays and will be used in the future not just in control engineering, but in many other engineering and scientific disciplines, like aerospace, motor industry, robotics, communications, signal processing, power systems, hydrology, computer science, bioengineering, chemical processes, economics, etc. In addition, showing how to use MATLAB extensively for "quick" control system design purposes makes the techniques presented in this book efficient design tools for modern control system practitioners.

The authors are thankful for support and contributions from Professor B. Lalević, Chairman of the Electrical and Computer Engineering Department at Rutgers University, Professors V. Kecman, M. Lim, P. Milojević, S. Orfanidis, B. Petrović, N. Puri, M. Qureshi, D. Skatarić, X. Shen, W. Su, our colleagues Drs Z. Aganović, I. Borno, graduate students R. Losada, V. Radisavljević, I. Seskar, and undergraduate students T. Carpenter, T. McCrimmon, and G. Topalović. For technical support we are indebted to N. Aganović, I. Lelić, and J. Li.

Dr M. Lelić is thankful to his wife Verica and daughters Dina and Merima for their love and support.

Last and the most we are thankful to our former teachers: Professor Z. Gajić to Professor H. Khalil from Michigan Sate University and Professor J. Medanić from University of Illinois at Urbana, and Professor M. Lelić to Professors P. Wellstead and M. Zarrop from Control Systems Centre, University of Manchester.

Z. Gajić
M. Lelić
February 1996
Piscataway, New Jersey, USA

Chapter One

Introduction

This book represents a modern treatment of classical control theory of continuous- and discrete-time linear systems. Classical control theory originated in the fifties and attained maturity in the sixties and seventies. During that time control theory and its applications were among the most challenging and interesting scientific and engineering areas. The success of the space program and the aircraft industry was heavily based on the power of classical control theory.

The rapid scientific development between 1960 and 1990 brought a tremendous number of new scientific results. Just within electrical engineering, we have witnessed the real explosion of the computer industry in the middle of the eighties, and the rapid development of signal processing, parallel computing, neural networks, and wireless communication theory and practice at the beginning of the nineties. In the years to come many scientific areas will evolve around vastly enhanced computers with the ability to solve by virtually brute force very complex problems, and many new scientific areas will open in that direction. The already established "information superhighway" is maybe just a synonym for the numerous possibilities for "informational breakthrough" in almost all scientific and engineering areas with the use of modern computers. Neural networks—dynamic systems able to process information through large number of inputs and outputs—will become specialized "dynamic" computers for solving specialized problems.

Where is the place of classical (and modern) control theory in contemporary scientific, industrial, and educational life? First of all, classical control theory values have to be preserved, properly placed, and incorporated into modern scientific knowledge of the nineties. Control theory will not get as much attention and recognition as it used to enjoy in the past. However, control theory is concerned with dynamic systems, and dynamics is present, and will be increasingly present, in almost all scientific and engineering disciplines. Even computers connected into networks can be studied as dynamic systems. Communication networks have long been recognized as dynamic systems, but their models are too complex to be studied without the use of powerful computers. Traffic highways of the future are already the subject of broad scale research as dynamic systems and an intensive search for the best optimal control of networks of highways is underway. Robotics, aerospace, chemical, and automotive industries are producing every day new and challenging models of dynamic systems which have to be optimized and controlled. Thus, there is plenty of room for further development of control theory applications, both behind or together with the "informational power" of modern computers.

Control theory must preserve its old values and incorporate them into modern scientific trends, which will be based on the already developed fast and reliable packages for scientific numerical computations, symbolic computations, and computer graphics. One of them, MATLAB, is already gaining broad recognition from the scientific community and academia. It represents an expert system for many control/system oriented problems and it is widely used in industry and academia either to solve new problems or to demonstrate the present state of scientific knowledge in control theory and its applications. The MATLAB package will be extensively used throughout of this book to solve many control theory problems and allow deeper understanding and analysis of problems that would not otherwise be solvable using only pen and paper.

Most contemporary control textbooks originated in the sixties or have kept the structure of the textbooks written in the sixties with a lot of emphasis on frequency domain techniques and a strong distinction between continuous- and discrete-time domains. At the present time, all undergraduate students in electrical engineering are exposed to discrete-time systems in their junior year while studying linear systems and signals and digital signal processing courses so that parallel treatment of continuous- and discrete-time systems saves time and space. The time domain techniques for system/control analysis and design are computationally more powerful than the frequency domain techniques. The time domain techniques are heavily based on differential/difference equations and linear algebra, which are very well developed areas of applied mathematics, for which efficient numerical methods and computer packages exist. In addition, the state space time domain method, to be presented in Chapter 3, is much more convenient for describing and studying high-order systems than the frequency domain method. Modern scientific problems to be addressed in the future will very often be of high dimensions.

In this book, the reader will find parallel treatment of continuous- and discrete-time systems with *emphasis on continuous-time control systems and on time domain techniques (state space method)* for analysis and design of linear control systems. However, all fundamental concepts known from the frequency domain approach will be presented in the book. Our goal is to present the essence, the fundamental concepts, of classic control theory—something that will be valuable and applicable for modern dynamic control systems.

The reader will find that some control concepts and techniques for discrete-time control systems are not fully explained in this book. The main reason for this omission is that those "untreated topics" can be simply obtained by extending the presented concepts and techniques given in detail for continuous-time control systems. Readers particularly interested in discrete-time control systems are referred to the specialized books on that topic (e.g. Ogata, 1987; Franklin *et al.*, 1990; Kuo, 1992; Phillips and Nagle, 1995). Instructors who are not enthusiastic about the simultaneous presentation of both continuous- and discrete-time control systems can completely omit the "discrete-time parts" of this book and give only continuous-time treatment of control systems. This book contains an introduction to discrete-time systems that naturally follows from their continuous-time counterparts, which historically are first considered, and which physically represent models of real-world systems.

Having in mind that this textbook will be used at a time when control theory is not at its peak, and is merging with other scientific fields dealing with dynamic systems, we have divided this book into two independent parts. In Chapters 2–5 we present *fundamental control theory methods and concepts*: transfer function method, state space method, system controllability and observability concepts, and system stability. In the next four chapters, we mostly deal with *applications* so that techniques useful for *design* of control systems are considered. In Chapter 10, an overview of modern control areas is given. A description of the topics considered in the introductory chapter of this book is given in the next paragraph.

Chapter Objectives

In the first chapter of this book, we introduce continuous- and discrete-time invariant linear control systems, and indicate the difference between open-loop and closed-loop (feedback) control. The two main techniques in control system analysis and design, i.e. state space and transfer function methods, are briefly discussed. Modeling of dynamic systems and linearization of nonlinear control systems are presented in detail. A few real-world control systems are given in order to demonstrate the system modeling and linearization. Several other models of real-world dynamic control systems will be considered in the following chapters of this book. In the concluding sections, we outline the book's structure and organization, and indicate the use of MATLAB and its CONTROL and SIMULINK toolboxes as teaching tools in computer control system analysis and design.

1.1 Continuous and Discrete Control Systems

Real-world systems are either static or dynamic. Static systems are represented by algebraic equations, and since not too many real physical systems are static they are of no interest to control engineers. Dynamic systems are described either by differential/difference equations (also known as *systems with concentrated* or *lumped parameters*) or by partial differential equations (known as *systems with distributed parameters*). Distributed parameter control systems are very hard to study from the control theory point of view since their analysis is based on very advanced mathematics, and hence will not be considered in this book. At some schools distributed parameter control systems are taught as advanced graduate courses. Thus, we will pay attention to concentrated parameter control systems, i.e. dynamic systems described by differential/difference equations. It is important to point out that many real physical systems belong to the category of concentrated parameter control systems and a large number of them will be encountered in this book.

Consider, for example, dynamic systems represented by scalar differential/difference equations

$$\dot{x}(t) = f_c(x(t)), \qquad x(t_0) = x_0 \tag{1.1}$$

$$x(k+1) = f_d(x(k)), \qquad x(k_0) = x_0 \tag{1.2}$$

where t stands for continuous-time, k represents discrete-time, subscript c indicates continuous-time and subscript d is used for discrete-time functions. By solving these equations we learn about the system's evolution in time (system response). If the system is under the influence of some external forces, or if we intend to change the system response intentionally by adding some external forces, then the corresponding system is represented by the so-called controlled differential/difference equation, that is

$$\dot{x}(t) = f_c(x(t), u(t)), \qquad x(t_0) = x_0 \tag{1.3}$$

$$x(k+1) = f_d(x(k), u(k)), \qquad x(k_0) = x_0 \tag{1.4}$$

where $u(t)$ and $u(k)$ play the role of control variables. By changing the control variable we hope that the system behavior can be changed in the desired direction, in other words, we intend to use the control variable such that the system response has the desired specifications. When we are able to achieve this goal, we are actually controlling the system behavior.

The general control problem can be formulated as follows: *find the control variable such that the solution of a controlled differential/difference equation has some prespecified characteristics.* This is a quite general definition. In order to be more specific, we have to precisely define the class of systems for which we are be able to solve the general control problem. Note that the differential/difference equations defined in (1.1)–(1.4) are nonlinear. In general, it is hard to deal with nonlinear systems. Nonlinear control systems are studied at the graduate level. In this undergraduate control course, we will study only *linear time invariant control systems.*

Continuous- and discrete-time *linear time invariant dynamic systems* are described, respectively, by linear differential and difference equations with constant coefficients. Mathematical models of such systems having one input and one output are given by

$$\frac{d^n y(t)}{dt^n} + a_{n-1} \frac{d^{n-1} y(t)}{dt^{n-1}} + \cdots + a_1 \frac{dy(t)}{dt} + a_0 y(t) = u(t) \tag{1.5}$$

and

$$y(k+n) + a_{n-1} y(k+n-1) + \cdots + a_1 y(k+1) + a_0 y(k) = u(k) \tag{1.6}$$

where n is the order of the system, y is the *system output* and u is the external forcing function representing the *system input*. In addition to the *external forcing function* the system is also driven by its *internal forces* coming from the *system initial conditions.* For continuous-time systems, the initial conditions are specified by known values of the system output derivatives up to the order of $n - 1$ at the initial time t_0, that is

$$y(t_0), \quad \frac{dy(t_0)}{dt}, \quad ..., \quad \frac{d^{n-1}y(t_0)}{dt^{n-1}} \tag{1.7}$$

In the discrete-time domain the initial conditions are specified by

$$y(k_0), \quad y(k_0 + 1), \quad ..., \quad y(k_0 + n - 1) \tag{1.8}$$

It is interesting to point out that in the discrete-time domain the initial conditions carry information about the evolution of the system output in time from k_0 to $k_0 + n - 1$. In this book, we study only time invariant continuous and discrete systems for which the coefficients $a_i, i = 0, 1, ..., n - 1$, are constants. A block diagram representation of such a system is given in Figure 1.1.

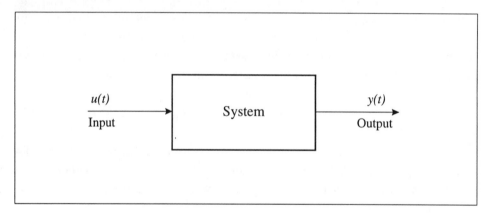

Figure 1.1: Input–output block diagram of a system

In general, the input function can be differentiated by the system so that the more general descriptions of time invariant continuous and discrete systems are given by

$$\frac{d^n y(t)}{dt^n} + a_{n-1}\frac{d^{n-1}y(t)}{dt^{n-1}} + \cdots + a_1\frac{dy(t)}{dt} + a_0 y(t)$$

$$= b_m\frac{d^m u(t)}{dt^m} + b_{m-1}\frac{d^{m-1}u(t)}{dt^{m-1}} + \cdots + b_1\frac{du(t)}{dt} + b_0 u(t) \tag{1.9}$$

and

$$y(k + n) + a_{n-1}y(k + n - 1) + \cdots + a_1y(k + 1) + a_0y(k)$$
$$= b_m u(k + m) + b_{m-1}u(k + m - 1) + \cdots + b_1u(k + 1) + b_0u(k) \tag{1.10}$$

where all coefficients $a_i, i = 1, 2, ..., n$, and $b_j, j = 0, 1, ..., m$, are constant.

The problem of obtaining differential (difference) equations that describe dynamics of real physical systems is known as *mathematical modeling*. In Sections 1.4 and 1.5 this problem will be addressed in detail and mathematical models for several real physical systems will be derived.

Basic Linear Control Problem

In summary, we outline the basic problem of the linear control theory. The problem of finding the system response for the given input function $u(t)$ or $u(k)$ is basically the straightforward problem of solving the corresponding linear differential or difference equation, (1.9) or (1.10). This problem can be solved by using standard knowledge from mathematical theory of linear differential and/or difference equations. However, *the linear control problem is much more challenging, namely the input function $u(t)$ has to be found such that the linear system response has the desired behavior*. A simplified version of the above basic linear control problem will be defined in Chapter 6 using the notion of system feedback. The basic linear control problem can be studied either in the time domain (state space approach) or in the frequency domain (transfer function approach). These two approaches will be presented in Chapters 2 and 3.

1.2 Open-Loop and Closed-Loop Control Systems

It seems that if an input function can be found such that the corresponding system has the desired response, then the control problem of interest is solved. This is true, but is it all that we need? Assume that $u(t)$ is such a function, which is apparently a function of time. Imagine that due to parameter variations or due to aging of the system components the system model is a little bit different than the original one or even worse that the coefficients $a_i, i = 0, 1, 2, ..., n - 1$; $b_j, j = 0, 1, 2, ..., m$, in equation (1.9) are not very precisely known. Then the function $u(t)$, given as a precomputed time function, might not produce a satisfactory solution (especially in the long run). One may try to solve the problem again and get a new expression for $u(t)$ at the expense of additional computation, which is fine if the system parameters are exactly known, but this approach will not bring any improvement in the case when the system coefficients, obtained either analytically or experimentally, are known only with certain accuracy. This *precomputed time function* $u(t)$ (or $u(k)$ in the discrete-time domain), which solves the control problem, is known as the *open-loop control*.

Imagine now that one is able, in an attempt to solve the basic linear control problem, to obtain the desired input function as a function of the system desired response, or even more precisely as a function of some essential system variables that completely determine the system dynamics. These essential system variables are called the *state space variables*. It is very natural to assume that for a system of order n, a collection of n such variables exist. Denote the state space variables by $x_1(t), x_2(t), ..., x_n(t)$. These state variables at any given time represent the actual state of the system. Even if some parameters of the system are changing in time or even if some coefficients in (1.9) are not precisely known, the state variables $x_1(t), x_2(t), ..., x_n(t)$ will reflect exactly the state of the system at any given time. The question now is: can we get the required control variable (system input) as a function of the state space variables? If the answer is yes, then the existence of such a $u(\mathbf{x}(t))$ indicates the existence of the so-called *state feedback control*. In some cases it is impossible to find the feedback control, but for the linear time invariant systems studied in this book, linear feedback control always exists. The linear feedback control is a linear function of the state space variables, that is

$$u(\mathbf{x}) = F_1 x_1 + F_2 x_2 + \cdots + F_n x_n = \mathbf{F}\mathbf{x}$$
$$\mathbf{F} = [F_1, F_2, ..., F_n] \tag{1.11}$$
$$\mathbf{x} = [x_1, x_2, ..., x_n]^T$$

where the coefficient matrix $\mathbf{F}$ is the *feedback gain* and the vector $\mathbf{x}$ is known as the *state space vector*. It is sometimes desirable (and possible) to achieve the goal by using instead of the state feedback control $u(\mathbf{x}(t))$, the so-called *output feedback control* given by $u(y(t)) = u(y(\mathbf{x}(t)))$, which in general does not contain all state variables, but only a certain collection of them. In Section 3.1 we will learn how to relate the output and state space variables.

Open-loop control, state feedback control, and output feedback control are represented schematically in Figure 1.2.

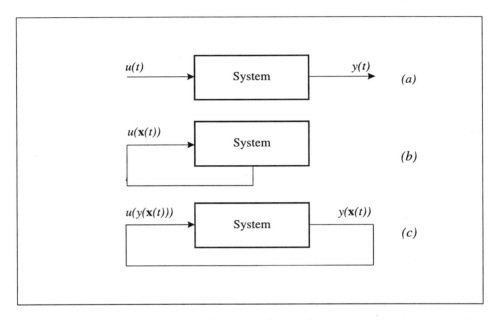

Figure 1.2: Open-loop (a), state feedback (b), and output feedback (c) controls

The system represented in Figure 1.1 and given in formulas (1.5)–(1.10) has only one input u and one output y. Such systems are known as *single-input single-output systems*. In general, systems have several inputs and several outputs, say r inputs and p outputs. In that case, we have

$$\mathbf{u} = [u_1, u_2, ..., u_r]^T, \quad \mathbf{y} = [y_1, y_2, ..., y_p]^T \qquad (1.12)$$

and the matrix $\mathbf{F}$ is of dimension $r \times n$. These systems are known as *multi-input multi-output systems*. They are also called *multivariable control systems*. A block diagram for a multi-input multi-output system is represented in Figure 1.3.

Feedback control is almost always desirable as a solution to the general control problem, and only in rare cases and in the cases when it is impossible to find the feedback control has one to stick with open-loop control. Throughout of book we will see and discuss many advantages of feedback control. *The main role of feedback is to stabilize the system under consideration.* The feedback also reduces the effect of uncertainties in the system model. In addition, it efficiently handles system parameter changes and external disturbances attacking the system, by simply reducing the system output sensitivity to all of these undesired phenomena. Most of these good feedback features will be analytically justified in the follow-up chapters.

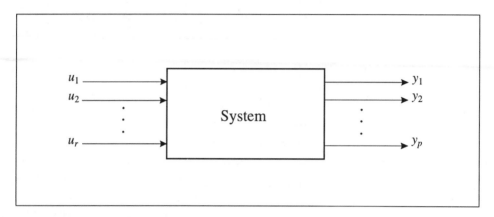

Figure 1.3: Block diagram of a multi-input multi-output system

However, one has to be careful when applying feedback since it changes the structure of the original system. Consider the system described by equation (1.3) under the state feedback control, that is

$$\dot{x}(t) = f_c(x(t), u(x(t))) = \mathcal{F}_c(x(t)), \qquad x(t_0) = x_0 \qquad (1.13)$$

Thus, after the feedback is applied, a new dynamic system is obtained, in other words, for different values of $u(x(t))$ we will get different expressions for the right-hand side of (1.13) so that equation (1.13) will represent different dynamic systems.

1.3 State Space and Transfer Functions

In analysis and design of linear time invariant systems two major approaches are available: the time domain state space approach and the frequency domain transfer function approach. Both approaches will be considered in detail in Chapters 2 and 3. Electrical engineering students are partially familiar with these methods and the corresponding terminology from undergraduate courses on linear systems and digital signal processing.

In this section only the main definitions are stated. This is necessary to maintain the continuum of presentation and to get mathematical models of some real physical systems in state space form. In addition, by being familiar with the notion of the state space, we will be able to present the linearization of nonlinear systems in the most general form.

In the previous section it is indicated that for an nth order system there are n essential state variables, the so-called state space variables, which form the corresponding state space vector $\mathbf{x}$. For linear time invariant systems, the vector of state space variables satisfies the linear time invariant vector differential (difference) equation known as the state space form. The state space equations of linear systems are defined by

$$\dot{\mathbf{x}}(t) = \mathbf{A}\mathbf{x}(t) + \mathbf{B}\mathbf{u}(t), \quad \mathbf{x}(0) = \mathbf{x}_o$$
$$\mathbf{y}(t) = \mathbf{C}\mathbf{x}(t) + \mathbf{D}\mathbf{u}(t) \tag{1.14}$$

and

$$\mathbf{x}(k+1) = \mathbf{A}\mathbf{x}(k) + \mathbf{B}\mathbf{u}(k), \quad \mathbf{x}(0) = \mathbf{x}_o$$
$$\mathbf{y}(k) = \mathbf{C}\mathbf{x}(k) + \mathbf{D}\mathbf{u}(k) \tag{1.15}$$

In view of the discussion in the previous section, the introduced constant matrices have the dimensions $\mathbf{A}^{n \times n}, \mathbf{B}^{n \times r}, \mathbf{C}^{p \times n}, \mathbf{D}^{p \times r}$. Of course, the given matrices for continuous and discrete systems have different entries. In formulas (1.14) and (1.15) the same notation is kept for both continuous and discrete systems for the sake of simplicity. In the case when a discrete-time system is obtained by sampling a continuous-time system, we will emphasize the corresponding difference by adding a subscript d for the discrete-time quantities, e.g. $\mathbf{A}_d, \mathbf{B}_d, \mathbf{C}_d, \mathbf{D}_d$.

The *system transfer function* for a single-input single-output time invariant continuous system is defined as *the ratio of the Laplace transform of the system output over the Laplace transform of the system input assuming that all initial conditions are zero.* This definition implies that the transfer function corresponding to (1.9) is

$$G(s) = \frac{b_m s^m + b_{m-1} s^{m-1} + \cdots + b_1 s + b_0}{s^n + a_{n-1} s^{n-1} + \cdots + a_1 s + a_0} \tag{1.16}$$

Polynomial exponents in transfer functions of real physical systems always satisfy $n \geq m$. In the discrete-time domain, the $\mathcal{Z}$-transform takes the role of the Laplace transform so that the discrete-time transform function is given by

$$G(z) = \frac{b_m z^m + b_{m-1} z^{m-1} + \cdots + b_1 z + b_0}{z^n + a_{n-1} z^{n-1} + \cdots + a_1 z + a_0} \tag{1.17}$$

For multi-input multi-output systems with r inputs and p outputs, transfer functions are matrices of order $p \times r$ whose entries are the corresponding transfer functions from the ith system input to the jth system output, $i = 1, 2, ..., r$; $j = 1, 2, ..., p$, that is

$$\mathbf{G}(s) = \begin{bmatrix} G_{11}(s) & G_{12}(s) & \cdots & \cdots & G_{1r}(s) \\ G_{21}(s) & \cdots & \cdots & \cdots & \cdots \\ \cdots & \cdots & G_{ji}(s) & \cdots & \cdots \\ \cdots & \cdots & \cdots & \cdots & \cdots \\ G_{p1}(s) & \cdots & \cdots & \cdots & G_{pr}(s) \end{bmatrix}^{p \times r} \tag{1.18}$$

Recall from basic circuits courses that while finding $G_{ji}(s) = Y_j(s)/U_i(s)$ *all other system inputs except for $U_i(s)$ must be set to zero.* Similarly, the discrete-time transfer function of a multi-input multi-output, time invariant, system is given by

$$\mathbf{G}(z) = \begin{bmatrix} G_{11}(z) & G_{12}(z) & \cdots & \cdots & G_{1r}(z) \\ G_{21}(z) & \cdots & \cdots & \cdots & \cdots \\ \cdots & \cdots & G_{ji}(z) & \cdots & \cdots \\ \cdots & \cdots & \cdots & \cdots & \cdots \\ G_{p1}(z) & \cdots & \cdots & \cdots & G_{pr}(z) \end{bmatrix}^{p \times r} \tag{1.19}$$

More will be said about the transfer function of multi-input multi-output systems in Chapter 2.

Since each entry in the matrix transfer functions given in (1.18) and (1.19) is a ratio of two polynomials with complex numbers, it is obvious that for high-order systems the required calculations in the frequency domain are mathematically very involved so that the state space system representation is simpler than the corresponding frequency domain representation. In addition, since the state space method is based on linear algebra, for which numerous efficient mathematical methods have already been developed, the state space method is also more convenient from the computational point of view than the frequency domain method. However, the importance of the frequency domain representation lies in the simplicity of presenting some basic concepts, and hence it very often gives a better understanding of the actual physical phenomena occurring within the system.

1.4 Mathematical Modeling of Real Physical Systems

Mathematical modeling of real-world physical systems is based on the application of known physical laws to the given systems, which leads to mathematical equations describing the behavior of systems under consideration. The equations obtained are either algebraic, ordinary differential or partial differential. Systems described by algebraic equations are of no interest for this course since they represent static phenomena. Dynamic systems mathematically described by partial differential equations are known as systems with distributed parameters. The study of distributed parameter systems is beyond the scope of this book. Thus, we will consider only systems described by ordinary differential equations. These systems are also known as systems with lumped (concentrated) parameters. Even the lumped parameter systems are, in general, too difficult from the point of view of solving the general control problem described in Section 1.1, so that we have to limit our attention to lumped parameter systems described by linear time invariant differential equations. Fortunately, many control systems do have this form. Even more, control systems described by nonlinear differential equations can very often be linearized in the neighborhood of their nominal (operating) trajectories and controls assuming that these quantities are known, which is very often the case, so that nonlinear systems can be studied as linear ones. The linearization procedure will be independently considered in Section 1.6.

In the following the modeling procedure is demonstrated on a simple RLC electrical network given in Figure 1.4. Assume that the initial values for the

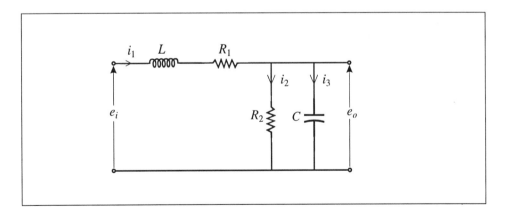

Figure 1.4: An RLC network

inductor current and capacitor voltage are zero. Applying the basic circuit laws (Thomas and Rosa, 1994) for voltages and currents, we get

$$e_i(t) = L\frac{di_1(t)}{dt} + R_1 i_1(t) + e_0(t) \tag{1.20}$$

$$e_0(t) = R_2 i_2(t) = \frac{1}{C} \int_0^t i_3(\tau) d\tau \quad \Rightarrow \quad i_3 = C\frac{de_0(t)}{dt} \tag{1.21}$$

$$i_1(t) = i_2(t) + i_3(t) \tag{1.22}$$

Using (1.21) in (1.22) produces

$$i_1(t) = \frac{1}{R_2} e_0(t) + C\frac{de_0(t)}{dt} \tag{1.23}$$

Taking the derivative of (1.23) and combining (1.20) and (1.23), we obtain the desired second-order differential equation, which relates the input and output of the system, and represents a mathematical model of the circuit given in Figure 1.4

$$\frac{d^2 e_0(t)}{dt^2} + \left(\frac{L + R_1 R_2 C}{R_2 LC}\right)\frac{de_0(t)}{dt} + \left(\frac{R_1 + R_2}{R_2 LC}\right)e_0(t) = \frac{1}{LC}e_i(t) \tag{1.24}$$

Note that in this mathematical model $e_i(t)$ represents the system input and $e_0(t)$ is the system output. However, any of the currents and any of the voltages can play the roles of either input or output variables.

Introducing the following change of variables

$$x_1(t) = e_o \quad \Rightarrow \quad \frac{dx_1(t)}{dt} = \frac{de_o(t)}{dt} = x_2(t)$$

$$x_2(t) = \frac{de_o(t)}{dt} \tag{1.25}$$

$$u(t) = e_i(t)$$

$$y(t) = e_o(t) \quad \Rightarrow \quad y(t) = x_1(t)$$

and combining it with (1.24) we get

$$\frac{dx_2(t)}{dt} + \left(\frac{L + R_1 R_2 C}{R_2 LC}\right)x_2(t) + \left(\frac{R_1 + R_2}{R_2 LC}\right)x_1(t) = \frac{1}{LC}u(t) \tag{1.26}$$

The first equation in (1.25) and equation (1.26) can be put into matrix form as

$$\begin{bmatrix} \dot{x}_1 \\ \dot{x}_2 \end{bmatrix} = \begin{bmatrix} 0 & 1 \\ -\frac{R_1+R_2}{R_2LC} & -\frac{L+R_1R_2C}{R_2LC} \end{bmatrix} \begin{bmatrix} x_1 \\ x_2 \end{bmatrix} + \begin{bmatrix} 0 \\ \frac{1}{LC} \end{bmatrix} u \qquad (1.27)$$

The last equation from (1.25), in matrix form, is written as

$$y(t) = \begin{bmatrix} 1 & 0 \end{bmatrix} \begin{bmatrix} x_1 \\ x_2 \end{bmatrix} \qquad (1.28)$$

Equations (1.27) and (1.28) represent the state space form for the system whose mathematical model is given by (1.24). The corresponding state space matrices for this system are given by

$$\mathbf{A} = \begin{bmatrix} 0 & 1 \\ -\frac{R_1+R_2}{R_2LC} & -\frac{L+R_1R_2C}{R_2LC} \end{bmatrix}, \quad \mathbf{B} = \begin{bmatrix} 0 \\ \frac{1}{LC} \end{bmatrix}, \quad \mathbf{C} = \begin{bmatrix} 1 & 0 \end{bmatrix}, \quad \mathbf{D} = 0 \quad (1.29)$$

The state space form of a system is not unique. Using another change of variables, we can get, for the same system, another state space form, which is demonstrated in Problem 1.1.

The transfer function of this single-input single-output system is easily obtained by taking the Laplace transform of (1.24), which leads to

$$G(s) = \frac{\mathcal{L}\{e_i(t)\}}{\mathcal{L}\{e_0(t)\}} = \frac{\frac{1}{LC}}{s^2 + \frac{L+R_1R_2C}{R_2LC}s + \frac{R_1+R_2}{R_2LC}} \qquad (1.30)$$

Note that a systematic approach for getting the state space form from differential (difference) equations will be given in detail in Chapter 3. In this chapter we present only the simplest cases. These cases are dictated by system physical structures described by a set of first- and second-order differential (difference) equations, which can be put in a straightforward way into matrix form, which in fact represents the desired state space form.

Another example, which demonstrates how to get a mathematical model for a real physical system, is taken from mechanical engineering.

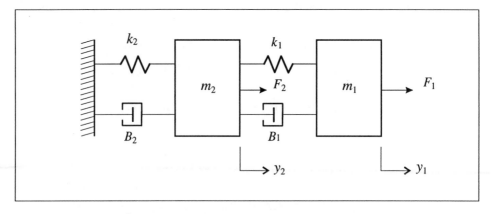

Figure 1.5: A translational mechanical system

A translational mechanical system is represented in Figure 1.5. The following two equations of motion for this system can be written by using the basic laws of dynamics (Greenwood, 1988)

$$F_1 = m_1 \frac{d^2 y_1}{dt^2} + B_1 \left(\frac{dy_1}{dt} - \frac{dy_2}{dt} \right) + k_1 (y_1 - y_2) \qquad (1.31)$$

and

$$F_2 = m_2 \frac{d^2 y_2}{dt^2} + B_2 \frac{dy_2}{dt} + k_2 y_2 - B_1 \left(\frac{dy_1}{dt} - \frac{dy_2}{dt} \right) - k_1 (y_1 - y_2) \qquad (1.32)$$

It can be seen that this system has two inputs, F_1 and F_2, and two outputs, y_1 and y_2. The rearranged form of equations (1.31) and (1.32) is given by

$$m_1 \frac{d^2 y_1}{dt^2} + B_1 \frac{dy_1}{dt} + k_1 y_1 - B_1 \frac{dy_2}{dt} - k_1 y_2 = F_1 \qquad (1.33)$$

and

$$-B_1 \frac{dy_1}{dt} - k_1 y_1 + m_2 \frac{d^2 y_2}{dt^2} + (B_1 + B_2) \frac{dy_2}{dt} + (k_1 + k_2) y_2 = F_2 \qquad (1.34)$$

From equations (1.33) and (1.34) the state space form can be obtained easily by choosing the following state space variables

$$x_1 = y_1, \quad x_2 = \frac{dy_1}{dt}, \quad x_3 = y_2, \quad x_4 = \frac{dy_2}{dt}$$

$$u_1 = F_1, \quad u_2 = F_2$$

$$(1.35)$$

The state space form of this two-input two-output system is given by

$$
\begin{bmatrix} \dot{x}_1 \\ \dot{x}_2 \\ \dot{x}_3 \\ \dot{x}_4 \end{bmatrix} = \begin{bmatrix} 0 & 1 & 0 & 0 \\ -\frac{k_1}{m_1} & -\frac{B_1}{m_1} & \frac{k_1}{m_1} & \frac{B_1}{m_1} \\ 0 & 0 & 0 & 1 \\ \frac{k_1}{m_2} & \frac{B_1}{m_2} & -\frac{k_1+k_2}{m_2} & -\frac{B_1+B_2}{m_2} \end{bmatrix} \begin{bmatrix} x_1 \\ x_2 \\ x_3 \\ x_4 \end{bmatrix} + \begin{bmatrix} 0 & 0 \\ \frac{1}{m_1} & 0 \\ 0 & 0 \\ 0 & \frac{1}{m_2} \end{bmatrix} \begin{bmatrix} u_1 \\ u_2 \end{bmatrix}
$$

(1.36)

and

$$
\begin{bmatrix} y_1 \\ y_2 \end{bmatrix} = \begin{bmatrix} 1 & 0 & 0 & 0 \\ 0 & 0 & 1 & 0 \end{bmatrix} \begin{bmatrix} x_1 \\ x_2 \\ x_3 \\ x_4 \end{bmatrix}
$$

(1.37)

It is interesting to find the transfer function for this multi-input multi-output system. Taking the Laplace transforms of (1.33) and (1.34), and assuming that all initial conditions are equal to zero, we get the scalar transfer functions from each input to each output. This is obtained by keeping the input under consideration different from zero and setting the other one to zero, that is

$$
G_{11}(s) = \left(\frac{Y_1(s)}{U_1(s)} \right)_{\big|_{U_2(s)=0}} = \frac{m_2 s^2 + (B_1 + B_2)s + (k_1 + k_2)}{a_4 s^4 + a_3 s^3 + a_2 s^2 + a_1 s + a_0}
$$

$$
G_{12}(s) = \left(\frac{Y_1(s)}{U_2(s)} \right)_{\big|_{U_1(s)=0}} = \frac{B_1 s + k_1}{a_4 s^4 + a_3 s^3 + a_2 s^2 + a_1 s + a_0}
$$

(1.38)

$$
G_{21}(s) = \left(\frac{Y_2(s)}{U_1(s)} \right)_{\big|_{U_2(s)=0}} = \frac{B_1 s + k_1}{a_4 s^4 + a_3 s^3 + a_2 s^2 + a_1 s + a_0}
$$

$$
G_{22}(s) = \left(\frac{Y_2(s)}{U_2(s)} \right)_{\big|_{U_1(s)=0}} = \frac{m_1 s^2 + B_1 s + k_1}{a_4 s^4 + a_3 s^3 + a_2 s^2 + a_1 s + a_0}
$$

where

$$
a_4 = m_1 m_2
$$
$$
a_3 = B_1(m_1 + m_2) + m_1 B_2
$$
$$
a_2 = B_1 B_2 + k_1(m_1 + m_2) + k_2 m_1
$$
$$
a_1 = k_1 B_2 + k_2 B_1
$$
$$
a_0 = k_1 k_2
$$

so that the system transfer function is given by

$$\mathbf{G}(s) = \begin{bmatrix} G_{11}(s) & G_{12}(s) \\ G_{21}(s) & G_{22}(s) \end{bmatrix} \tag{1.39}$$

Sometimes due to the complexity of dynamic systems it is not possible to establish mathematical relations describing the dynamical behavior of the systems under consideration. In those cases one has to use experimentation in order to get data that can be used in establishing some mathematical relations caused (induced) by system dynamics. The experimental way of getting system models is the subject of the area of control systems known as *system identification*. More about system identification can be found in Chapter 10. A classic textbook on system identification is given in the list of references (Ljung, 1987).

The reader particularly interested in mathematical modelling of real physical systems is referred to Wellstead (1979) and Kecman (1988).

1.5 Models of Some Control Systems

In this section we consider the modeling of two common physical control systems. In that direction modeling of an inverted pendulum and a complex electrical network is presented. Mathematical models of a DC motor will be derived in Section 2.2.1. DC motors are integral parts of several control system schemes. Mathematical models of many other interesting control systems can be found in Wellstead (1979), Kecman (1988), and Dorf (1992).

Inverted Pendulum

The inverted pendulum is a familiar dynamic system used very often in textbooks (Kwakernaak and Sivan, 1972; Kamen, 1990). Here, we follow derivations of Kecman (1988) for an idealized pendulum of length l whose mass, m_1, is concentrated at its end. It is assumed that a cart of mass m_2 is subjected to an external force F, which, as a control variable, has to keep the pendulum upright. Cart displacement is denoted by x and pendulum displacement is represented by angle θ (see Figure 1.6).

Using elementary knowledge from college physics (see for example Serway 1992), the equation of motion for translation in the direction of x axis is obtained by applying Newton's law

$$m_2 \frac{d^2 x(t)}{dt^2} + m_1 \frac{d^2(x + l \sin \theta)}{dt^2} = F \tag{1.40}$$

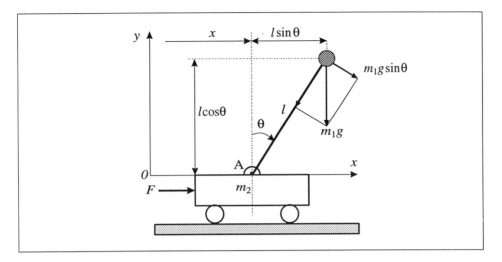

Figure 1.6: Inverted pendulum

The conservation of momentum equation with respect to rotation about point A implies

$$m_1 l \cos \theta \frac{d^2}{dt^2}(x + l \sin \theta) - m_1 l \sin \theta \frac{d^2}{dt^2}(l \cos \theta) = m_1 g l \sin \theta \qquad (1.41)$$

where $g = 9.8 \, \text{m/s}^2$ is the gravitational constant. Bearing in mind that

$$\frac{d}{dt} \sin \theta = \cos \theta \frac{d\theta}{dt}, \qquad \frac{d}{dt} \cos \theta = -\sin \theta \frac{d\theta}{dt}$$

$$\frac{d^2}{dt^2} \sin \theta = -\sin \theta \left(\frac{d\theta}{dt}\right)^2 + \cos \theta \frac{d^2\theta}{dt^2} \qquad (1.42)$$

$$\frac{d^2}{dt^2} \cos \theta = -\cos \theta \left(\frac{d\theta}{dt}\right)^2 - \sin \theta \frac{d^2\theta}{dt^2}$$

we get a system of two second-order differential equations

$$(m_1 + m_2)\frac{d^2 x}{dt^2} - m_1 l \sin \theta \left(\frac{d\theta}{dt}\right)^2 + m_1 l \cos \theta \frac{d^2\theta}{dt^2} = F$$

$$\cos \theta \frac{d^2 x}{dt^2} + l \frac{d^2\theta}{dt^2} = g \sin \theta$$

$$(1.43)$$

Equations (1.43) represent the desired mathematical model of an inverted pendulum.

Complex Electrical Network

Complex electrical networks are obtained by connecting basic electrical elements: resistors, inductors, capacitors (passive elements) and voltage and current sources (active elements). Of course, other electrical engineering elements like diodes and transistors can be present, but since in this course we study only the linear time invariant networks and since this textbook is intended for all engineering students we will limit our attention to basic electrical elements. The complexity of the network will be manifested by a large number of passive and active elements and large number of loops. Such a network is represented in Figure 1.7.

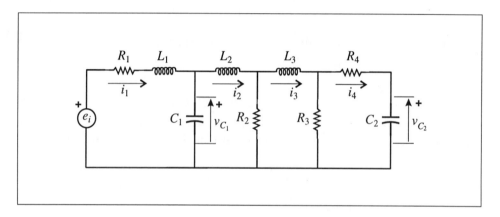

Figure 1.7: Complex electrical network

In electrical networks composed of inductors and capacitors *the total number of inductors and capacitors indicates the order of the dynamic system.* In this particular example, we have three inductors and two capacitors, i.e. the order of this dynamical system is $n = 5$. Having a dynamic system of order $n = 5$ indicates that the required number of first-order differential equations to be set is five. If one sets more than n differential equations, for a system of order n, some of them are redundant. Redundant equations have to be eliminated since they do not carry any new information; they are just linear combinations of the remaining equations. Using basic laws for currents and voltages, we can easily set up five first-order differential equations; for voltages around loops

$$e_i - L_1 \frac{di_1}{dt} - R_1 i_1 - v_{c_1} = 0$$

$$L_2 \frac{di_2}{dt} - v_{c_1} + R_2(i_2 - i_3) = 0 \tag{1.44}$$

$$L_3 \frac{di_3}{dt} + R_3(i_3 - i_4) - R_2(i_2 - i_3) = 0$$

and for currents

$$i_1 - i_2 - C_1 \frac{dv_{c_1}}{dt} = 0$$

$$i_4 - C_2 \frac{dv_{c_2}}{dt} = 0 \tag{1.45}$$

We have set up five equations for four currents and two voltages. Current i_4 can be eliminated by using the following algebraic voltage balance equation, which is valid for the last loop

$$R_3(i_3 - i_4) = R_4 i_4 + v_{c_2} \tag{1.46}$$

from which the current i_4 can be expressed as

$$i_4 = \frac{R_3}{R_3 + R_4} i_3 - \frac{1}{R_3 + R_4} v_{c_2} \tag{1.47}$$

Replacing current i_4 in (1.44) and (1.45) by the expression obtained in (1.47), the following five first-order differential equations are obtained

$$\frac{di_1}{dt} = -\frac{R_1}{L_1} i_1 - \frac{1}{L_1} v_{c_1} + \frac{1}{L_1} e_i$$

$$\frac{di_2}{dt} = -\frac{R_2}{L_2} i_2 + \frac{R_2}{L_3} i_3 + \frac{1}{L_2} v_{c_1}$$

$$\frac{di_3}{dt} = \frac{R_2}{L_3} i_2 - \frac{R_2 R_3 + R_2 R_4 + R_3 R_4}{L_3(R_3 + R_4)} i_3 - \frac{R_3}{L_3(R_3 + R_4)} v_{c_2} \tag{1.48}$$

$$\frac{dv_{c_1}}{dt} = \frac{1}{C_1} i_1 - \frac{1}{C_1} i_2$$

$$\frac{dv_{c_2}}{dt} = \frac{R_3}{C_2(R_3 + R_4)} i_3 - \frac{1}{C_2(R_3 + R_4)} v_{c_2}$$

The matrix form of this system of first-order differential equations represents the system state space form. Take $x_1 = i_1, x_2 = i_2, x_3 = i_3, x_4 = v_{c_1}, x_5 = v_{c_2}$, $u = e_i$, then the state space form is given by

$$
\begin{bmatrix} \dot{x}_1 \\ \dot{x}_2 \\ \dot{x}_3 \\ \dot{x}_4 \\ \dot{x}_5 \end{bmatrix} = \begin{bmatrix} -\frac{R_1}{L_1} & 0 & 0 & -\frac{1}{L_1} & 0 \\ 0 & -\frac{R_2}{L_2} & \frac{R_2}{L_2} & \frac{1}{L_2} & 0 \\ 0 & \frac{R_2}{L_3} & a_{33} & 0 & -\frac{R_3}{L_3(R_3+R_4)} \\ \frac{1}{C_1} & -\frac{1}{C_1} & 0 & 0 & 0 \\ 0 & 0 & \frac{R_3}{C_2(R_3+R_4)} & 0 & -\frac{1}{C_2(R_3+R_4)} \end{bmatrix} \begin{bmatrix} x_1 \\ x_2 \\ x_3 \\ x_4 \\ x_5 \end{bmatrix} + \begin{bmatrix} \frac{1}{L_1} \\ 0 \\ 0 \\ 0 \\ 0 \end{bmatrix} u
$$

$$(1.49)$$

where $a_{33} = -\frac{R_2 R_3 + R_2 R_4 + R_3 R_4}{L_3(R_3+R_4)}$.

Note that while modeling electrical networks, it is advisable to use the currents through inductors and the voltages on capacitors for the state space variables, which, in fact, is done in this example.

We would like to point out that in this chapter we have presented only modeling and mathematical models for continuous-time real dynamic physical systems. The reason for this is twofold: (1) there are no real-world discrete-time physical dynamic systems; (2) discrete-time models obtained from social, economic, hydrological, and meteorological sciences are usually of no interest to control engineers. However, discrete-time models obtained by discretization of continuous-time systems will be treated in detail in Chapter 3.

1.6 Linearization of Nonlinear Systems

We have mentioned before that in this book we study only time invariant linear control systems and that the study of nonlinear control systems is rather difficult. However, in some cases it is possible to linearize nonlinear control systems and study them as linear ones. In this section we show how to perform linearization of control systems described by nonlinear differential equations. The procedure introduced is based on the Taylor series expansion and on the knowledge of nominal (operating) system trajectories and nominal control inputs. Readers particularly interested in the study of nonlinear systems are referred to a comprehensive book by Khalil (1992).

We will start with a simple scalar first-order nonlinear dynamic system represented by

$$\dot{x}(t) = f(x(t), u(t)) \tag{1.50}$$

Assume that under usual working circumstances this system operates along the trajectory $x_n(t)$ while it is driven by the control input $u_n(t)$. We call $x_n(t)$ and $u_n(t)$, respectively, the *nominal system trajectory* and the *nominal control input*. On the nominal trajectory the following differential equation is satisfied

$$\dot{x}_n(t) = f(x_n(t), u_n(t)) \tag{1.51}$$

Now assume that the motion of the nonlinear system (1.50) is in the neighborhood of the nominal system trajectory and that the distance from the nominal trajectory is small, that is

$$x(t) = x_n(t) + \Delta x(t) \tag{1.52}$$

where $\Delta x(t)$ represents a small quantity. It is natural to assume that the system motion in close proximity to the nominal trajectory will be sustained by a control input which is obtained by adding a small quantity to the nominal control input, that is

$$u(t) = u_n(t) + \Delta u(t) \tag{1.53}$$

For the system motion in close proximity to the nominal trajectory, from equations (1.50), (1.52), and (1.53), we have

$$\dot{x}_n(t) + \Delta \dot{x}(t) = f(x_n(t) + \Delta x(t), u_n(t) + \Delta u(t)) \tag{1.54}$$

Since $\Delta x(t)$ and $\Delta u(t)$ are small quantities, the right-hand side of (1.54) can be expanded into a Taylor series about the nominal trajectory and control, which produces

$$\dot{x}_n(t) + \Delta \dot{x}(t) = f(x_n(t), u_n(t)) + \frac{\partial f}{\partial x}(x_n, u_n)\Delta x(t) + \frac{\partial f}{\partial u}(x_n, u_n)\Delta u(t)$$
$$+ \text{ high–order terms}$$
$$\tag{1.55}$$

Using (1.51) and canceling high-order terms (which contain very small quantities $\Delta x^2, \Delta u^2, \Delta x \Delta u, \Delta x^3, ...$), the following linear differential equation is obtained

$$\Delta \dot{x}(t) = \frac{\partial f}{\partial x}(x_n, u_n)\Delta x(t) + \frac{\partial f}{\partial u}(x_n, u_n)\Delta u(t) \tag{1.56}$$

whose solution represents a valid approximation for $\Delta x(t)$. Note that the *partial derivatives in the linearization procedure are evaluated at the nominal points.* Introducing the notation

$$a_0(t) = -\frac{\partial f}{\partial x}(x_n, u_n), \qquad b_0 = \frac{\partial f}{\partial u}(x_n, u_n) \tag{1.57}$$

the linear system (1.56) can be represented as

$$\Delta \dot{x}(t) + a_0(t)\Delta x(t) = b_0(t)\Delta u(t) \tag{1.58}$$

In general, the obtained linear system is time varying. Since in this course we study only time invariant systems, we will consider only those examples for which the linearization procedure produces time invariant systems. It remains to find the initial condition for the linearized system, which can be obtained from (1.52) as

$$\Delta x(t_0) = x(t_0) - x_n(t_0) \tag{1.59}$$

Similarly, we can linearize the second-order nonlinear dynamic system

$$\ddot{x} = f(x, \dot{x}, u, \dot{u}) \tag{1.60}$$

by assuming that

$$\begin{aligned} x(t) &= x_n(t) + \Delta x(t), \quad \dot{x}(t) = \dot{x}_n(t) + \Delta \dot{x}(t) \\ u(t) &= u_n(t) + \Delta u(t), \quad \dot{u}(t) = \dot{u}_n(t) + \Delta \dot{u}(t) \end{aligned} \tag{1.61}$$

and expanding

$$\ddot{x}_n + \Delta \ddot{x} = f(x_n + \Delta x_n, \dot{x}_n + \Delta \dot{x}, u_n + \Delta u, \dot{u}_n + \Delta \dot{u}) \tag{1.62}$$

into a Taylor series about nominal points $x_n, \dot{x}_n, u_n, \dot{u}_n$, which leads to

$$\Delta \ddot{x}(t) + a_1 \Delta \dot{x}(t) + a_0 \Delta x(t) = b_1 \Delta \dot{u}(t) + b_0 \Delta u(t) \tag{1.63}$$

where the corresponding coefficients are evaluated at the nominal points as

$$a_1 = -\frac{\partial f}{\partial \dot{x}}(x_n, \dot{x}_n, u_n, \dot{u}_n), \quad a_0 = -\frac{\partial f}{\partial x}(x_n, \dot{x}_n, u_n, \dot{u}_n)$$

$$(1.64)$$

$$b_1 = \frac{\partial f}{\partial \dot{u}}(x_n, \dot{x}_n, u_n, \dot{u}_n), \quad b_0 = \frac{\partial f}{\partial u}(x_n, \dot{x}_n, u_n, \dot{u}_n)$$

The initial conditions for the second-order linearized system are easily obtained from (1.61)

$$\Delta x(t_0) = x(t_0) - x_n(t_0), \quad \Delta \dot{x}(t_0) = \dot{x}(t_0) - \dot{x}_n(t_0) \qquad (1.65)$$

Example 1.1: The mathematical model of a stick-balancing problem is given in Sontag (1990) by

$$\ddot{\theta} = \sin \theta - u \cos \theta = f(\theta, u)$$

where u is the horizontal force of a finger and θ represents the stick's angular displacement from the vertical. This second-order dynamic system is linearized at the nominal points $(\dot{\theta}_n(t) = \theta_n(t) = 0, u_n(t) = 0)$ by using formulas (1.64), which produces

$$a_1 = -\frac{\partial f}{\partial \dot{\theta}} = 0, \quad a_0 = -\left(\frac{\partial f}{\partial \theta}\right)_{|_n} = -(\cos \theta + u \sin \theta)_{|_{\substack{\theta_n(t)=0 \\ u_n(t)=0}}} = -1$$

$$b_1 = \frac{\partial f}{\partial \dot{u}} = 0, \quad b_0 = \left(\frac{\partial f}{\partial u}\right)_{|_n} = -(\cos \theta)_{|_{\theta_n(t)=0}} = -1$$

The linearized equation is given by

$$\ddot{\theta}(t) - \theta(t) = -u(t)$$

Note that $\Delta\theta(t) = \theta(t), \Delta u(t) = u(t)$ since $\theta_n(t) = 0, u_n(t) = 0$. It is important to point out that the same linearized model could have been obtained by setting $\sin\theta(t) \approx \theta(t)$, $\cos\theta(t) \approx 1$, which is valid for small values of $\theta(t)$.

$\diamond$

Of course, we can extend the presented linearization procedure to an n-order nonlinear dynamic system with one input and one output in a straightforward way. However, for multi-input multi-output systems this procedure becomes cumbersome. Using the state space model, the linearization procedure for the multi-input multi-output case is quite simple.

Consider now the general nonlinear dynamic control system in matrix form represented by

$$\frac{d}{dt}\mathbf{x}(t) = \mathcal{F}(\mathbf{x}(t), \mathbf{u}(t)) \tag{1.66}$$

where $\mathbf{x}(t)$, $\mathbf{u}(t)$, and $\mathcal{F}$ are, respectively, the n-dimensional state space vector, the r-dimensional control vector, and the n-dimensional vector function. Assume that the nominal (operating) system trajectory $\mathbf{x}_n(t)$ is known and that the nominal control that keeps the system on the nominal trajectory is given by $\mathbf{u}_n(t)$. Using the same logic as for the scalar case, we can assume that the actual system dynamics in the immediate proximity of the system nominal trajectories can be approximated by the first terms of the Taylor series. That is, starting with

$$\mathbf{x}(t) = \mathbf{x}_n(t) + \Delta\mathbf{x}(t), \quad \mathbf{u}(t) = \mathbf{u}_n(t) + \Delta\mathbf{u}(t) \tag{1.67}$$

and

$$\frac{d}{dt}\mathbf{x}_n(t) = \mathcal{F}(\mathbf{x}_n(t), \mathbf{u}_n(t)) \tag{1.68}$$

we expand equation (1.66) as follows

$$\frac{d}{dt}\mathbf{x}_n + \frac{d}{dt}\Delta\mathbf{x} = \mathcal{F}(\mathbf{x}_n + \Delta\mathbf{x}, \mathbf{u}_n + \Delta\mathbf{u})$$

$$= \mathcal{F}(\mathbf{x}_n, \mathbf{u}_n) + \left(\frac{\partial \mathcal{F}}{\partial \mathbf{x}}\right)_{\substack{|\mathbf{x}_n(t) \\ |\mathbf{u}_n(t)}} \Delta\mathbf{x} + \left(\frac{\partial \mathcal{F}}{\partial \mathbf{u}}\right)_{\substack{|\mathbf{x}_n(t) \\ |\mathbf{u}_n(t)}} \Delta\mathbf{u} + \text{high-order terms} \tag{1.69}$$

High-order terms contain at least quadratic quantities of $\Delta\mathbf{x}$ and $\Delta\mathbf{u}$. Since $\Delta\mathbf{x}$ and $\Delta\mathbf{u}$ are small their squares are even smaller, and hence the high-order terms can be neglected. Using (1.67) and neglecting high-order terms, an approximation is obtained

$$\frac{d}{dt}\Delta\mathbf{x}(t) = \left(\frac{\partial \mathcal{F}}{\partial \mathbf{x}}\right)_{\substack{|\mathbf{x}_n(t) \\ |\mathbf{u}_n(t)}} \Delta\mathbf{x}(t) + \left(\frac{\partial \mathcal{F}}{\partial \mathbf{u}}\right)_{\substack{|\mathbf{x}_n(t) \\ |\mathbf{u}_n(t)}} \Delta\mathbf{u}(t) \tag{1.70}$$

Partial derivatives in (1.70) represent the Jacobian matrices given by

$$
\left(\frac{\partial \mathcal{F}}{\partial \mathbf{x}}\right)_{\Big|\begin{subarray}{l}\mathbf{x}_n(t)\\ \mathbf{u}_n(t)\end{subarray}} = \mathbf{A}^{n\times n} =
\begin{bmatrix}
\frac{\partial \mathcal{F}_1}{\partial x_1} & \frac{\partial \mathcal{F}_1}{\partial x_2} & \cdots & \cdots & \frac{\partial \mathcal{F}_1}{\partial x_n} \\
\frac{\partial \mathcal{F}_2}{\partial x_1} & \cdots & \cdots & \cdots & \frac{\partial \mathcal{F}_2}{\partial x_n} \\
\cdots & \cdots & \frac{\partial \mathcal{F}_i}{\partial x_j} & \cdots & \cdots \\
\cdots & \cdots & \cdots & \cdots & \cdots \\
\frac{\partial \mathcal{F}_n}{\partial x_1} & \frac{\partial \mathcal{F}_n}{\partial x_2} & \cdots & \cdots & \frac{\partial \mathcal{F}_n}{\partial x_n}
\end{bmatrix}_{\Big|\begin{subarray}{l}\mathbf{x}_n(t)\\ \mathbf{u}_n(t)\end{subarray}}
\tag{1.71a}
$$

$$
\left(\frac{\partial \mathcal{F}}{\partial \mathbf{u}}\right)_{\Big|\begin{subarray}{l}\mathbf{x}_n(t)\\ \mathbf{u}_n(t)\end{subarray}} = \mathbf{B}^{n\times r} =
\begin{bmatrix}
\frac{\partial \mathcal{F}_1}{\partial u_1} & \frac{\partial \mathcal{F}_1}{\partial u_2} & \cdots & \cdots & \frac{\partial \mathcal{F}_1}{\partial u_r} \\
\frac{\partial \mathcal{F}_2}{\partial u_1} & \cdots & \cdots & \cdots & \frac{\partial \mathcal{F}_2}{\partial u_r} \\
\cdots & \cdots & \frac{\partial \mathcal{F}_i}{\partial u_j} & \cdots & \cdots \\
\cdots & \cdots & \cdots & \cdots & \cdots \\
\frac{\partial \mathcal{F}_n}{\partial u_1} & \frac{\partial \mathcal{F}_n}{\partial u_2} & \cdots & \cdots & \frac{\partial \mathcal{F}_n}{\partial u_r}
\end{bmatrix}_{\Big|\begin{subarray}{l}\mathbf{x}_n(t)\\ \mathbf{u}_n(t)\end{subarray}}
\tag{1.71b}
$$

Note that the Jacobian matrices have to be evaluated at the nominal points, i.e. at $\mathbf{x}_n(t)$ and $\mathbf{u}_n(t)$. With this notation, the linearized system (1.70) has the form

$$
\frac{d}{dt}\Delta\mathbf{x}(t) = \mathbf{A}\Delta\mathbf{x}(t) + \mathbf{B}\Delta\mathbf{u}(t), \quad \Delta\mathbf{x}(t_0) = \mathbf{x}(t_0) - \mathbf{x}_n(t_0)
\tag{1.72}
$$

The output of a nonlinear system, in general, satisfies a nonlinear algebraic equation, that is

$$
\mathbf{y}(t) = \mathcal{G}(\mathbf{x}(t), \mathbf{u}(t))
\tag{1.73}
$$

This equation can be also linearized by expanding its right-hand side into a Taylor series about nominal points $\mathbf{x}_n(t)$ and $\mathbf{u}_n(t)$. This leads to

$$
\mathbf{y}_n + \Delta\mathbf{y} = \mathcal{G}(\mathbf{x}_n, \mathbf{u}_n) + \left(\frac{\partial \mathcal{G}}{\partial \mathbf{x}}\right)_{\Big|\begin{subarray}{l}\mathbf{x}_n(t)\\ \mathbf{u}_n(t)\end{subarray}}\Delta\mathbf{x} + \left(\frac{\partial \mathcal{G}}{\partial \mathbf{u}}\right)_{\Big|\begin{subarray}{l}\mathbf{x}_n(t)\\ \mathbf{u}_n(t)\end{subarray}}\Delta\mathbf{u}
$$
$$
+ \text{ high–order terms}
\tag{1.74}
$$

Note that $\mathbf{y}_n$ cancels term $\mathcal{G}(\mathbf{x}_n, \mathbf{y}_n)$. By neglecting high-order terms in (1.74), the linearized part of the output equation is given by

$$
\Delta\mathbf{y}(t) = \mathbf{C}\Delta\mathbf{x}(t) + \mathbf{D}\Delta\mathbf{u}(t)
\tag{1.75}
$$

where the Jacobian matrices $\mathbf{C}$ and $\mathbf{D}$ satisfy

$$\mathbf{C}^{p \times n} = \left(\frac{\partial \mathcal{G}}{\partial \mathbf{x}}\right)_{\substack{\mathbf{x}_n(t) \\ \mathbf{u}_n(t)}} = \begin{bmatrix} \frac{\partial \mathcal{G}_1}{\partial x_1} & \frac{\partial \mathcal{G}_1}{\partial x_2} & \cdots & \cdots & \frac{\partial \mathcal{G}_1}{\partial x_n} \\ \frac{\partial \mathcal{G}_2}{\partial x_1} & \cdots & \cdots & \cdots & \frac{\partial \mathcal{G}_2}{\partial x_n} \\ \cdots & \cdots & \frac{\partial \mathcal{G}_i}{\partial x_j} & \cdots & \cdots \\ \cdots & \cdots & \cdots & \cdots & \cdots \\ \frac{\partial \mathcal{G}_p}{\partial x_1} & \frac{\partial \mathcal{G}_p}{\partial x_2} & \cdots & \cdots & \frac{\partial \mathcal{G}_p}{\partial x_n} \end{bmatrix}_{\substack{\mathbf{x}_n(t) \\ \mathbf{u}_n(t)}} \tag{1.76a}$$

$$\mathbf{D}^{p \times r} = \left(\frac{\partial \mathcal{G}}{\partial \mathbf{u}}\right)_{\substack{\mathbf{x}_n(t) \\ \mathbf{u}_n(t)}} = \begin{bmatrix} \frac{\partial \mathcal{G}_1}{\partial u_1} & \frac{\partial \mathcal{G}_1}{\partial u_2} & \cdots & \cdots & \frac{\partial \mathcal{G}_1}{\partial u_r} \\ \frac{\partial \mathcal{G}_2}{\partial u_1} & \cdots & \cdots & \cdots & \frac{\partial \mathcal{G}_2}{\partial u_r} \\ \cdots & \cdots & \frac{\partial \mathcal{G}_i}{\partial u_j} & \cdots & \cdots \\ \cdots & \cdots & \cdots & \cdots & \cdots \\ \frac{\partial \mathcal{G}_p}{\partial u_1} & \frac{\partial \mathcal{G}_p}{\partial u_2} & \cdots & \cdots & \frac{\partial \mathcal{G}_p}{\partial u_r} \end{bmatrix}_{\substack{\mathbf{x}_n(t) \\ \mathbf{u}_n(t)}} \tag{1.76b}$$

Example 1.2: Let a nonlinear system be represented by

$$\frac{dx_1}{dt} = x_1 \sin x_2 + x_2 u$$

$$\frac{dx_2}{dt} = x_1 e^{-x_2} + u^2$$

$$y = 2x_1 x_2 + x_2^2$$

Assume that the values for the system nominal trajectories and control are known and given by x_{1n}, x_{2n}, and u_n. The linearized state space equation of the above nonlinear system is obtained as

$$\begin{bmatrix} \Delta \dot{x}_1(t) \\ \Delta \dot{x}_2(t) \end{bmatrix} = \begin{bmatrix} \sin x_{2n} & x_{1n} \cos x_{2n} + u_n \\ e^{-x_{2n}} & -x_{1n} e^{-x_{2n}} \end{bmatrix} \begin{bmatrix} \Delta x_1(t) \\ \Delta x_2(t) \end{bmatrix} + \begin{bmatrix} x_{2n} \\ 2u_n \end{bmatrix} \Delta u(t)$$

$$\Delta y(t) = \begin{bmatrix} 2x_{2n} & 2x_{1n} + 2x_{2n} \end{bmatrix} \begin{bmatrix} \Delta x_1(t) \\ \Delta x_2(t) \end{bmatrix} + 0 \Delta u(t)$$

Having obtained the solution of this linearized system under the given control input $\Delta u(t)$, the corresponding approximation of the nonlinear system trajectories is

$$\mathbf{x}_n(t) + \Delta \mathbf{x}(t) = \begin{bmatrix} x_{1n}(t) \\ x_{2n}(t) \end{bmatrix} + \begin{bmatrix} \Delta x_1(t) \\ \Delta x_2(t) \end{bmatrix}$$

$\diamond$

Example 1.3: Consider the mathematical model of a single-link robotic manipulator with a flexible joint (Spong and Vidyasagar, 1989)

$$I\ddot{\theta}_1 + mgl \sin\theta_1 + k(\theta_1 - \theta_2) = 0$$

$$J\ddot{\theta}_2 - k(\theta_1 - \theta_2) = u$$

where θ_1, θ_2 are angular positions, I, J are moments of inertia, m and l are, respectively, the link's mass and length, and k is the link's spring constant. Introducing the change of variables as

$$x_1 = \theta_1, \ x_2 = \dot{\theta}_1, \ x_3 = \theta_2, \ x_4 = \dot{\theta}_2$$

the manipulator's state space nonlinear model equivalent to (1.66) is given by

$$\dot{x}_1 = x_2$$

$$\dot{x}_2 = -\frac{mgl}{I}\sin x_1 - \frac{k}{I}(x_1 - x_3)$$

$$\dot{x}_3 = x_4$$

$$\dot{x}_4 = \frac{k}{J}(x_1 - x_3) + \frac{1}{J}u$$

Take the nominal points as $(x_{1n}, x_{2n}, x_{3n}, x_{4n}, u_n)$, then the matrices $\mathbf{A}$ and $\mathbf{B}$ defined in (1.71) are given by

$$\mathbf{A} = \begin{bmatrix} 0 & 1 & 0 & 0 \\ -\frac{k+mgl\cos x_{1n}}{I} & 0 & \frac{k}{I} & 0 \\ 0 & 0 & 0 & 1 \\ \frac{k}{J} & 0 & -\frac{k}{J} & 0 \end{bmatrix}, \quad \mathbf{B} = \begin{bmatrix} 0 \\ 0 \\ 0 \\ \frac{1}{J} \end{bmatrix}$$

In Spong (1995) the following numerical values are used for system parameters: $mgl = 5, \ I = J = 1, \ k = 0.08$.

Assuming that the output variable is equal to the link's angular position, that is

$$y = x_1$$

the matrices $\mathbf{C}$ and $\mathbf{D}$, defined in (1.76), are given by

$$\mathbf{C} = \begin{bmatrix} 1 & 0 & 0 & 0 \end{bmatrix}, \quad \mathbf{D} = 0$$

$\diamond$

In the next example, we give state space matrices for two linearized models of an F-15 aircraft obtained by linearizing nonlinear equations for two sets of operating points.

Example 1.4: F-15 Aircraft

The longitudinal dynamics of an F-15 aircraft can be represented by a fourth-order mathematical model. For two operating conditions (subsonic and supersonic) two linear mathematical models have been derived (Brumbaugh, 1994; Schomig *et al.*, 1995). The corresponding state space models are given by

$$
\begin{bmatrix} \dot{x}_1 \\ \dot{x}_2 \\ \dot{x}_3 \\ \dot{x}_4 \end{bmatrix} = \begin{bmatrix} -0.00819 & -25.70839 & 0 & -32.17095 \\ -0.00019 & -1.27626 & 1.0000 & 0 \\ 0.00069 & 1.02176 & -2.40523 & 0 \\ 0 & 0 & 1.0000 & 0 \end{bmatrix} \begin{bmatrix} x_1 \\ x_2 \\ x_3 \\ x_4 \end{bmatrix}
$$

$$
+ \begin{bmatrix} -6.80939 \\ -0.14968 \\ -14.06111 \\ 0 \end{bmatrix} u, \qquad y = x
$$

for subsonic flight conditions, and

$$
\begin{bmatrix} \dot{x}_1 \\ \dot{x}_2 \\ \dot{x}_3 \\ \dot{x}_4 \end{bmatrix} = \begin{bmatrix} -0.01172 & -95.91071 & 0 & -32.11294 \\ -0.00011 & -1.87942 & 1.0000 & 0 \\ 0.00056 & -3.61627 & -3.44478 & 0 \\ 0 & 0 & 1.0000 & 0 \end{bmatrix} \begin{bmatrix} x_1 \\ x_2 \\ x_3 \\ x_4 \end{bmatrix}
$$

$$
+ \begin{bmatrix} -25.40405 \\ -0.22042 \\ -53.42460 \\ 0 \end{bmatrix} u, \qquad y = x
$$

for supersonic flight conditions.

Model derivations are beyond the scope of this book. The state space variables represent: $x_1(t)$—velocity in feet per second, $x_2(t)$—angle of attack in radians, $x_3(t)$—pitch rate in radians per second, and $x_4(t)$—pitch attitude in radians. The control input $u(t)$ represents the elevator control in radians.

◇

Linearization of an Inverted Pendulum

Sometimes is not necessary to go through the entire linearization procedure. It is possible to simplify and linearize mathematical equations describing a given dynamic system by using simple mathematics. This will be demonstrated on an example of the inverted pendulum considered in Section 1.5. A linearized model of the inverted pendulum can be obtained from equations (1.43) by assuming that in the normal operating position (pendulum in an upright position) the pendulum displacement θ is very small so that the following approximations are valid $\sin\theta(t) \approx \theta(t)$, $\cos\theta(t) \approx 1$, $\theta(t)(d\theta(t)/dt)^2 \approx 0$, $\forall t$. Then, from (1.43), the linearized model of the inverted pendulum is obtained as

$$(m_1 + m_2)\frac{d^2x}{dt^2} + m_1 l\frac{d^2\theta}{dt^2} = F$$
$$\frac{d^2x}{dt^2} + l\frac{d^2\theta}{dt^2} = g\theta \tag{1.77}$$

This model can easily be put in the state space form equivalent to (1.14) by introducing the following change of variables

$$x_1 = x \Rightarrow \dot{x}_1 = x_2$$
$$x_2 = \frac{dx}{dt} = \dot{x}$$
$$x_3 = \theta \Rightarrow \dot{x}_3 = x_4 \tag{1.78}$$
$$x_4 = \frac{d\theta}{dt}$$
$$u = F$$

With this change of variables equations (1.77) imply

$$\dot{x}_2 = -\frac{m_1 g}{m_2}x_3 + \frac{1}{m_2}u$$
$$\dot{x}_4 = \frac{(m_1 + m_2)g}{m_2 l}x_3 - \frac{1}{m_2 l}u \tag{1.79}$$

From (1.78) and (1.79) the state space form of the inverted pendulum is given by

$$\begin{bmatrix} \dot{x}_1 \\ \dot{x}_2 \\ \dot{x}_3 \\ \dot{x}_4 \end{bmatrix} = \begin{bmatrix} 0 & 1 & 0 & 0 \\ 0 & 0 & -\frac{m_1 g}{m_2} & 0 \\ 0 & 0 & 0 & 1 \\ 0 & 0 & \frac{(m_1 + m_2)g}{m_2 l} & 0 \end{bmatrix} \begin{bmatrix} x_1 \\ x_2 \\ x_3 \\ x_4 \end{bmatrix} + \begin{bmatrix} 0 \\ \frac{1}{m_2} \\ 0 \\ -\frac{1}{m_2 l} \end{bmatrix} u \tag{1.80}$$

The output equation can be chosen such that it gives information about the cart's horizontal displacement x_1 and the link's angular position x_3. In that case a possible choice for the output equation is

$$y = \begin{bmatrix} 1 & 0 & 1 & 0 \end{bmatrix} \mathbf{x} \tag{1.81}$$

The same state variables will appear directly on the output if the following output equations are used

$$\mathbf{y} = \begin{bmatrix} 1 & 0 & 0 & 0 \\ 0 & 0 & 1 & 0 \end{bmatrix} \mathbf{x} \tag{1.82}$$

$$\mathbf{y} = \begin{bmatrix} 0 & 0 & 1 & 0 \\ 1 & 0 & 0 & 0 \end{bmatrix} \mathbf{x} \tag{1.83}$$

The main difference between (1.81) and (1.82)–(1.83) is that in output equations (1.82) and (1.83) we have two channels, each of which independently produces information about the particular state variable.

Finally, we would like to point out that the SIMULINK package is very convenient for simulation of nonlinear systems. It can also be used to obtain linearized models of nonlinear systems around given operating points (nominal system trajectories and controls).

1.7 MATLAB Computer Analysis and Design

MATLAB is a very advanced and reliable computer package, which can be used for computer-aided control system analysis and design. In addition to handling standard linear algebra problems, it has several specialized control theory and application toolboxes. One of them, the CONTROL toolbox, will be extensively used in this book. At some places in the book we also refer to the SIMULINK package, which is very convenient for simulation (finding system responses due to given inputs) of linear and nonlinear systems. MATLAB is user friendly. It takes only a few hours to master all of its functions. MATLAB will help students obtain a deeper understanding of the main control theory concepts and techniques, and to study high-order real physical control systems, which would be impossible using only pencil and paper. Many MATLAB problems, laboratory experiments, and case studies will be encountered in this book. More about MATLAB and its CONTROL toolbox can be found in Appendix D.

1.8 Book Organization

This book has two parts: Part I, titled *Methods and Concepts*, Chapters 2–5, and Part II, titled *Analysis and Design*, Chapters 6–9. In the first part of the book, in Chapters 2 and 3, two main techniques of control theory—the transfer function approach and the state space method—are discussed in detail. In Chapter 4 we consider the concepts of system controllability and observability, and in Chapter 5 the stability concept of time-invariant continuous and discrete systems is presented.

In this introductory chapter we have defined the general control problem. The main control system characteristics and control objectives will be presented in the second part of this book starting with Chapter 6. In Chapter 6, the control system specifications relating to a system's transient behavior and steady state properties will be considered. Since the emphasis in this book is on the time domain controller design (based on the root locus technique), the corresponding technique is presented in detail in Chapter 7. Design of controllers that solve specific control problems will be presented in Chapters 8 and 9. In Chapter 10 we give an overview of modern control theory, which can serve as an introduction for further studies of control theory and its applications. The presentation of Chapter 10 to undergraduate students can either be completely omitted or even expanded at schools that have strong programs in control systems.

1.9 References

Brumbaugh, R., "An aircraft model for the AIAA controls design challenge," *Journal of Guidance, Control, and Dynamics*, vol. 17, 747–752, 1995.

Dorf, R., *Modern Control Systems*, Addison Wesley, Reading, Massachusetts, 1992.

Franklin, G., J. Powel, and M. Workman, *Digital Control of Dynamic Systems*, Addison Wesley, Reading, Massachusetts, 1990.

Greenwood, D., *Principles of Dynamics*, Prentice Hall, Englewood Cliffs, New Jersey, 1988.

Kamen, E., *Introduction to Signals and Systems*, Macmillan, New York, 1990.

Kecman, V., *State Space Models of Lumped and Distributed Systems*, Springer-Verlag, Berlin, 1988.

Khalil, H., *Nonlinear Systems*, Macmillan, New York, 1992.

Kuo, B., *Digital Control Systems*, Saunders College Publishing, New York, 1992.

Kwakernaak, H. and R. Sivan, *Linear Optimal Control Systems*, Wiley, New York, 1972.

Lewis, F., *Applied Optimal Control and Estimation: Digital Design and Implementation*, Prentice Hall, Englewood Cliffs, New Jersey, 1992.

Ljung, L., *System Identification: Theory for the User*, Prentice Hall, Englewood Cliffs, New Jersey, 1987.

Ogata, K., *Discrete-Time Control Systems,* Prentice Hall, Englewood Cliffs, New Jersey, 1987.

Phillips, C. and H. Nagle, *Digital Control System Analysis and Design*, Prentice Hall, Englewood Cliffs, New Jersey, 1995.

Schomig, E., M. Sznaier, and U. Ly, "Mixed H_2/H_∞ control of multimodel plants," *Journal of Guidance, Control, and Dynamics*, vol. 18, 525–531, 1995.

Serway, R., *Physics for Scientists and Engineers*, H.B. Jovanovich College Publishing, Orlando, Florida, 1992.

Sontag, E., *Mathematical Control Theory*, Springer-Verlag, New York, 1990.

Spong, M., "Adaptive control of flexible joint manipulators: comments on two papers," *Automatica*, vol. 31, 585–590, 1995.

Spong, M. and M. Vidyasagar, *Robot Dynamics and Control*, Wiley, New York, 1989.

Thomas, R. and A. Rosa, *Analysis and Design of Linear Circuits*, Prentice Hall, Englewood Cliffs, New Jersey, 1994.

Wellstead, P., *Physical System Modelling*, Academic Press, London, 1979.

1.10 Problems

1.1 Find the state space form for the electrical circuit whose mathematical model is given in (1.20)–(1.22) by taking for the state space variables the input current and output voltage, i.e. by choosing $x_1 = i_1, x_2 = e_o$. In addition, take $y = x_2$ and $u = e_i$.

1.2 Find a mathematical model for the inverted pendulum, assuming that its mass is concentrated at the center of gravity, and linearize the obtained nonlinear system at $\dot{\theta}_n(t) = \theta_n(t) = \dot{x}_n(t) = x_n(t) = 0$ (Kwakernaak and Sivan, 1972; Lewis, 1992). Compare the linear model obtained with model (1.80) derived under the assumption that the pendulum mass is concentrated at its end.

1.3 Verify the expressions given in (1.38) for the transfer function of a two-input two-output translational mechanical system.

1.4 Find the mathematical model, transfer function, and state space form of the electrical network given in Figure 1.8.

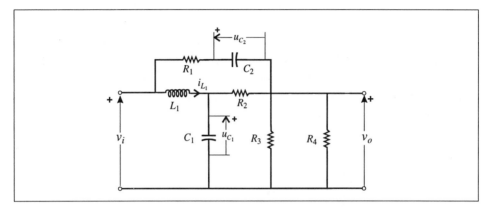

Figure 1.8: An electrical network

1.5 Linearize a scalar system represented by the first-order differential equation

$$\frac{dx(t)}{dt} = x(t)u(t)e^{-u(t)}, \qquad x(0) = 0.9$$

at a nominal point given by $(x_n(t), u_n(t)) = (1, 0)$.

1.6 Consider a nonlinear continuous-time system given by

$$\frac{d^2x(t)}{dt^2} = -2\frac{dx(t)}{dt}\cos u(t) - (1 + u(t))x(t) + 1, \quad x(0) = 1.1, \quad \frac{dx(0)}{dt} = 0.1$$

Derive its linearized equation with respect to a nominal point defined by $(x_n(t), u_n(t)) = (1, 0)$. Find the linearized system response due to $\Delta u(t) = e^{-2t}$.

1.7 For a nonlinear system

$$\frac{d^2x(t)}{dt^2} + 2\frac{dx(t)}{dt}u(t) + (1 - u(t))x(t) = u^2(t) + 1, \quad x(0) = 0, \quad \frac{dx(0)}{dt} = 1$$

find the nominal system response on the nominal system trajectory defined by $u_n(t) = 1$, subject to $x_n(0) = 0$ and $dx_n(0)/dt = 1.1$. Find the linearized state space equation and its initial conditions.

1.8 The mathematical model of a simple pendulum is given by (see for example Kamen, 1990)

$$I\frac{d^2\theta}{dt^2} + mgl\sin\theta = lu(t), \quad \theta(t_0) = \theta_0, \quad \dot{\theta}(t_0) = \omega_0$$

where I is the moment of inertia, l, m are pendulum length and mass, respectively, and $u(t)$ is an external tangential force. Assume that $\theta_n(t) = 0$, $u_n(t) = 0, \theta_n(t_0) = 0, \dot{\theta}_n(t_0) = 0$, and $\theta_0, \dot{\theta}_0$ are small. Find the linearized equation for this pendulum by using formulas (1.64). Determine the initial conditions.

1.9 Linearize the given system at a nominal point $(x_{1n}, x_{2n}, x_{3n}) = (0, 1, 1)$

$$\dot{x}_1 = x_1 x_2 - \sin x_1$$
$$\dot{x}_2 = 1 - 3x_2 e^{-x_1}$$
$$\dot{x}_3 = x_1 x_2 x_3$$

1.10 Linearize a nonlinear control system represented by

$$\dot{x}_1 = u\ln x_1 + x_2 e^{-u}$$
$$\dot{x}_2 = x_1 \sin u - \sin x_2$$
$$y = \sin x_1$$

Assume that x_{1n}, x_{2n}, and u_n are known. Find the transfer function of the linearized model obtained.

1.11 Linearize the Volterra predator–prey mathematical model

$$\dot{x}_1 = -x_1 + x_1 x_2$$
$$\dot{x}_2 = x_2 - x_1 x_2$$

at a nominal point given by $(x_{1n}, x_{2n}) = (0, 0)$.

1.12 A linearized model of a single-link manipulator with a flexible joint is given by (Spong and Vidyasagar, 1989)

$$J_l \ddot{\theta}_l + B_l \dot{\theta}_l + k(\theta_l - \theta_m) = 0$$
$$J_m \ddot{\theta}_m + B_m \dot{\theta}_m - k(\theta_l - \theta_m) = u(t)$$

where J_l, J_m are moments of inertia, B_l, B_m are damping factors, k is the spring constant, $u(t)$ is the input torque, and $\theta_m(t), \theta_l(t)$ are angular positions. Write the state space form for this manipulator by taking the following change of variables: $x_1 = \theta_l, x_2 = \dot{\theta}_l, x_3 = \theta_m, x_4 = \dot{\theta}_m$.

Remark: Note that the SIMULINK package can be used for linearization of nonlinear systems. Students may check most of the linearization problems given in this section by using the `linmod` function of SIMULINK.

Part I

METHODS AND CONCEPTS

Chapter Two

Transfer Function Approach

In the previous chapter it has been indicated that modeling, analysis, and design of control systems can be performed in two domains, namely in the time and frequency domains. In this chapter we will consider the frequency (complex) domain technique, also known as the transfer function method. Our main goal is to present methods for finding the system transfer function. This is particularly important for systems composed of many blocks, where each block represents an internal transfer function. In Chapter 9, the frequency domain approach will be used to design controllers for linear time invariant systems.

Modern control theory has its foundation in the state space approach; classical control theory is based on the transfer function approach. The state space method is widely used in modern control theory and practice due to the extensive support from modern packages for computer-aided control system analysis and design. The state space method will be considered in detail in Chapter 3. The transfer function approach is based on the Laplace and $\mathcal{Z}$-transforms and their time derivative properties, which convert differential/difference equations into algebraic equations with complex coefficients. The algebraic equations obtained are frequency domain representations of the considered dynamic systems. The basics of the Laplace and $\mathcal{Z}$-transforms are reviewed in Appendices A and B.

In classical control theory, it is desirable to have a tool that permits analysis and design of control systems, especially when instead of knowing the internal state of the system, we just need to know the relationship between the system inputs and outputs. This can be facilitated if the system model is transformed from the time domain into the frequency domain. The transfer function—the main concept of the frequency domain technique—is considered for both continuous- and discrete-time systems in Section 2.1. This book emphasizes continuous-time systems since many discrete-time results are easily derived by analogy from the corresponding continuous-time results.

A conventional way of representing linear time invariant systems is via block diagrams. This provides a pictorial view of a control system. Block diagrams are considered in Section 2.2. Block diagram algebra is introduced in Section 2.3 as a suitable tool for obtaining transfer functions of systems whose block diagrams are known. The use of block diagram algebra to find the system transfer function is advisable for simple systems, but for complex systems it gets quite involved.

The signal flow graph technique is employed in Section 2.4 as an alternative to the block diagram system representation. Mason's gain rule, the main result of the signal flow graph technique, is an elegant way of finding transfer functions, especially for complex and high-dimensional systems. Several examples are given in order to demonstrate the power of Mason's rule.

In Section 2.5 we present specialized methods for finding transfer functions of sampled data control systems obtained by sampling continuous-time systems.

At the end of this chapter, in Section 2.6, a laboratory MATLAB experiment on the system transfer function is designed.

Chapter Study Guide and Objectives

Students not completely familiar with the Laplace and $\mathcal{Z}$-transforms should first read Appendices A and B. Instructors not interested in teaching transfer functions of sampled data control systems may skip Section 2.5 without loss of continuity. Sections 2.1–2.4 represent the core of the chapter. The main objective of this chapter is that students master a technique for finding transfer functions of any time invariant linear control system by using either the block diagram algebra or Mason's rule.

2.1 Frequency Domain Representations

Real dynamic systems operate in real continuous time so that it is natural to describe and study their dynamical behavior and evolution in continuous time. This is done by using differential equations to model them. Some artificial dynamic systems operate in discrete time so that their models are represented by difference equations. In addition, discrete-time systems can be obtained by discretizing continuous-time systems so that the obtained sampled data systems are also described by difference equations. The study of dynamic systems in both continuous and discrete time will be presented in detail in Chapter 3.

Another way of studying continuous- and discrete-time systems is the frequency domain approach. This approach is performed in the space of complex numbers: by using the Laplace and $\mathcal{Z}$-transforms, the differential/difference equations are transformed into linear algebraic equations with complex coefficients. In general, it is easier to solve linear algebraic equations than linear differential/difference equations, and hence the frequency domain approach seems to be very attractive. The frequency domain approach is often called the complex domain approach. Since all calculations have to be performed in the complex domain, and since the complex numbers $s = \sigma + j\omega$ and $z = e^{-sT}$ (where T stands for a sampling period) are also known in engineering as complex frequencies, the common name for these methods is the frequency domain methods. The importance of such a representation of a system is especially emphasized in classical control system theory.

We would like to point out that the frequency domain very often gives a better understanding of the actual control system phenomena than the time domain, but from the computational point of view the frequency domain is inferior to the time domain state space approach, especially for high-order dimensional systems.

2.1.1 System Transfer Functions

The system transfer function relates to the frequency domain system outputs and inputs. In other words, the system transfer function gives what is in between the system inputs and outputs, i.e. it indicates what kind of dynamic elements input signals have to face before they appear on the system outputs. This pictorial definition can be put in rigorous mathematical form by using the Laplace and $\mathcal{Z}$-transforms.

In the first part of this section we present the transfer functions for single-input single-output systems, and in the second part we study the general case of multi-input multi-output systems.

Definition 2.1: The transfer function of a *continuous-time single-input single-output system* is defined as the ratio of the *Laplace transform* of the system output over the *Laplace transform* of the system input, when *all initial conditions are zero*.

Definition 2.2: The transfer function of a *discrete-time single-input single-output system* is defined as the ratio of the $\mathcal{Z}$-*transform* of the system output over the $\mathcal{Z}$-*transform* of the system input, when *all initial conditions are zero*.

Note that there are several other ways to introduce the definition of the system transfer function (Franklin *et al.*, 1990; Kuo, 1995), but all of them are basically the same.

Some preliminary results on system transfer functions have been presented in Section 1.3. Since the presentation of the discrete-time transfer function parallels that for continuous time, we will mostly present the results for continuous-time transfer functions and give only the final results for discrete-time transfer functions. In Section 2.5 we will pay special attention to the transfer functions of sampled data systems.

Consider a single-input single-output control system represented by an n-order differential equation, that is

$$\frac{d^n y(t)}{dt^n} + a_{n-1}\frac{d^{n-1}y(t)}{dt^{n-1}} + \cdots + a_1\frac{dy(t)}{dt} + a_0 y(t)$$
$$= b_m\frac{d^m u(t)}{dt^m} + b_{m-1}\frac{d^{m-1}u(t)}{dt^{m-1}} + \cdots + b_1\frac{du(t)}{dt} + b_0 u(t) \tag{2.1}$$

with $n \geq m$. If the initial conditions are zero, its complex counterpart is obtained simply by a substitution of d^i/dt^i by s^i and $y(t) \to Y(s)$, $u(t) \to U(s)$ (see (a.4) in Appendix A), to give

$$\left(s^n + a_{n-1}s^{n-1} + \cdots + a_1 s + a_0\right)Y(s)$$
$$= \left(b_m s^m + b_{m-1}s^{m-1} + \cdots + b_1 s + b_0\right)U(s) \tag{2.2}$$

Hence, the transfer function of this system is

$$G(s) = \frac{Y(s)}{U(s)} = \frac{b_m s^m + b_{m-1}s^{m-1} + \cdots + b_1 s + b_0}{s^n + a_{n-1}s^{n-1} + \cdots + a_1 s + a_0} = \frac{N_m(s)}{D_n(s)} \tag{2.3}$$

Similarly, for discrete-time systems the transfer function is obtained by applying the $\mathcal{Z}$-transform to the difference equation describing system dynamics

$$y(k+n) + a_{n-1}y(k+n-1) + \cdots + a_1 y(k+1) + a_0 y(k)$$
$$= b_m u(k+m) + b_{m-1}u(k+m-1) + \cdots + b_1 u(k+1) + b_0 u(k) \tag{2.4}$$

This yields the discrete-time transfer function of the form

$$G(z) = \frac{Y(z)}{U(z)} = \frac{b_m z^m + b_{m-1} z^{m-1} + \cdots + b_1 z + b_0}{z^n + a_{n-1} z^{n-1} + \cdots + a_1 z + a_0} = \frac{N_m(z)}{D_n(z)} \qquad (2.5)$$

Polynomials N_m and D_n (with the s and z arguments dropped) have real coefficients, and for the so-called proper systems (real physical systems or causal systems) it must be satisfied that $m \leq n$. The meaning of a proper system is that the system cannot respond before an input to the system is applied (system causality, see for example, Kamen, 1990).

The polynomial in the denominator of a single-input single-output system transfer function, D_n, is called the *characteristic polynomial,* and its roots are known as the *system poles.* At any of these n roots the denominator polynomial D_n is zero, so that the overall transfer system function becomes infinite. The roots of the numerator polynomial N_m are called the *system zeros* since at these m values both the numerator polynomial and the system transfer function are zero. If the system poles and zeros are known, the transfer function $G(s)$ can be recorded in *pole-zero* form as

$$G(s) = K \frac{(s + z_1)(s + z_2) \cdots (s + z_m)}{(s + p_1)(s + p_2) \cdots (s + p_n)} \qquad (2.6)$$

or in *time constant* form

$$G(s) = K_\tau \frac{(\tau_{b_1} s + 1)(\tau_{b_2} s + 1) \cdots (\tau_{b_m} s + 1)}{(\tau_{a_1} s + 1)(\tau_{a_2} s + 1) \cdots (\tau_{a_n} s + 1)} \qquad (2.7)$$

From (2.3) and (2.6)–(2.7) we have

$$K = b_m, \qquad K_\tau = K \frac{z_1 z_2 \cdots z_m}{p_1 p_2 \cdots p_n}$$

and

$$\tau_{a_i} = \frac{1}{p_i}, \; i = 1, 2, ..., n; \qquad \tau_{b_i} = \frac{1}{z_i}, \; i = 1, 2, ..., m$$

Discrete-time transfer functions can be represented by the forms identical to (2.6) and (2.7) with the complex frequency s replaced by the complex frequency z.

Example 2.1: The discrete transfer function of the following system

$$y(k + 4) + 3y(k + 2) - y(k + 1) + 5y(k) = u(k + 1) + 2u(k)$$

is obtained by assuming that all initial conditions are equal to zero and by applying the derivative (left shift in time) property of the $\mathcal{Z}$-transform, which leads to

$$\left(z^4 + 3z^2 - z + 5\right)Y(z) = (z + 2)U(z)$$

so that

$$G(z) = \frac{Y(z)}{U(z)} = \frac{z + 2}{z^4 + 3z^2 - z + 5}$$

◇

Example 2.2: For the transfer function

$$G(s) = \frac{s^3 + 0.4s^2 - 0.95s - 0.45}{s^5 + 8.3s^4 + 23.1s^3 + 26.2s^2 + 10.4s}$$

we find the pole-zero form by using MATLAB function `tf2zp`. This produces

$$G(s) = \frac{(s - 1)(s + 0.5)(s + 0.9)}{s(s + 1)(s + 1.3)(s + 2)(s + 4)}$$

◇

Example 2.3: The transfer function for the linearized system from Example 1.1 is easily obtained by setting all initial conditions to zero and taking the Laplace transform, which leads to

$$G(s) = \frac{\Theta(s)}{U(s)} = \frac{-1}{s^2 - 1}$$

◇

It is important to point out that the system transfer function carries information about the *system impulse response*. In general, we have

$$Y(s) = G(s)U(s)$$
$$Y(z) = G(z)U(z) \tag{2.8}$$

Using impulse delta functions as inputs $(U(s) = 1, U(z) = 1)$, we get

$$Y_{impulse}(s) = G(s), \quad Y_{impulse}(z) = G(z) \tag{2.9}$$

so that, in the time domain

$$g(t) = \mathcal{L}^{-1}(Y_{impulse}(s)) = \mathcal{L}^{-1}\{G(s)\}$$
$$g(k) = \mathcal{Z}^{-1}\{Y_{impulse}(z)\} = \mathcal{Z}^{-1}\{G(z)\} \tag{2.10}$$

The impulse response is obtained simply by finding the inverse transformation (from the frequency domain to the time domain) of the corresponding system transfer function.

For multi-input multi-output (multivariable) systems the definition of the transfer function is more general since vectors and matrices are involved in the system transfer function description.

Definition 2.3: Transfer functions of *multivariable systems* relate the frequency representation of system vector inputs and system vector outputs assuming that all initial conditions are equal to zero, that is

$$\mathbf{Y}(s) = \mathbf{G}(s)\mathbf{U}(s)$$
$$\mathbf{Y}(z) = \mathbf{G}(z)\mathbf{U}(z) \tag{2.11}$$

Note that $\mathbf{G}(s)$ and $\mathbf{G}(z)$ are matrices of dimensions $p \times r$ since we have assumed that the number of inputs is r and the number of outputs is p, so that vectors $\mathbf{U}(s), \mathbf{U}(z)$ are of dimensions $r \times 1$ and vectors $\mathbf{Y}(s), \mathbf{Y}(z)$ have dimensions $p \times 1$. Due to this "matrix" nature, one has to be careful while relating inputs and outputs for multivariable systems. For single-input single-output systems, one can write

$$Y(s) = G(s)U(s) = U(s)G(s)$$
$$Y(z) = G(z)U(z) = U(z)G(z) \tag{2.12}$$

However, for multivariable systems this commutativity does not hold since in that case we are dealing with vectors and matrices.

We have seen in Section 1.3 that the transfer matrix of a system with r inputs and p outputs has the form

$$\mathbf{G}^{p \times r}(s) = \begin{bmatrix} G_{11}(s) & G_{12}(s) & \cdots & G_{1r}(s) \\ G_{21}(s) & G_{22}(s) & \cdots & G_{2r}(s) \\ \vdots & \vdots & \ddots & \vdots \\ G_{p1}(s) & G_{p2}(s) & \cdots & G_{pr}(s) \end{bmatrix}^{p \times r} \tag{2.13}$$

where the coefficients G_{ij} denote the transfer functions between the jth input and the ith output, when all other inputs are zero, that is

$$G_{ij}(s) = \frac{Y_i(s)}{U_j(s)}\Big|_{all\ inputs\ except\ jth\ are\ set\ to\ zero} \tag{2.14}$$

The following example illustrates the procedure for finding the transfer function for an inverted pendulum, which can be viewed as a multivariable system with two outputs and one input.

Example 2.4: Recall a linearized model of the inverted pendulum, given by (1.77), that is

$$(m_1 + m_2)\frac{d^2x(t)}{dt^2} + m_1 l\frac{d^2\theta(t)}{dt^2} = F(t)$$

$$\frac{d^2x(t)}{dt^2} + l\frac{d^2\theta(t)}{dt^2} = g\theta(t)$$

Let us first find the transfer function $\Theta(s)/F(s)$. The above system of two equations can be simplified by eliminating d^2x/dt^2, which produces

$$m_2 l\frac{d^2\theta(t)}{dt^2} - (m_1 + m_2)g\theta(t) = -F(t)$$

After taking the Laplace transform of this equation, the transfer function is obtained as

$$G_{11}(s) = \frac{\Theta(s)}{F(s)} = \frac{-\frac{1}{m_2 l}}{s^2 - \left(1 + \frac{m_1}{m_2}\right)\frac{g}{l}}$$

The second transfer function, $X(s)/F(s)$, is obtained by taking the Laplace transform of the second equation, that is

$$s^2 X(s) + l s^2 \Theta(s) = g\Theta(s)$$

so that

$$X(s) = -\frac{l s^2 - g}{s^2}\Theta(s) = -\frac{l s^2 - g}{s^2}G_{11}(s)F(s)$$

which implies

$$G_{21}(s) = -\frac{l s^2 - g}{s^2}G_{11}(s)$$

The same result could have been obtained by taking simultaneously the Laplace transform of both the equations

$$\begin{bmatrix} (m_1 + m_2)s^2 & m_1 l s^2 \\ s^2 & l s^2 - g \end{bmatrix} \begin{bmatrix} X(s) \\ \Omega(s) \end{bmatrix} = \begin{bmatrix} F(s) \\ 0 \end{bmatrix}$$

and then solving this system of algebraic equations with respect to $X(s)$ and $\Theta(s)$, that is

$$\begin{bmatrix} X(s) \\ \Omega(s) \end{bmatrix} = \begin{bmatrix} (m_1 + m_2)s^2 & m_1 l s^2 \\ s^2 & l s^2 - g \end{bmatrix}^{-1} \begin{bmatrix} F(s) \\ 0 \end{bmatrix} = \begin{bmatrix} G_{11}(s) \\ G_{21}(s) \end{bmatrix} F(s)$$

◇

Sometimes systems are so complex that playing with algebraic equations in the complex domain in order to obtain the system transfer function(s) is mathematically very involved. A graphical system representation in terms of either block diagrams or signal flow graphs will help us to develop systematic methods for finding the system transfer function(s).

2.2 Block Diagrams

A pictorial description is a very convenient way of representing dynamic systems. It gives a clear picture of all components of the control system and the flow of information (signals) in the system. Such a representation is called the *system block diagram*. In the following, we show how to use block diagrams in order to obtain information about input and output variables, the relationships between these variables, and how to get the transfer function(s). In some cases the block diagram is used just to represent the composition and interconnections of a system.

The simplest possible block diagram of a single-input single-output system is represented in Figure 2.1, where $U(s)$ and $Y(s)$ are, respectively, the Laplace transforms of the input and output signals, and $G(s)$ is the block transfer function. This is consistent with the transfer function definition given in (2.3). The rule that connects (2.3) and the block diagram in Figure 2.1, and which is also valid for all block diagrams, can be formulated as follows.

Main Block Diagram Rule: *The output signal is a product of the transfer function of the given block and block's input signal.*

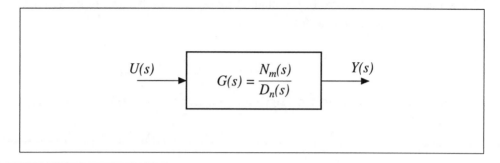

Figure 2.1: Block diagram of a general linear system defined in (2.3)

The basic structure of a single-input single-output *closed-loop system* with a non-unit feedback is presented in Figure 2.2a, and with a unit feedback is given in Figure 2.2b. Applying the main block diagram rule and taking into account the directions of the flow of signals in the system as indicated by arrows, we have

$$Y(s) = G(s)E(s)$$
$$E(s) = U(s) - H(s)Y(s)$$

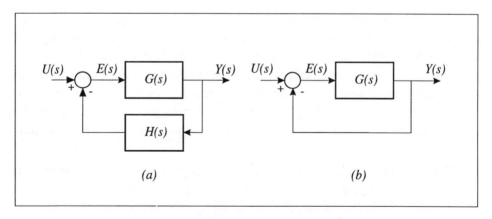

Figure 2.2: Simple feedback structures

Eliminating $E(s)$ from the last equation, it follows

$$Y(s) = G(s)U(s) - G(s)H(s)Y(s)$$

so that for a single-input single-output control system the *closed-loop transfer function*, denoted by $M(s)$, is given by

$$M(s) = \frac{Y(s)}{U(s)} = \frac{G(s)}{1 + G(s)H(s)} \qquad (2.15)$$

Note that if the system loop is open, which is the case for $H(s) = 0$, we have the *open-loop system transfer function*

$$Y(s) = G(s)E(s) = G(s)U(s) \Rightarrow \frac{Y(s)}{U(s)} = G(s) \qquad (2.16)$$

A more general block diagram of a closed-loop control system is given in Figure 2.3.

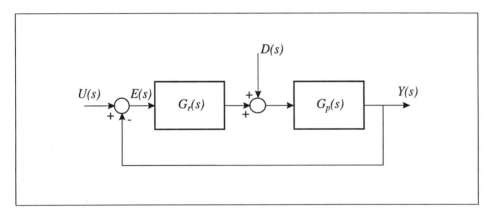

Figure 2.3: Basic structure of a control system

The two principal components (blocks) in this diagram are the *plant* and the *controller*. The plant has its own dynamics represented by the transfer function $G_p(s)$. The purpose of the controller (often called the regulator) $G_r(s)$ is to "reshape" the dynamics of the plant, such that the overall system transfer function has the desired form. In Figure 2.3, the signal $D(s)$ represents the potential system disturbance. The main roles of the controllers (regulators) are system stabilization, improvement of the system transient response, reduction of steady state errors, disturbance rejection, etc. These issues will be addressed in the subsequent chapters. Here, we study only the problem of obtaining the system transfer function from the block diagrams.

In this block diagram we have two inputs, $U(s)$ and $D(s)$, and one output, $Y(s)$. Setting $D(s) = 0$, we get

$$Y(s) = G_p(s)G_r(s)E(s)$$
$$E(s) = U(s) - Y(s)$$

that is

$$Y(s) = G_p(s)G_r(s)U(s) - G_p(s)G_r(s)Y(s)$$

so that the system closed-loop transfer function is given by

$$M(s) = \frac{Y(s)}{U(s)} = \frac{G_p(s)G_r(s)}{1 + G_p(s)G_r(s)} \tag{2.17}$$

It is interesting to find the transfer function from the system disturbance $D(s)$ to the system output $Y(s)$. By setting $U(s) = 0$, we get

$$E(s) = -Y(s)$$
$$Y(s) = G_p(s)[D(s) + G_r(s)E(s)] = G_p(s)D(s) - G_p(s)G_r(s)Y(s)$$

so that

$$\frac{Y(s)}{D(s)} = \frac{G_p(s)}{1 + G_p(s)G_r(s)} \tag{2.18}$$

Since it is not desirable for the disturbance to affect the system output, the magnitude of the corresponding transfer function

$$\left| \frac{G_p(s)}{1 + G_p(s)G_r(s)} \right| \tag{2.19}$$

should be minimized as much as possible by a proper choice of the controller $G_r(s)$.

In the case of *multivariable systems*, the closed-loop transfer function can be found using the same technique, bearing in mind that the corresponding quantities are vectors and matrices. The closed-loop block diagram of a multivariable system is given in Figure 2.4.

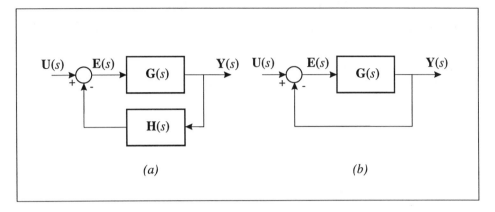

Figure 2.4: Multivariable feedback system

From this block diagram by using the main block diagram rule, *to be applied strictly in the order: output = transfer function × input*, it follows

$$\mathbf{Y}(s) = \mathbf{G}(s)\mathbf{E}(s)$$
$$\mathbf{E}(s) = \mathbf{U}(s) - \mathbf{H}(s)\mathbf{Y}(s)$$

so that

$$\mathbf{Y}(s) = \mathbf{G}(s)\mathbf{U}(s) - \mathbf{G}(s)\mathbf{H}(s)\mathbf{Y}(s)$$

which implies

$$\mathbf{Y}(s) = [\mathbf{I} + \mathbf{G}(s)\mathbf{H}(s)]^{-1}\mathbf{G}(s)\mathbf{U}(s) \qquad (2.20)$$

The *closed-loop multivariable control system transfer function* relates the frequency representations of the system vector input and vector output. From (2.20) it is given by

$$\mathbf{M}(s) = [\mathbf{I} + \mathbf{G}(s)\mathbf{H}(s)]^{-1}\mathbf{G}(s) \qquad (2.21)$$

For $\mathbf{H}(s) = 0$, we have the open-loop transfer function of a multivariable control system obtained from (2.20) as $\mathbf{G}(s)$.

In the next subsection we show how to perform modeling and construct a block diagram from mathematical equations describing system dynamics. We consider a model of a DC motor, which is frequently used in control systems.

2.2.1 Modeling and Block Diagrams of a DC Motor

A DC motor is an electromechanical energy converter which converts electrical energy into mechanical energy. It is often used as an actuator in control systems. Figure 2.5 illustrates such a motor schematically. In this section we present the modeling of a DC motor and draw two block diagrams corresponding to two different working conditions.

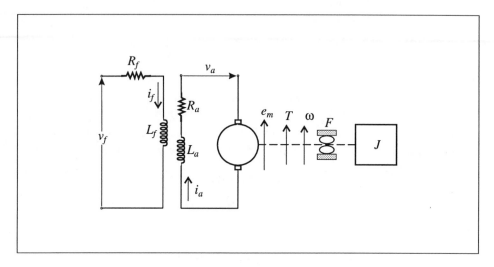

Figure 2.5: Schematic of a DC motor

The basic equations of a DC motor (electric part) are obtained from Maxwell's electromagnetic theory. The *magnetic flux* $\phi(t)$ is proportional to the field current $i_f(t)$

$$\phi(t) = k_1 i_f(t) \tag{2.22}$$

The *torque* produced by a motor is proportional to the product of the magnetic flux $\phi(t)$ and the armature current $i_a(t)$

$$T(t) = k_2 \phi(t) i_a(t) = k_1 k_2 i_f(t) i_a(t) \tag{2.23}$$

The motor's electromotive force (induced voltage), denoted by $e_m(t)$, is proportional to the product of the magnetic flux and the rotor shaft rotational speed $\omega(t)$

$$e_m(t) = k_3 \phi(t) \omega(t) = k_1 k_3 i_f(t) \omega(t) \tag{2.24}$$

The last three equations are valid if the values of $\phi(t)$, $T(t)$, and $e_m(t)$ are close to their nominal values.

On the mechanical side, the torque $T(t)$, developed by the motor, is balanced by the load and disturbance torques. $T(t)$ is also related to the rotational speed $\omega(t)$ by the differential equation

$$T(t) = T_l(t) + T_d(t) = J\frac{d\omega(t)}{dt} + F\omega(t) + T_d(t) \qquad (2.25)$$

where J is the combined load and armature mass moment, and F is the viscous friction coefficient. $T_l(t)$ represents the load torque and $T_d(t)$ is a disturbance torque (T_d is frequently negligible).

Balancing the voltages in the field and armature windings, we obtain

$$v_f(t) = L_f\frac{di_f(t)}{dt} + R_f i_f(t) \qquad (2.26)$$

$$\begin{aligned}v_a(t) &= L_a\frac{di_a(t)}{dt} + R_a i_a(t) + e_m(t) \\ &= L_a\frac{di_a(t)}{dt} + R_a i_a(t) + k_1 k_3 i_f(t)\omega(t)\end{aligned} \qquad (2.27)$$

The above set of equations is nonlinear due to the presence of the products $i_f(t)i_a(t)$ and $i_f(t)\omega(t)$. However, usually one of the currents is kept constant. For constant $i_f(t)$, we have the so-called *armature-controlled* DC motor; if $i_a(t)$ is constant, then the motor is said to be *field-controlled*. Mathematical models for these two regimes are different. They are presented below.

Armature-Controlled DC Motor

In this case $i_f(t) = I_{fo} = const$, so that $v_f(t) = R_f I_{fo}$. Balancing the voltages and torques, we obtain from (2.23), (2.25), and (2.27)

$$v_a(t) = L_a\frac{di_a(t)}{dt} + R_a i_a(t) + k_4\omega(t) \qquad (2.28)$$

$$k_5 i_a(t) = J\frac{d\omega(t)}{dt} + F\omega(t) + T_d(t) \qquad (2.29)$$

where

$$k_4 = k_1 k_3 I_{fo}, \qquad k_5 = k_1 k_2 I_{fo}$$

Quantities $\tau_a = L_a/R_a$ and $\tau_m = J/F$ are usually called the system time constants. The Laplace transform of the above system of equations produces

$$I_a(s) = \frac{1}{(L_a s + R_a)}[V_a(s) - k_4\Omega(s)] \tag{2.30}$$

$$\Omega(s) = \frac{1}{(Js + F)}[k_5 I_a(s) - T_d(s)] \tag{2.31}$$

The block diagram for this system is easily drawn by looking at equations (2.30) and (2.31) and using the *main block diagram rule*. This is shown in Figure 2.6.

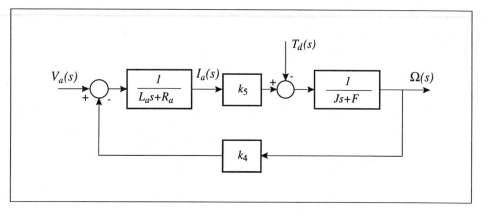

Figure 2.6: Block diagram for an armature-controlled DC motor

Field-Controlled DC Motor

In this case $i_a = I_{ao} = const$, the following differential equation is obtained from (2.23) and (2.25)

$$k_6 i_f(t) = J\frac{d\omega(t)}{dt} + F\omega(t) + T_d(t), \quad k_6 = k_1 k_2 I_{ao} \tag{2.32}$$

Taking the Laplace transforms of (2.26) and (2.32), we get

$$V_f(s) = R_f(\tau_f s + 1)I_f(s), \quad \tau_f = \frac{L_f}{R_f} \tag{2.33}$$

$$k_6 I_f(s) = F(\tau_m s + 1)\Omega(s) + T_d(s)$$

where τ_f is a time constant. Rearranging these equations in the form

$$I_f(s) = \frac{1}{R_f(\tau_f s + 1)}V_f(s)$$

$$\Omega(s) = \frac{1}{F(\tau_m s + 1)}[k_6 I_f(s) - T_d(s)] \tag{2.34}$$

and using the *main block diagram rule*, the block diagram is easily drawn and is represented in Figure 2.7.

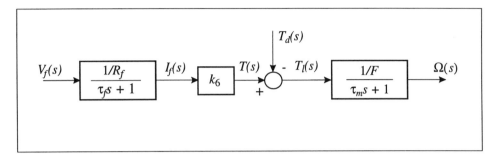

Figure 2.7: Block diagram for a field-controlled DC motor

Real control systems can have complex structures including several local feedback loops, and many inputs and outputs. Local feedback loops can arise for two reasons: they are either the result of the physical nature of the specific element in the system or the system as an entity, or they can be intentionally built in, with the aim of achieving a desirable performance for the system. No matter how complex the block diagram of a system is, it can be reduced to one of the basic structures given in Figures 2.1–2.3 by using the block diagram algebra rules. In the following we present the main results of the block diagram algebra.

2.3 Block Diagram Algebra

Block diagram algebra is a set of rules that facilitates modification and simplification of block diagrams. The rules of block diagram algebra for continuous-time systems are quite simple. They are based on simple principles of algebra that are used for writing input–output relations for the specific blocks in the block diagram.

Cascade (serial) connection: The transfer function equivalent to a serial connection of n blocks with transfer functions $G_1(s), G_2(s), ..., G_n(s)$, represented in Figure 2.8, is given by

$$G(s) = G_1(s)G_2(s)\cdots G_n(s) = \prod_{i=1}^{n} G_i(s) \qquad (2.35)$$

This rule is obtained by generalizing the main block diagram rule.

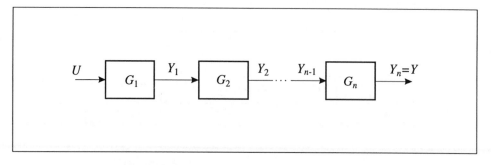

Figure 2.8 A serial connection of n blocks

Parallel (tandem) connection: The equivalent transfer function for such a connection representing a summation of signals, given in Figure 2.9, is obtained as

$$G(s) = G_1(s) + G_2(s) + \cdots + G_n(s) = \sum_{i=1}^{n} G_i(s) \qquad (2.36)$$

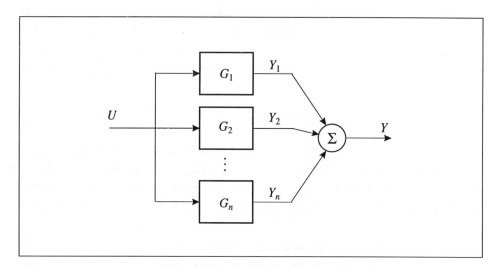

Figure 2.9: Parallel connection of n blocks

Feedback connection: The simplest form of a feedback control system is given in Figure 2.2. For such a system connection the transfer function is given by (2.15).

Example 2.5: In this example we demonstrate the procedure for obtaining the transfer functions for cascade, parallel, and feedback connections by using MATLAB. Consider the transfer functions

$$G_1(s) = \frac{5}{s(s+1)(s+2)}, \quad G_2(s) = \frac{s+4}{s+5}$$

We obtain the *cascade* (series) transfer function by the following sequence of MATLAB operators

```
% define G1(s)
z1=[inf;inf;inf];
% all three zeros of G1(s) are at infinity
p1=[0;-1;-2];
k1=5;
[n1,d1]=zp2tf(z1,p1,k1);
% zp2tf maps zero-pole transfer function into
          numerator-denominator transfer function
% define G2(s)
n2=[1  4];
d2=[1  5];
% find the series connection
[ns,ds]=series(n1,d1,n2,d2);
% print
printsys(ns,ds,'s')
```

Execution of these operators produces the following result

$$G_1(s)G_2(s) = \frac{5s + 20}{s^4 + 8s^3 + 17s^2 + 10s}$$

The transfer function for *parallel* connection is obtained by using

```
[np,dp]=parallel(n1,d1,n2,d2);
printsys(np,dp,'s')
```

This produces

$$G_1(s) + G_2(s) = \frac{s^4 + 7s^3 + 14s^2 + 13s + 25}{s^4 + 8s^3 + 17s^2 + 10s}$$

The *feedback* connection is executed by

```
[nf,df]=feedback(n1,d1,n2,d2,-1);
% -1 indicate negative feedback
printsys(nf,df,'s')
```

which leads to

$$\frac{G_1(s)}{1 + G_1(s)G_2(s)} = \frac{5s + 25}{s^4 + 8s^3 + 17s^2 + 10s}$$

◇

In addition to algebraic formulas (2.35) and (2.36), block diagram algebra is complemented by several "geometric" rules. Two of them are given below.

Moving Pick-Off Point: In some cases it is desirable to move a pick-off point in front or behind a block in a block diagram, such that the terminal signals do not change their values. Figure 2.10 shows the equivalent block diagrams for the cases before and after replacement of the pick-off points.

Moving Summing Point: An adder or subtracter may be moved from one side of a block to another as Figure 2.11 illustrates. It is easy to show that the diagrams on the left-hand and right-hand sides are equivalent. We leave the proofs to the reader.

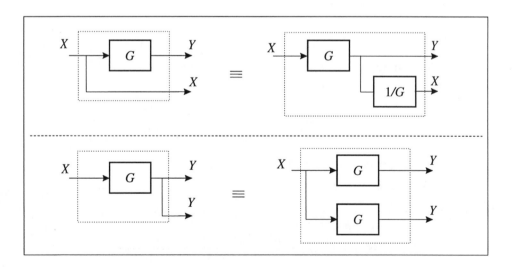

Figure 2.10: Moving pick-off point transformation

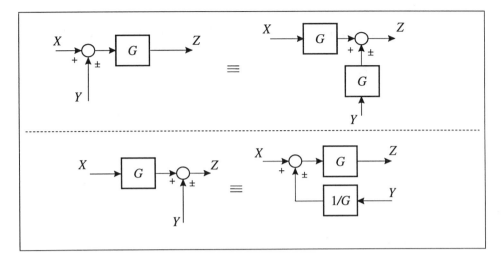

Figure 2.11: Moving adder/subtracter transformation

Using formulas (2.35) and (2.36) and moving pick-off and summing point rules, additional block diagram algebra rules can be derived. A summary of some additional useful block diagram algebra "geometric" rules is given in Table 2.1.

Next we solve two simple examples in order to demonstrate the procedure for finding the system transfer function from block diagrams.

Example 2.6: The original block diagram, presented in Figure 2.12a, is first simplified by moving the adder in front of the block G_1 (Figure 2.12b), then by interchanging adder and adder/subtracter (Figure 2.12c), and finally by finding the corresponding closed-loop transfer functions (Figure 2.12d).

⋄

Example 2.7: The block diagram from Figure 2.13a is redrawn in Figure 2.13b in order to explicitly indicate block connections and signal flows. In the next step, presented in Figure 2.13c, the closed-loop transfer function of blocks G_1 and H_2 is found and the pick-off point is moved in front of block G_2. Finally, two closed-loop transfer functions are found (Figure 2.13d) and their cascaded connection is evaluated (Figure 2.13e).

⋄

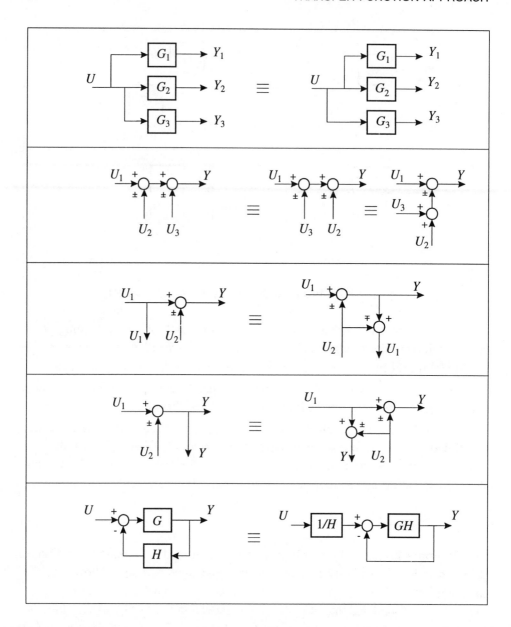

Table 2.1: Block diagram algebra rules

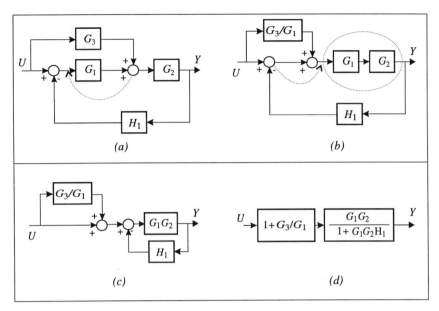

Figure 2.12: Simplification of the block diagram in Example 2.6

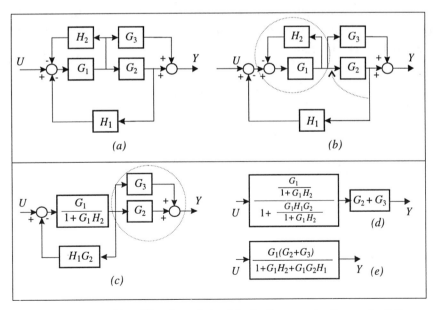

Figure 2.13: Simplification of the block diagram in Example 2.7

Using the rules for simplifying the block diagrams presented in formulas (2.35) and (2.36), Figures 2.10 and 2.11, and Table 2.1, and finding the corresponding transfer functions is relatively easy for simple systems. However, for complex systems the procedure can be quite involved since it requires drawing of many intermediate block diagrams before the final (simple feedback form) is reached. Example 2.8 demonstrates the required procedure for a complex feedback system.

Example 2.8: The reduction of a complex block diagram for a system shown in Figure 2.14a is illustrated in Figures 2.14b–i. The process of reduction is pretty much self-explanatory from the corresponding Figures 2.14b–i. The above simplification is primarily done by using the established rules, but in addition, one has to use common sense, as was done in going from Figure 2.14b to Figure 2.14c and from Figure 2.14c to Figure 2.14d (see Problems 2.12 and 2.13).

The final expression for the transfer function is given by

$$\frac{Y(s)}{U(s)} = \frac{G_1 G_2 G_4 (G_3 + G_5)}{1 + G_2(H_1 G_1 + H_2 G_3) + H_3 G_1 G_3 G_4 (G_2 + G_5) - H_1 H_2 G_1 G_2 G_3 G_5}$$

It can be seen from this particular example that for complex systems the block diagram algebra produces the required answer after many redrawings of the original block diagram.

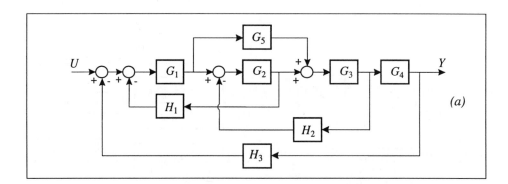

Figure 2.14a: Block diagram of the control system for Example 2.8

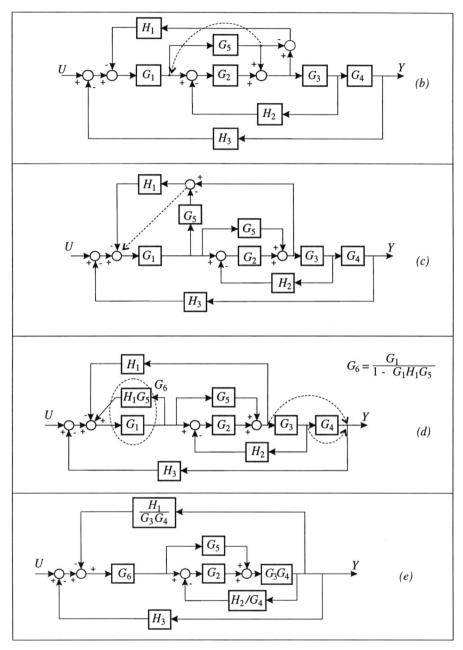

$$G_6 = \frac{G_1}{1 - G_1 H_1 G_5}$$

Figure 2.14b–e: Simplification of the block diagram from Figure 2.14a

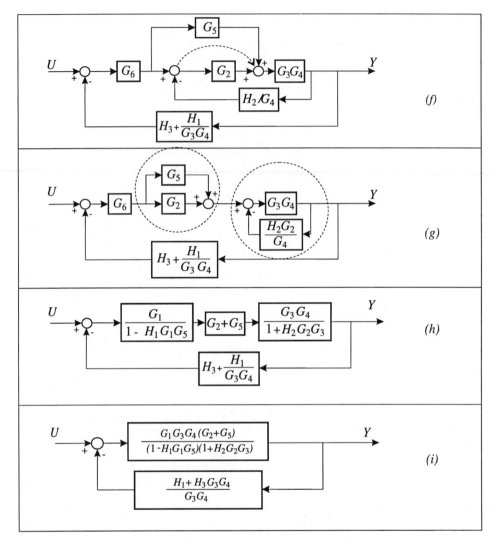

Figure 2.14f–i: Simplification of the block diagram from Figure 2.14a

◇

In the next section we present another method, based on signal flow graph theory, known as Mason's rule, which is particularly efficient for high-dimensional and complex systems.

2.4 Signal Flow Graphs and Mason's Rule

Another way of finding transfer functions of linear time invariant systems represented by their block diagrams is the so-called Mason's rule, which is based on signal flow graph theory (Mason, 1953, 1956). Signal flow graph techniques are also used in several other areas of engineering and sciences (Robichaud *et al.*, 1962; Rao and Koshy, 1991).

There is an analogy between block diagrams and signal flow graphs. The main elements of the signal flow graph technique are *nodes* and *branches*, with the nodes being connected by branches. *A branch is equivalent to a block in the block diagram and represents the transfer function between the nodes.* A branch consists of input node, output node, and an arrow showing the signal flow direction. A transfer function is associated with each branch. A branch in a signal flow graph and its transfer function counterpart are represented in Figure 2.15.

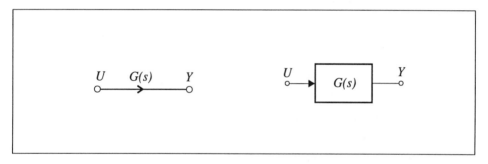

Figure 2.15: Equivalent elements in a block diagram and a signal flow graph

A node represents a signal. The basic rule for nodes is that *a signal at a node is equal to the sum of signals coming into the node from branches.* Note that signals leaving a node do not count. It is only important to pay attention to the signals coming into a node. *A signal entering a node from a branch is equal to the signal from the input node of that branch multiplied by the branch transfer function.*

In Figure 2.16 a simple feedback block diagram and its signal flow graph are given. It can be seen from this figure that the expressions for two signals at two nodes are given by

$$E = 1 \times U - HY$$
$$Y = GE$$

$$(2.37)$$

which produces our familiar closed-loop result

$$Y = GU - GHY \;\Rightarrow\; Y = \frac{G}{1 + GH}U \qquad\qquad (2.38)$$

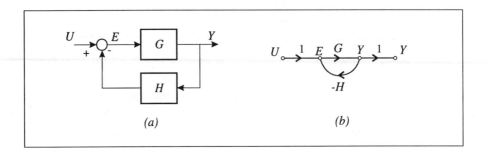

Figure 2.16: Equivalence between closed-loop systems

In order to be able to present the method for finding transfer functions by the signal flow graph technique, we need to introduce the following terminology.

Source node — a node at which signals flow *only away* from the node. Input signals are represented by such nodes.

Sink node — a node at which signals flow *only towards* the node. These nodes represent output signals. It is customary to extract the inputs and outputs out of a signal flow graph by using additional branches whose transfer functions are equal to 1 (see Figure 2.16b).

Path — a succession of branches from a source node (input) to a sink node (output) with all arrowheads in the same direction *which does not pass any node more than once. The path gain is the product of all transfer functions in the path.*

Loop — a closed path of branches with all arrowheads in the same direction in which *no node is encountered more than once.* A source node cannot be a part of a loop since each node in the loop must have at least one branch into the node and at least one branch out of it. *The loop gain is the product of transfer functions of the branches comprising the loop.*

Nontouching loops — two loops are nontouching if they have no common node.

The above notions are demonstrated on a signal flow graph presented in Figure 2.17.

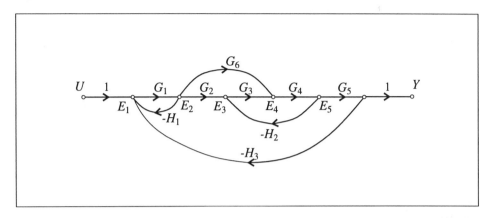

Figure 2.17: Example of a signal flow graph

In this figure the input node is U and the output node is Y. There are two paths connecting the input U and the output Y. One of them has a path gain equal to $G_1G_2G_3G_4G_5$, and the other has a path gain of $G_1G_6G_4G_5$. There are four loops in this signal flow graph with gains $-G_1H_1$, $-G_3G_4H_2$, $-G_1G_2G_3G_4G_5H_3$, and $-G_1G_6G_4G_5H_3$. Two loops, $-G_1H_1$ and $-G_3G_4H_2$, do not touch each other, i.e. they represent nontouching loops.

Note that the signal flow graph contains fewer elements than the corresponding block diagram. Signal flow graphs can be simplified by employing similar rules to those we have been using in block diagram algebra in order to determine system transfer functions. However, we have seen from the examples in the previous section that the simplification procedure based on block diagram algebra is quite lengthy.

In Mason (1953, 1956) an elegant and powerful formula for finding the transfer function between input and output nodes was derived. That formula is known as Mason's gain formula and is given by

$$G(s) = \frac{1}{\Delta} \sum_{k=1}^{N} P_k \Delta_k = \frac{1}{\Delta}(P_1\Delta_1 + P_2\Delta_2 + \cdots + P_N\Delta_N) \qquad (2.39)$$

where P_k is the path gain for the kth path, N stands for the number of paths, Δ is the *determinant* of the signal flow graph, and Δ_k is the cofactor of path k.

Δ and Δ_k are computed as follows:

$\Delta = 1 - $ (sum of all loop gains) $ + $ (sum of products of the loop gains of all possible combinations of nontouching loops taken two at a time) $ - $ (sum of products of the loop gains of all possible combinations of nontouching loops taken three at a time) $ + \cdots$.

$\Delta_k = $ value of Δ for the part of the flow graph not touching the kth forward path.

The application of formula (2.39) is demonstrated on the signal flow graph presented in Figure 2.17. We have already found that there are two paths with the corresponding gains

$$P_1 = G_1 G_2 G_3 G_4 G_5, \qquad P_2 = G_1 G_4 G_5 G_6$$

and four loops whose loop gains are

$$L_1 = -G_1 H_1, \quad L_2 = -G_3 G_4 H_2$$
$$L_3 = -G_1 G_2 G_3 G_4 G_5 H_3, \quad L_4 = -G_1 G_4 G_5 G_6 H_3$$

There are also two nontouching loops with gains L_1 and L_2. Then

$$\Delta = 1 - (L_1 + L_2 + L_3 + L_4) + L_1 L_2$$

Apparently, if we eliminate the path $G_1 G_2 G_3 G_4 G_5$, the remaining signal flow graph will have no loops left, so that $\Delta_1 = 1$. The same conclusion is obtained if we eliminate the path $G_1 G_6 G_4 G_5$ leading to $\Delta_2 = 1$. Thus, the transfer function of the considered signal flow graph, according to formula (2.39), is given by

$$\frac{Y(s)}{U(s)} = G(s) = \frac{P_1 \times 1 + P_2 \times 1}{1 - (L_1 + L_2 + L_3 + L_4) + L_1 L_2} =$$

$$\frac{G_1 G_2 G_3 G_4 G_5 + G_1 G_4 G_5 G_6}{1 + G_1 H_1 + G_3 G_4 H_2 + G_1 G_2 G_3 G_4 G_5 H_3 + G_1 G_4 G_5 G_6 H_3 - G_1 G_3 G_4 H_1 H_2}$$

In the above expression $G_i, i = 1, ..., 6$, and $H_j, j = 1, 2, 3$, are either constants or functions of the complex frequency s.

Note that all signals in a signal flow graph are mutually related by linear algebraic equations. For example, in the case of the signal flow graph given in

Figure 2.17 we have

$$E_1 = U - H_1 E_2 - H_3 Y$$
$$E_2 = G_1 E_1$$
$$E_3 = G_2 E_2 - H_2 E_5$$
$$E_4 = G_6 E_2 + G_3 E_3$$
$$E_5 = G_4 E_4$$
$$Y = G_5 E_5$$

where $G_i, i = 1, ..., 6$, and $H_j, j = 1, 2, 3$, are coefficients. By playing simple algebra with the above system of linear equations one is able to obtain the required relationship between Y and U, i.e. the required transfer function. However, that approach is not systematic. Mason's formula (2.39) is derived by using Kramer's determinant method for solving systems of linear algebraic equations, which are obtained by relating signals in a signal flow graph. The complete proof of formula (2.39) is beyond the scope of this textbook; it can be found in Mason (1956) and Younger (1963).

Next, we give an example to find the system transfer function by using both block diagram algebra and Mason's rule.

Example 2.9: Consider the block diagram given in Figure 2.18.

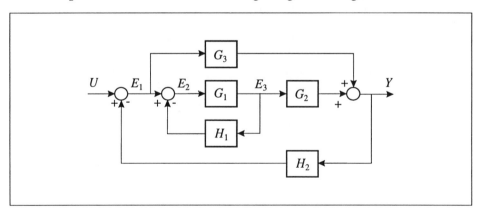

Figure 2.18: Block diagram of a feedback control system

The block transfer functions are given by

$$G_1(s) = \frac{5}{s(s + 1)}, \quad G_2(s) = \frac{2}{s}, \quad G_3(s) = 2$$

$$H_1(s) = \frac{s}{s+4}, \quad H_2(s) = \frac{5s}{s+2}$$

(a) Using block diagram algebra rules, this block diagram is simplified as shown in Figure 2.19. The required transfer function is given in Figure 2.19d.

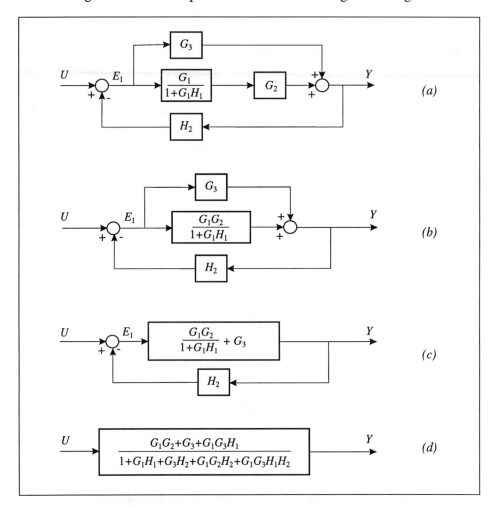

Figure 2.19: Block diagram simplification using block diagram algebra rules

(b) The signal flow graph for this example is represented in Figure 2.20. It contains two paths and three loops. In addition the loops G_1H_1 and G_3H_2 are nontouching ones. Expressions for path and loop gains are

$$P_1 = G_1 G_2, \quad P_2 = G_3, \quad L_1 = -G_1 H_1, \quad L_2 = -G_3 H_2, \quad L_3 = -G_1 G_2 H_2$$

The graph's determinant and path's cofactors are obtained as

$$\Delta = 1 - (L_1 + L_2 + L_3) + L_1 L_2, \quad \Delta_1 = 1, \quad \Delta_2 = 1 - L_1$$

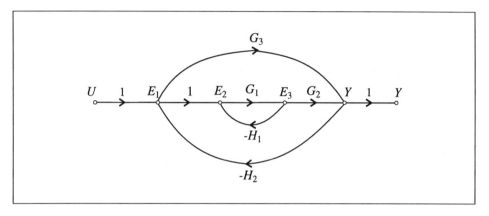

Figure 2.20: Signal flow graph for the system given in Figure 2.18

The required closed-loop transfer function, according to Mason's formula (2.39), is given by

$$\frac{Y(s)}{U(s)} = \frac{P_1 + P_2(1 - L_1)}{1 - (L_1 + L_2 + L_3) + L_1 L_2}$$

$$= \frac{G_1 G_2 + G_3 + G_1 G_3 H_1}{1 + G_1 H_1 + G_3 H_2 + G_1 G_2 H_2 + G_1 G_3 H_1 H_2}$$

After substitution of the given values for $G_i(s), i = 1, 2, 3,$ and $H_j(s), j = 1, 2,$ the transfer function is obtained as a ratio of two polynomials with respect to the complex frequency s. The final result for the transfer function is given in Part (c) of this example.

(c) The same problem can be solved by MATLAB using its functions for feedback, parallel, and series connections. Note that the procedure given below cannot be applied to any signal flow graph. It can be applied only to those with explicitly distinguished feedback loops, series, and parallel connections. However, using the SIMULINK package one is able to obtain the transfer function for any block diagram and the corresponding signal flow graph. The transfer function of the feedback control system given in Figure 2.18 is found by using the following sequence of MATLAB instructions

```
% feedback configuration of G1 and H1
[n,d]=feedback([0 0 5],[1 1 0],[1 0],[1 4],-1);
% cascade connection to G2
[n,d]=series(n,d,[0 2],[1 0]);
% parallel connection to G3
[n,d]=parallel(n,d,[2],[1]);
% feedback connection with H2
[n,d]=feedback(n,d,[5 0],[1 2],-1);
printsys(n,d,'s')
```

This MATLAB program produces the following result

$$\frac{Y(s)}{U(s)} = \frac{2s^5 + 14s^4 + 38s^3 + 46s^2 + 60s + 80}{11s^5 + 57s^4 + 109s^3 + 68s^2 + 200s}$$

◇

Example 2.10: Consider the block diagram given in Figure 2.21.

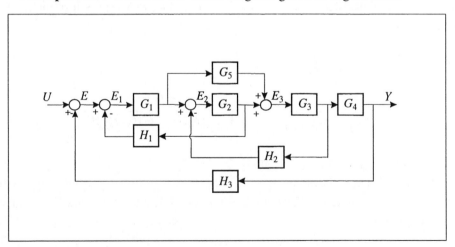

Figure 2.21: Block diagram of a feedback control system

The corresponding signal flow graph is presented in Figure 2.22. In this example we have two paths and five loops. There are no nontouching loops.

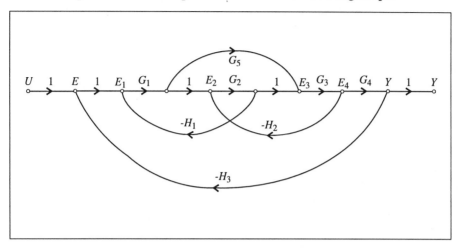

Figure 2.22: Signal flow graph for the system given in Figure 2.21

The corresponding path gains $P_i, i = 1, 2$, loop gains $L_j, j = 1, 2, ..., 5$, signal flow graph determinant, and cofactors are given by

$$P_1 = G_1 G_2 G_3 G_4, \quad P_2 = G_1 G_5 G_3 G_4$$

$$L_1 = -G_1 G_2 H_1, \quad L_2 = -G_2 G_3 H_2, \quad L_3 = G_1 G_5 G_3 H_2 G_2 H_1$$

$$L_4 = -G_1 G_5 G_3 G_4 H_3, \quad L_5 = -G_1 G_2 G_3 G_4 H_3$$

$$\Delta_1 = 1, \quad \Delta_2 = 1$$

$$\Delta = 1 + G_1 G_2 H_1 + G_2 G_3 H_2 + G_1 G_2 G_3 G_4 H_3 + G_1 G_5 G_4 H_3 - G_1 G_5 G_3 H_2 G_2 H_1$$

so that the closed-loop transfer function for this system is obtained as

$$\frac{Y(s)}{U(s)} =$$

$$\frac{G_1 G_2 G_3 G_4 + G_1 G_3 G_4 G_5}{1 + G_1 G_2 H_1 + G_2 G_3 H_2 + G_1 G_2 G_3 G_4 H_3 + G_1 G_3 G_4 G_5 H_3 - G_1 G_2 G_3 G_5 H_1 H_2}$$

$\diamond$

Note that the same problem is studied in Example 2.8 using block diagram algebra. Comparing the required calculations done in Examples 2.8 and 2.10, it is obvious that for complex systems Mason's rule is much more efficient than the block diagram algebra approach.

It should be pointed out that Mason's rule is also applicable to the signal flow graphs corresponding to multi-input multi-output linear systems (see Problem 2.17d). Mason's rule can be also used for finding transfer functions from any two nodes in the signal flow graph (see Problem 2.14).

Having obtained the system transfer functions, we will be able to design controllers in the frequency domain such that feedback systems satisfy certain specifications, like desired transient and steady state responses. A frequency domain controller design technique based on Bode diagrams will be presented in Chapter 9. Note that Bode diagrams, in fact, represent the frequency plots, for $s = j\omega$, of the magnitude and phase of the system transfer function.

For discrete-time systems that are inherently discrete, duality can be employed and the same rules for finding discrete transfer functions as for continuous-time systems are valid. However, there are some differences in the case of discrete-time systems obtained through sampling (sampled data systems). Transfer functions of sampled data systems are considered in the next section.

2.5 Sampled Data Control Systems[1]

In determining discrete transfer functions of sampled data systems the procedure is a little bit more complex. In some cases, the corresponding transfer function even does not exist since it is impossible to find a linear relationship in the frequency domain between output and input signals.

In Sections 2.5.1 and 2.5.2 we present procedures for finding the basic transfer functions of open-loop and closed-loop sampled data control systems. Section 2.5.3 studies the closed-loop transfer function for a special class of sampled data control systems known as digital computer controlled systems. Here, we present only the basics. For more information about the sampled data control systems, the reader is referred to specialized books (e.g. Astrom and Wittenmark, 1990; Ogata, 1987; Franklin et al., 1990; Kuo, 1992; Phillips and Nagle, 1995).

---

[1]   This section may be skipped without loss of continuity.

2.5.1 Open-Loop Transfer Functions

Depending on where the samplers are positioned, even for very simple block diagrams, several interesting cases may arise. These are illustrated in Figure 2.23, where T stands for the sampling period.

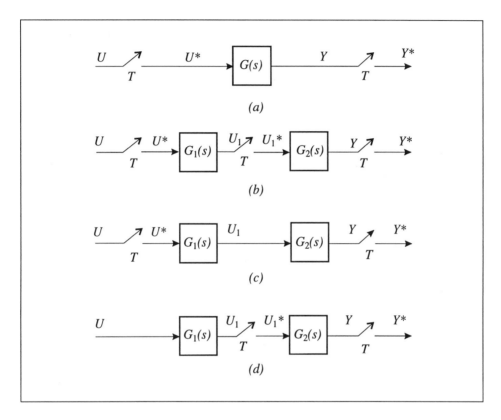

Figure 2.23: Possible cascade connections in a sampled data system

From Figure 2.23a, it follows that

$$Y(s) = G(s)U^*(s)$$
$$Y^*(s) = [G(s)U^*(s)]^* = G^*(s)U^*(s)$$
(2.40)

Formula (2.40) indicates one of the main properties of the starred Laplace transform defined in Appendix B in (b.11), see also (b.16). Since the starred

Laplace transform of a signal is equal to its $\mathcal{Z}$-transform (see (b.12) in Appendix B), it follows from the above equation that

$$Y(z) = G(z)U(z)$$

$$\frac{Y(z)}{U(z)} = G(z) = G^*(s)|_{s=\frac{1}{T}\ln z} \tag{2.41}$$

where T is the sampling period, and $G(z)$ the required transfer function.

Serial connection of two cascaded blocks with samplers, presented in Figure 2.23b, can be described by the following set of equations

$$U_1(s) = G_1(s)U^*(s)$$
$$Y(s) = G_2(s)U_1^*(s) \tag{2.42}$$

Using the same procedure as in the previous example, we have

$$U_1^*(s) = G_1^*(s)U^*(s)$$
$$Y^*(s) = G_2^*(s)U_1^*(s) \tag{2.43}$$

and

$$U_1(z) = G_1(z)U(z)$$
$$Y(z) = G_2(z)U_1(z) \tag{2.44}$$

which yields

$$Y(z) = G_2(z)G_1(z)U(z) \tag{2.45}$$

so that the transfer function for this system is

$$\frac{Y(z)}{U(z)} = G_1(z)G_2(z) \tag{2.46}$$

Note that in this case the transfer function is equal to the product of the transfer functions of each block, as in the case of two cascaded blocks representing continuous-time systems.

A sampled data system with cascaded elements and no sampler in between is given in Figure 2.23c. Here, we have

$$Y(s) = G_1(s)G_2(s)U^*(s) \tag{2.47}$$

The starred Laplace transform of (2.47) gives

$$Y^*(s) = G_1 G_2^*(s) U^*(s) \tag{2.48}$$

where $G_1 G_2^*(s)$ stands for $[G_1(s)G_2(s)]^*$. Then, it follows

$$Y(z) = G_1 G_2(z) U(z)$$
$$\frac{Y(z)}{U(z)} = G_1 G_2(z) \tag{2.49}$$

It is important to note that the transfer functions in (2.46) and (2.49) are not the same, i.e. in general

$$G_1(z)G_2(z) \neq G_1 G_2(z) \tag{2.50}$$

The last case, given in Figure 2.23d, is a serial connection of two elements with a sampler in between. For such a structure, we have

$$U_1(s) = G_1(s)U(s) \;\Rightarrow\; U_1^*(s) = G_1 U^*(s)$$
$$Y(s) = G_2(s)U_1^*(s) \tag{2.51}$$

Equation (2.51) gives, after $Y(s)$ is starred

$$Y^*(s) = G_2^*(s)G_1 U^*(s) \tag{2.52}$$

so that

$$Y(z) = G_2(z)G_1 U_1(z) \tag{2.53}$$

It can be seen from (2.53) that, in this case, we are not able to identify the quantity that relates the system input and output in the frequency domain, in other words, *for this particular open-loop sampled data structure the transfer function does not exist.*

The above discussion suggests that similar rules for continuous- and discrete-time block diagrams of the same structure are valid if there exists a sampler in front of each block of a sampled-data system (see Figure 2.23b).

2.5.2 Closed-Loop Transfer Functions

Typical structures for closed-loop sampled data control systems are given in Figure 2.24. We will find that in two cases the closed-loop transfer function

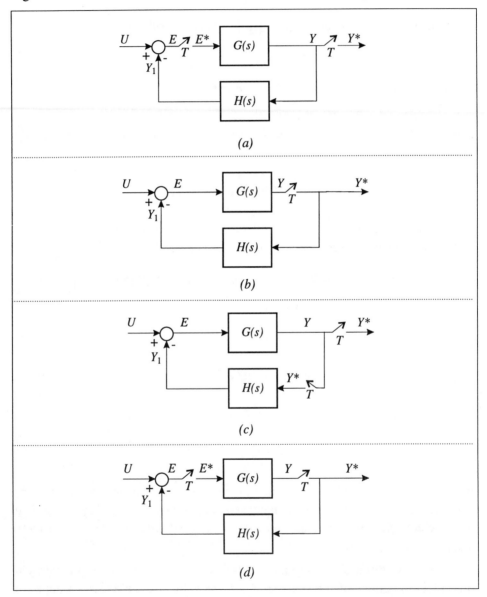

(a)

(b)

(c)

(d)

Figure 2.24: Four possible positions of a sampler in a closed-loop system

does not exist, and that in one case it has the same form as for the closed-loop continuous-time system given by (2.15). However, in this case, the transfer function depends on the complex frequency z.

According to Figure 2.24a, we have

$$Y(s) = G(s)E^*(s)$$
$$E(s) = U(s) - Y_1(s) = U(s) - G(s)H(s)E^*(s)$$

(2.54)

By taking the starred Laplace transform, we obtain

$$E^*(s) = U^*(s) - GH^*(s)E^*(s)$$

(2.55)

or

$$E^*(s) = \frac{U^*(s)}{1 + GH^*(s)}$$

(2.56)

Since

$$Y^*(s) = G^*(s)E^*(s)$$

(2.57)

it follows

$$Y^*(s) = \frac{G^*(s)}{1 + GH^*(s)}U^*(s)$$

(2.58)

In terms of the $\mathcal{Z}$-transform notation, the output is given by

$$Y(z) = \frac{G(z)}{1 + GH(z)}U(z)$$

(2.59)

and the transfer function for the closed-loop system in Figure 2.24a is

$$\frac{Y(z)}{U(z)} = \frac{G(z)}{1 + GH(z)}$$

(2.60)

In a similar way, we obtain for the system in Figure 2.24b

$$Y(z) = \frac{GU(z)}{1 + GH(z)}$$

(2.61)

which means that *it is not possible to determine the transfer function for this sampled data feedback system configuration.* The same expression for the output as the one in (2.61) is obtained for the system given in Figure 2.24c.

Using the above procedure it is easy to find that the closed-loop system in Figure 2.24d has the transfer function given by

$$\frac{Y(z)}{U(z)} = \frac{G(z)}{1 + G(z)H(z)} \tag{2.62}$$

which corresponds to (2.15), obtained for continuous-time closed-loop systems.

The study of this section shows that *the position of the sampler has a very important role*, because it determines whether or not the input signal $U(s)$ can be separated from the system dynamics. Therefore, we have seen from these examples that the transfer function exists if the sampler is just behind the comparator (subtracter), and conversely, does not exist if the sampler is in any other place in the system.

It is important to note that in the case of sampled data systems, it is forbidden for a sampler and a continuous-time system element in cascade to mutually interchange their positions, because in that case the performance of the control system is changed. In other words, the *commutative law* is not, in general, applicable for such systems.

2.5.3 Transfer Functions of Digital Control Systems

A typical block diagram of a digital control system (digital computer controlled) is shown in Figure 2.25. One part of this configuration is a plant representing a continuous-time process that has to be controlled. Another part is a digital

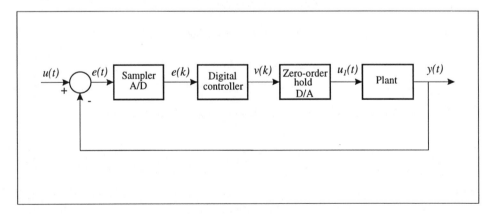

Figure 2.25: Block diagram of a digital control system

controller, usually a digital computer. The interfacing of these two parts is realized by a sampler (A/D converter), which converts analog signals into digital ones, and a zero-order hold (D/A converter), which converts digital signals into continuous-time signals. The continuous-time signals obtained are fed to the plant as control inputs. The block diagram showing transfer functions of all blocks in the above digital control system is given in Figure 2.26.

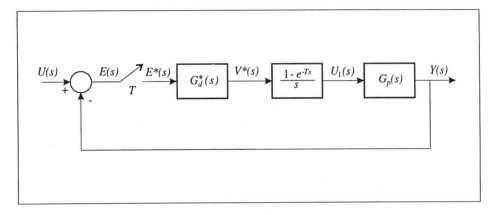

Figure 2.26: Transfer functions in a digital control system

In this system the error signal $e(t) = u(t) - y(t)$ is sampled so that the analog signal $e(t)$ is converted into a digital signal by an A/D device. The digital form of the error signal $e(k)$ is fed to the digital controller, whose transfer function is $G_d(z)$. After the controller has solved the difference equation described by this transfer function, the control signal $v(k)$ is fed to a zero-order hold. The zero-order hold in time domain is represented by a unit pulse of duration T so that its the transfer function is

$$G_h(s) = \mathcal{L}\{h(t) - h(t - T)\} = \frac{1}{s} - \frac{1}{s}e^{-Ts} = \frac{1}{s}\left(1 - e^{-sT}\right) \qquad (2.63)$$

The plant transfer function is $G_p(s)$. From Figure 2.26 it follows

$$Y(s) = G_h(s)G_p(s)G_d^*(s)E^*(s) = G(s)G_d^*(s)E^*(s), \quad G(s) = G_h(s)G_p(s) \qquad (2.64)$$

By the property of the starred Laplace transform given in (2.40), we have

$$Y^*(s) = G^*(s)G_d^*(s)E^*(s) \qquad (2.65)$$

which in the z-domain produces

$$Y(z) = G(z)G_d(z)E(z) \qquad (2.66)$$

Since

$$E(z) = U(z) - Y(z) \qquad (2.67)$$

the last two equations give the required closed-loop transfer function of the system in Figure 2.26 as

$$\frac{Y(z)}{U(z)} = \frac{G_d(z)G(z)}{1 + G_d(z)G(z)} \qquad (2.68)$$

Note that in this introductory control theory course we will not pay attention to sampled data control systems. Material presented in this section is used to demonstrate the procedures for finding transfer functions of a class of linear discrete-time control systems known as sampled-data control systems and to indicate that special care has to be taken while finding the corresponding transfer functions. However, in this book we will study fundamental concepts and methods for discrete-time linear control systems mostly by using dualities with continuous-time linear control systems.

2.6 Transfer Function MATLAB Laboratory Experiment

Part 1. Consider the continuous-time system represented by its transfer function

$$G(s) = \frac{s+1}{s^2 + 5s + 6}$$

Using MATLAB plot:

(a) The impulse response of the system. Use impulse(num,den,t) with t=0:0.1:5.
(b) The step response of the system. Use the function step(num,den,t) with t=0:0.1:5.
(c) The system output response due to the input $\sin(2t)$. Use the function lsim(num,den,u,t) with t=0:0.2:20 and u=sin(2*t).
(d) The system output response due to the input e^{-t}. Use the function lsim(num,den,u,t) with t=0:0.1:5 and u=exp(-t).

Part 2. A discrete-time system is described by the transfer function

$$G(z) = \frac{z - 2}{z^2 - 2.5z + 1}$$

Using MATLAB plot:

(a) The impulse response of the system. Use `dimpulse(num,den)` and `axis([0 10 0 1.5])`.
(b) The step response of the system. Use `dstep(num,den)` and `axis([0 10 0 3])`.
(c) The response due to the input $\sin(2k)$. Use `dlsim(num,den,u)` with `k=0:1:50` and `u=sin(2*k)`.
(d) The steady state response due to the input 2^{-k}. Use `dlsim(num,den,u)` with `u=2.^(-k)` and `axis([0 11 0 1.1])`. Note that "." after 2 in MATLAB indicates a pointwise operation.

Part 3. Consider a flexible beam system (Qiu and Davison, 1993) whose linearized model has the transfer function

$$G(s) = \frac{1.65s^4 - 0.331s^3 - 576s^2 + 90.6s + 19080}{s^6 + 0.996s^5 + 463s^4 + 97.8s^3 + 12131s^2 + 8.11s}$$

Use MATLAB in order to find:

(a) The system open-loop poles and zeros. Use the functions `roots(num)` and `roots(den)`.
(b) The system closed-loop transfer function assuming unity negative feedback. Use `[numc,denc]=cloop(num,den,-1)`.
(c) The closed-loop poles and zeros and compare them to the open-loop poles and zeros found in (a). Use `roots(numc)` and `roots(denc)`.

Part 4. The block diagram of a simple positioning control system using a field-controlled DC motor is shown in Figure 2.27.

(a) Using the rules of block diagram algebra reduce this system to the basic feedback system shown in Figure 2.2a, and find the system transfer function $Y(s)/U(s)$.
(b) Find the transfer function of this system using Mason's gain formula.
(c) If $K = 2$, $k_1 = -0.05$, $k_2 = 0.16$, and $k_3 = 0.24$, write the MATLAB script to find the transfer function of this system (see Example 2.5).

(d) Using MATLAB functions, find the poles of the closed-loop system.

(e) For the given system poles find the partial fraction expansion using the MATLAB function `residue`. From the partial fraction expansion find analytically the system response to a unit step input. Check the obtained results using the MATLAB function `step`.

(f) Plot the unit step response of the system for the time interval of 10 seconds.

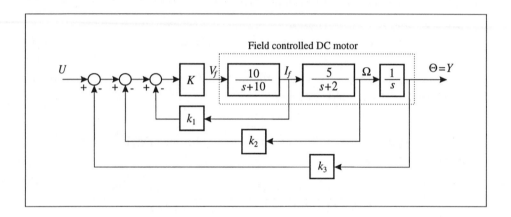

Figure 2.27: Position control system

2.7 References

Astrom, K. and B. Wittenmark, *Computer Controlled Systems: Theory and Design*, Prentice Hall, Englewood Cliffs, New Jersey, 1990.

Franklin, G., J. Powel, and M. Workman, *Digital Control of Dynamic Systems*, Addison-Wesley, Reading, Massachusetts, 1990.

Kamen, E., *Introduction to Signals and Systems*, Macmillan, New York, 1990.

Kuo, B., *Digital Control Systems*, Saunders College Publishing, New York, 1992.

Kuo, B., *Automatic Control Systems*, Prentice Hall, Englewood Cliffs, New Jersey, 1995.

Mason, S., "Feedback theory—some properties of signal flow graphs," *Proc. IRE*, vol. 41, 1144–1156, 1953.

Mason, S., "Feedback theory—further properties of signal flow graphs," *Proc. IRE*, vol. 44, 920–926, 1956.

Ogata, K., *Discrete-Time Control Systems*, Prentice Hall, Englewood Cliffs, New Jersey, 1987.

Phillips, C. and H. Nagle, *Digital Control System Analysis and Design*, Prentice Hall, Englewood Cliffs, New Jersey, 1995.

Qiu, L. and E. Davison, "Performance limitations of non-minimum phase systems in the servomechanism problem," *Automatica*, vol. 29, 337–349, 1993.

Rao, M. and G. Koshy, "Mason's formula: A new and simplified approach," *Control—Theory and Advanced Technology*, vol. 7, 355–363, 1991.

Robichaud, L., M. Boisvert, and J. Robert, *Signal Flow Graphs and Applications*, Prentice Hall, Englewood Cliffs, New Jersey, 1962.

Younger, D., "A simple derivation of Mason's gain formula," *Proc. IEEE*, vol. 51, 1043–1044, 1963.

2.8 Problems

2.1 A control system is described by the following set of equations

$$\frac{d^2 y(t)}{dt^2} + 2\frac{dy(t)}{dt} = e(t), \quad e(t) = u(t) - y(t)$$

(a) Find the transfer function of this system and its impulse response.

(b) Using the inverse Laplace transform, find the response of the system to a unit step input and zero initial conditions.

(c) Using the final value theorem, calculate the steady-state error due to a unit step input, i.e. find $e(t)$ for $t \to \infty$.

(d) Plot the unit step response of this system by using the MATLAB function `step(num,den,t)` for the time interval of 10 seconds. Take `t=0:0.1:10`.

2.2 Consider the electrical circuit given in Figure 2.28.

(a) Find the voltage transfer function $V_0(s)/V_i(s)$.

(b) Suppose that an inductor L_1 is connected in parallel to resistor R_2. Find the voltage transfer function for the modified circuit.

(c) If a constant input voltage $v_i(t) = 5\,\mathrm{V}$ is applied to the circuit, find the output voltages $v_o(t)$ for the circuits given in (a) and (b) by using the inverse Laplace transform.

(d) Using the final value theorem of the Laplace transform, find the steady state values of the output voltages in cases (a) and (b).

(e) If $R_1 = 1, C_1 = 1, R_2 = 2, L_1 = 1$, plot the output $v_o(t)$ for the circuits given in (a) and (b) by using MATLAB.

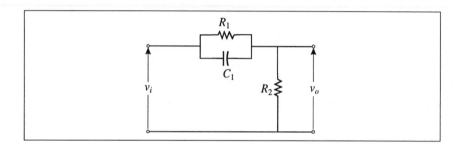

Figure 2.28: An RC network

2.3 Find the transfer function of the following continuous-time system

$$y^{(5)} + 3y^{(4)} + 2y^{(3)} + y^{(2)} + 5y^{(1)} + y = 3u^{(2)} + 2u^{(1)} + u$$

2.4 Consider the continuous-time linear system represented by the transfer function

$$G(s) = \frac{2s^5 + s^3 - 3s^2 + s + 4}{5s^8 + 2s^7 - s^6 - 3s^5 + 5s^4 + 2s^3 - 4s^2 + 2s - 1}$$

Use MATLAB to find:

(a) The zeros and poles of the system.

(b) The inverse Laplace transform of the transfer function. Use the `residue` function.

(c) The system closed-loop transfer function assuming a negative unit feedback and find the corresponding closed-loop poles.

2.5 Consider the control system given in Figure 2.29.

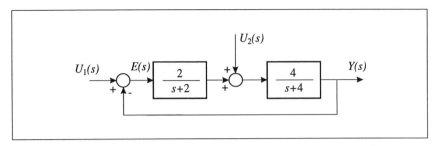

Figure 2.29: A control system with two inputs

(a) Assuming that the initial conditions are zero, find the Laplace transform of the system output due to inputs: $u_1(t) = 2e^{-t}$, $u_2(t) = 4h(t)$, where $h(t)$ is a unit step function.

(b) Find $y(t)$ by using the partial fraction expansion and the inverse Laplace transform.

(c) Using MATLAB, find $y_1(t) = y(t)$ for $u_2(t) = 0$, and $y_2(t) = y(t)$ for $u_1(t) = 0$. Then, using superposition, find $y(t)$ as a response to both $u_1(t)$ and $u_2(t)$. Plot the outputs, $y_1(t), y_2(t)$, and $y(t)$.

2.6 Using the Laplace transform, find the transfer function of the electric network given in Figure 1.7, i.e. find $V_{c_2}(s)/E_i(s)$.

2.7 Find the transfer functions $\Omega(s)/V_a(s)$ and $\Omega(s)/T_d(s)$ of the armature-controlled DC motor given in Figure 2.6.

2.8 Use formula (2.21) to find the closed-loop transfer function of a multivariable system represented by

$$\mathbf{G}(s) = \begin{bmatrix} \frac{s+1}{s^2+6s+8} & \frac{s}{s^2+6s+8} \\ \frac{-s}{s^2+6s+8} & \frac{s+3}{s^2+6s+8} \end{bmatrix}, \quad \mathbf{H}(s) = \begin{bmatrix} \frac{1}{s+1} & 0 \\ 0 & \frac{1}{s+2} \end{bmatrix}$$

2.9 The block diagram of a control system is shown in Figure 2.30.

(a) Reduce this system using block diagram algebra rules and find its transfer function $Y(s)/U(s)$.

(b) Draw the signal flow diagram of the system and find the transfer function by Mason's rule.

(c) Find the transfer function of the system using the MATLAB functions series and feedback repeatedly.

(d) Find the unit step response of this system using MATLAB.

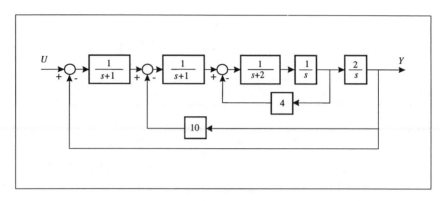

Figure 2.30: Block diagram of a control system

2.10 Using block diagram algebra rules, simplify the block diagrams shown in Figure 2.31 and find their transfer functions (the matrix transfer function for multi-input multi-output case).

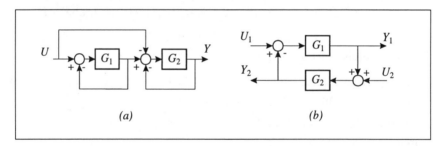

Figure 2.31: Block diagrams

(a) Repeat the procedure using Mason's gain formula.

(b) Find the transfer functions using MATLAB with

$$G_1(s) = \frac{3}{s(s+1)}, \quad G_2(s) = \frac{5}{s^2+2}$$

2.11 Verify that the block diagrams shown in Figure 2.32 are equivalent, i.e. show that they have identical transfer functions $Y(s)/U(s)$ and $Y(s)/D(s)$.

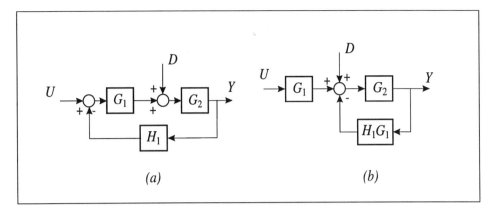

Figure 2.32: Equivalent block diagrams

2.12 Show that the transfer functions of the block diagrams given in Figures 2.33a and 2.33b are identical, i.e. conclude that the block diagrams are equivalent. Note that this equivalence has been used to move from Figure 2.14b to Figure 2.14c in Example 2.8.

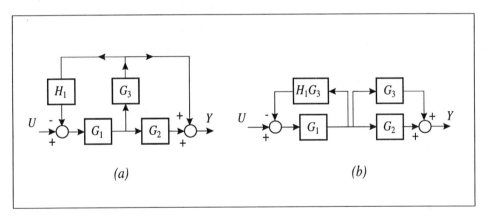

Figure 2.33: Equivalent block diagrams

2.13 Show that both transfer functions $Y(s)/U_1(s)$ and $Y(s)/U_2(s)$ in the block diagrams given in Figures 2.34a and 2.34b are identical, i.e. conclude that the block diagrams are equivalent. Note that this equivalence has been used to move from Figure 2.14c to Figure 2.14d in Example 2.8.

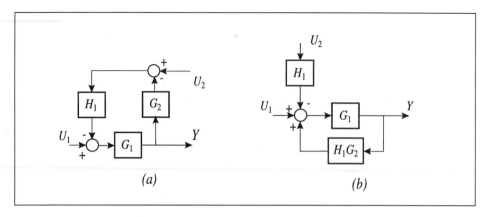

Figure 2.34: Equivalent block diagrams

2.14 Find the transfer function between the nodes E_2 and E_5, i.e. find the ratio $E_5(s)/E_2(s)$ for the signal flow graph given in Figure 2.17.

2.15 SIMULINK can be used for drawing block diagrams. Once a block diagram has been drawn and a name to it given, we can obtain the state space model by invoking the SIMULINK function `linmod('block diagram name')`. From the state space representation, the MATLAB function `ss2tf` produces the corresponding transfer function. Find the transfer functions for the block diagrams considered in Problems 2.9–2.13 by using the SIMULINK package.

2.16 Omit the branch containing block G_5 in the block diagram presented in Figure 2.14a and find the system transfer function by using:

(a) Mason's rule.
(b) Block diagram algebra.

2.17 Using Mason's gain formula, find the transfer functions of the systems whose signal flow graphs are shown in Figure 2.35.

2.18 Consider the discrete-time linear system represented by the transfer function

$$G(z) = \frac{z^4 - 3z^3 + 5z^2 + 2z}{2z^7 + 5z^5 - 3z^4 + z^3 - 2z^2 + 3z - 1}$$

Use MATLAB in order to find:

(a) The zeros and poles of the system.

(b) The inverse $\mathcal{Z}$-transform of the transfer function. Hint: Use the `residue` function.

(c) The closed-loop zeros and poles.

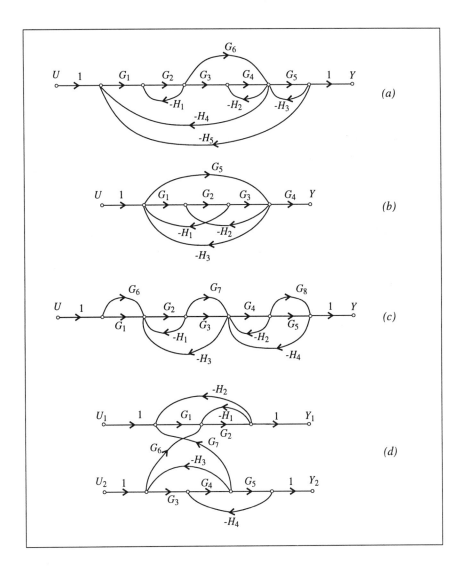

Figure 2.35: Signal flow graphs of control systems

2.19 Find the inverse $\mathcal{Z}$-transform for the following discrete transfer functions

$$H(z) = \frac{3z^2 - z}{z^2 + 2z + 5}, \quad H(z) = \frac{2z^2 + z}{(z + 0.4)(z - 1)(z - 0.2)}$$

Using MATLAB, find the first 10 samples of system outputs $y(k)$ if the system input is $u(k) = 1, k = 0, 1, 2, ...,$ and $u(k) = 0$ for $k < 0$.

2.20 For the digital control system shown in Figure 2.26 the plant transfer function is

$$G(s) = \frac{1}{s(s + 1)}$$

(a) Find the discrete transfer function of the closed-loop system if the sampling interval is $T = 0.5\,\text{s}$, and the digital controller is

$$G_d(z) = K_p + \frac{K_i}{1 - z^{-1}}$$

(b) Using MATLAB, find the poles and zeros of the closed-loop system if $K_p = 1, K_i = 1$. Plot the output response to a unit step input.

Chapter Three

State Space Approach

The state space approach has been introduced in Section 1.3. Due to its fundamental importance for control systems, the state space technique will be considered thoroughly in this chapter. Both continuous- and discrete-time linear time invariant systems will be presented. It has already been pointed out that the state space technique represents the modern approach to control system theory and its applications. The state space approach is very convenient for representation of high-order dimensional and complex systems, and extremely efficient for numerical calculations since many efficient and reliable numerical algorithms developed in mathematics, especially within the area of numerical linear algebra, can be used directly. In addition, the state space form is the basis for introducing system controllability and observability concepts and many modern control theory techniques.

The state space model of a continuous-time linear system is represented by a system of n linear differential equations. In matrix form, it is given by

$$\frac{d}{dt}\mathbf{x}(t) = \dot{\mathbf{x}}(t) = \mathbf{A}\mathbf{x}(t) + \mathbf{B}\mathbf{u}(t), \quad \mathbf{x}(0) = \mathbf{x}_o \tag{3.1}$$

$$\mathbf{y}(t) = \mathbf{C}\mathbf{x}(t) + \mathbf{D}\mathbf{u}(t) \tag{3.2}$$

where $\mathbf{x} \in \Re^n$, $\mathbf{u} \in \Re^r$, and $\mathbf{y} \in \Re^p$ are, respectively, vectors of system states, control inputs, and system outputs. The matrix $\mathbf{A}^{n \times n}$ describes the *internal* behavior of the system, while matrices $\mathbf{B}^{n \times r}$, $\mathbf{C}^{p \times n}$, and $\mathbf{D}^{p \times r}$ represent connections between the *external world* and the system. If there are no direct paths between inputs and outputs, which is often the case, the matrix $\mathbf{D}^{p \times m}$ is

95

zero. It is assumed in this book that all matrices in (3.1) and (3.2) are time invariant. Studying linear control systems with time varying coefficient matrices requires knowledge of some advanced topics in mathematics (see for example Chen, 1984; see also Section 10.1).

The state space model for linear discrete-time control systems has exactly the same form as (3.1) and (3.2) with differential equations replaced by difference equations, that is

$$\mathbf{x}(k+1) = \mathbf{A}_d\mathbf{x}(k) + \mathbf{B}_d\mathbf{u}(k), \quad \mathbf{x}(0) = \mathbf{x}_0 \tag{3.3}$$

$$\mathbf{y}(k) = \mathbf{C}_d\mathbf{x}(k) + \mathbf{D}_d\mathbf{u}(k) \tag{3.4}$$

All vectors and matrices defined in (3.3) and (3.4) have the same dimensions as corresponding ones given in (3.1) and (3.2). In this chapter, we present and derive in detail the main state space concepts for continuous-time linear control systems and then give the corresponding interpretations in the discrete-time domain.

The chapter is organized as follows. In Section 3.1 several systematic methods for obtaining the state space form from differential equations and transfer functions are developed. The time response of linear systems given in the state form is considered in Section 3.2. The corresponding results for discrete-time systems, and the procedure for discretization of continuous-time systems leading to discrete-time models, are given in Section 3.3. The concepts of the system characteristic equation, eigenvalues, and eigenvectors and their use in control system theory are presented in Section 3.4. At the end of the chapter, in Section 3.5, three MATLAB laboratory experiments are outlined.

Chapter Objectives

The dynamical systems considered in this book are either described by differential/difference equations or given in the form of system transfer functions. One of the goals is to present procedures for obtaining the state space forms either from differential/difference equations or from transfer functions. In that respect students will be exposed to four standard state space forms, known as canonical forms: the phase variable form or controller form, the observer form, the modal form, and the Jordan form.

Another important objective is to show students how to analyze linear systems given in the state space form, i.e. how to find responses (state variables and outputs) of the corresponding state space models to any input signal (control input). A working knowledge of undergraduate linear algebra and the basic theory of differential equations is helpful for complete understanding of this chapter. Some useful results on linear algebra are given in Appendix C. Students without a strong background in these topics may consult any undergraduate text, (for example, Fraleigh and Beauregard, 1990; Boyce and DiPrima, 1992).

3.1 State Space Models

In this section we study state space models of continuous-time linear systems. The corresponding results for discrete-time systems, obtained via duality with the continuous-time models, are given in Section 3.3.

The state space model of a continuous-time dynamic system can be derived either from the system model given in the time domain by a differential equation or from its transfer function representation. Both cases will be considered in this section. Four state space forms—the phase variable form (controller form), the observer form, the modal form, and the Jordan form—which are often used in modern control theory and practice, are presented.

3.1.1 The State Space Model and Differential Equations

Consider a general nth-order model of a dynamic system represented by an nth-order differential equation

$$\frac{d^n y(t)}{dt^n} + a_{n-1}\frac{d^{n-1}y(t)}{dt^{n-1}} + \cdots + a_1\frac{dy(t)}{dt} + a_0 y(t)$$
$$= b_n\frac{d^n u(t)}{dt^n} + b_{n-1}\frac{d^{n-1}u(t)}{dt^{n-1}} + \cdots + b_1\frac{du(t)}{dt} + b_0 u(t) \tag{3.5}$$

At this point we assume that all initial conditions for the above differential equation, i.e. $y(0^-), dy(0^-)/dt, ..., d^{n-1}y(0^-)/dt^{n-1}$, are equal to zero. We will show later how to take into account the effect of initial conditions.

In order to derive a systematic procedure that transforms a differential equation of order n to a state space form representing a system of n first-order differential equations, we first start with a simplified version of (3.5), namely we study the case when no derivatives with respect to the input are present

$$\frac{d^n y(t)}{dt^n} + a_{n-1}\frac{d^{n-1}y(t)}{dt^{n-1}} + \cdots + a_1\frac{dy(t)}{dt} + a_0 y(t) = u(t) \tag{3.6}$$

Introduce the following (easy to remember) change of variables

$$x_1(t) = y(t)$$

$$x_2(t) = \frac{dy(t)}{dt}$$

$$x_3(t) = \frac{d^2 y(t)}{dt^2} \tag{3.7}$$

$$\vdots$$

$$x_n(t) = \frac{d^{n-1} y(t)}{dt^{n-1}}$$

which after taking derivatives leads to

$$\frac{dx_1(t)}{dt} = \dot{x}_1 = \frac{dy(t)}{dt} = x_2(t)$$

$$\frac{dx_2(t)}{dt} = \dot{x}_2 = \frac{d^2 y(t)}{dt^2} = x_3(t)$$

$$\frac{dx_3(t)}{dt} = \dot{x}_3 = \frac{d^3 y(t)}{dt^3} = x_4(t)$$

$$\vdots \tag{3.8}$$

$$\frac{dx_n(t)}{dt} = \dot{x}_n = \frac{d^n y(t)}{dt^n}$$

$$= -a_0 y(t) - a_1 \frac{dy(t)}{dt} - a_2 \frac{d^2 y(t)}{dt^2} - \cdots - a_{n-1} \frac{d^{n-1} y(t)}{dt^{n-1}} + u(t)$$

$$= -a_0 x_1(t) - a_1 x_2(t) - \cdots - a_2 x_3(t) - \cdots - a_{n-1} x_n(t) + u(t)$$

The state space form of (3.8) is given by

$$
\begin{bmatrix} \dot{x}_1 \\ \dot{x}_2 \\ \vdots \\ \vdots \\ \dot{x}_{n-1} \\ \dot{x}_n \end{bmatrix} = \begin{bmatrix} 0 & 1 & 0 & \cdots & \cdots & 0 \\ 0 & 0 & 1 & 0 & \cdots & 0 \\ \vdots & \vdots & \ddots & \ddots & \ddots & \vdots \\ \vdots & \vdots & \cdots & \ddots & \ddots & 0 \\ 0 & 0 & \cdots & \cdots & 0 & 1 \\ -a_0 & -a_1 & -a_2 & \cdots & \cdots & -a_{n-1} \end{bmatrix} \begin{bmatrix} x_1(t) \\ x_2(t) \\ \vdots \\ \vdots \\ x_{n-1}(t) \\ x_n(t) \end{bmatrix} + \begin{bmatrix} 0 \\ 0 \\ \vdots \\ \vdots \\ 0 \\ 1 \end{bmatrix} u(t)
$$

$$\tag{3.9}$$

with the corresponding output equation obtained from (3.7) as

$$y(t) = \begin{bmatrix} 1 & 0 & \cdots & 0 \end{bmatrix} \begin{bmatrix} x_1(t) \\ x_2(t) \\ \vdots \\ x_{n-1}(t) \\ x_n(t) \end{bmatrix} \tag{3.10}$$

The state space form (3.9) and (3.10) is known in the literature as the *phase variable canonical form*.

In order to extend this technique to the general case defined by (3.5), which includes derivatives with respect to the input, we form an auxiliary differential equation of (3.5) having the form of (3.6) as

$$\frac{d^n \xi(t)}{dt^n} + a_{n-1} \frac{d^{n-1} \xi(t)}{dt^{n-1}} + \cdots + a_1 \frac{d\xi(t)}{dt} + a_0 \xi(t) = u(t) \tag{3.11}$$

for which the change of variables (3.7) is applicable

$$\begin{aligned} x_1(t) &= \xi(t) \\ x_2(t) &= \frac{d\xi(t)}{dt} \\ x_3(t) &= \frac{d^2 \xi(t)}{dt^2} \\ &\vdots \\ x_n(t) &= \frac{d^{n-1} \xi(t)}{dt^{n-1}} \end{aligned} \tag{3.12}$$

and then apply the superposition principle to (3.5) and (3.11). Since $\xi(t)$ is the response of (3.11), then by the superposition property the response of (3.5) is given by

$$y(t) = b_0 \xi(t) + b_1 \frac{d\xi(t)}{dt} + b_2 \frac{d^2 \xi(t)}{dt^2} + \cdots + b_n \frac{d^n \xi(t)}{dt^n} \tag{3.13}$$

Equations (3.12) produce the state space equations in the form already given by (3.9). The output equation can be obtained by eliminating $d^n \xi(t)/dt^n$ from (3.13), by using (3.11), that is

$$\frac{d^n \xi(t)}{dt^n} = u(t) - a_{n-1} x_n(t) - \cdots - a_1 x_2(t) - a_0 x_1(t)$$

This leads to the output equation

$$y(t) = [(b_0 - a_0 b_n) \quad (b_1 - a_1 b_n) \quad \cdots \quad (b_{n-1} - a_{n-1} b_n)] \begin{bmatrix} x_1(t) \\ x_2(t) \\ \vdots \\ x_n(t) \end{bmatrix} + b_n u(t)$$

(3.14)

It is interesting to point out that for $b_n = 0$, which is almost always the case, the output equation also has an easy-to-remember form given by

$$y(t) = [b_0 \quad b_1 \quad \cdots \quad b_{n-1}] \begin{bmatrix} x_1(t) \\ x_2(t) \\ \vdots \\ x_n(t) \end{bmatrix}$$

(3.15)

Thus, in summary, for a given dynamic system modeled by differential equation (3.5), one is able to write immediately its state space form, given by (3.9) and (3.15), just by identifying coefficients a_i and b_i, $i = 0, 1, 2, ..., n - 1$, and using them to form the corresponding entries in matrices $\mathbf{A}$ and $\mathbf{C}$.

Example 3.1: Consider a dynamical system represented by the following differential equation

$$y^{(6)} + 6y^{(5)} - 2y^{(4)} + y^{(2)} - 5y^{(1)} + 3y = 7u^{(3)} + u^{(1)} + 4u$$

where $y^{(i)}$ stands for the ith derivative, i.e. $y^{(i)} = d^i y/dt^i$. According to (3.9) and (3.14), the state space model of the above system is described by the following matrices

$$\mathbf{A} = \begin{bmatrix} 0 & 1 & 0 & 0 & 0 & 0 \\ 0 & 0 & 1 & 0 & 0 & 0 \\ 0 & 0 & 0 & 1 & 0 & 0 \\ 0 & 0 & 0 & 0 & 1 & 0 \\ 0 & 0 & 0 & 0 & 0 & 1 \\ -3 & 5 & -1 & 0 & 2 & -6 \end{bmatrix}, \quad \mathbf{B} = \begin{bmatrix} 0 \\ 0 \\ 0 \\ 0 \\ 0 \\ 1 \end{bmatrix}$$

$$\mathbf{C} = [4 \quad 1 \quad 0 \quad 7 \quad 0 \quad 0], \quad \mathbf{D} = 0$$

◇

3.1.2 State Space Variables from Transfer Functions

In this section, we present two methods, known as direct and parallel programming techniques, which can be used for obtaining state space models from system transfer functions. For simplicity, like in the previous subsection, we consider only single-input single-output systems.

The resulting state space models may or may not contain all the modes of the original transfer function, where by transfer function modes we mean poles of the original transfer function (before zero-pole cancellation, if any, takes place). If some zeros and poles in the transfer function are cancelled, then the resulting state space model will be of reduced order and the corresponding modes will not appear in the state space model. This problem of system reducibility will be addressed in detail in Chapter 5 after we have introduced the system controllability and observability concepts.

In the following, we first use direct programming techniques to derive the state space forms known as the controller canonical form and the observer canonical form; then, by the method of parallel programing, the state space forms known as modal canonical form and Jordan canonical form are obtained.

The Direct Programming Technique and Controller Canonical Form

This technique is convenient in the case when the plant transfer function is given in a nonfactorized polynomial form

$$\frac{Y(s)}{U(s)} = \frac{b_n s^n + b_{n-1} s^{n-1} + \cdots + b_1 s + b_0}{s^n + a_{n-1} s^{n-1} + \cdots + a_1 s + a_0} \qquad (3.16)$$

For this system an auxiliary variable $V(s)$ is introduced such that the transfer function is split as

$$\frac{V(s)}{U(s)} = \frac{1}{s^n + a_{n-1} s^{n-1} + \cdots + a_1 s + a_0} \qquad (3.17a)$$

$$\frac{Y(s)}{V(s)} = b_n s^n + b_{n-1} s^{n-1} + \cdots + b_1 s + b_0 \qquad (3.17b)$$

The block diagram for this decomposition is given in Figure 3.1.

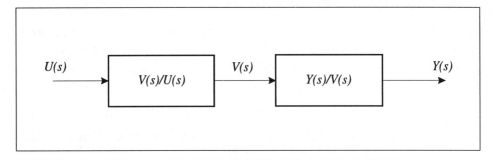

Figure 3.1: Block diagram representation for (3.17)

Equation (3.17a) has the same structure as (3.6), after the Laplace transformation is applied, which directly produces the state space system equation identical to (3.9). It remains to find matrices for the output equation (3.2). Equation (3.17b) can be rewritten as

$$Y(s) = b_n s^n V(s) + b_{n-1} s^{n-1} V(s) + \cdots + b_1 s V(s) + b_0 V(s) \qquad (3.18)$$

indicating that $y(t)$ is just a superposition of $v(t)$ and its derivatives. Note that (3.17) may be considered as a differential equation in the operator form for zero initial conditions, where $s \equiv d/dt$. In that case, $V(s)$, $Y(s)$, and $U(s)$ are simply replaced with $v(t)$, $y(t)$, and $u(t)$, respectively.

The common procedure for obtaining state space models from transfer functions is performed with help of the so-called transfer function *simulation diagrams*. In the case of continuous-time systems, the simulation diagrams are elementary analog computers that solve differential equations describing systems dynamics. They are composed of integrators, adders, subtracters, and multipliers, which are physically realized by using operational amplifiers. In addition, function generators are used to generate input signals. The number of integrators in a simulation diagram is equal to the order of the differential equation under consideration. It is relatively easy to draw (design) a simulation diagram. There are many ways to draw a simulation diagram for a given dynamic system, and there are also many ways to obtain the state space form from the given simulation diagram.

The simulation diagram for the system (3.17) can be obtained by direct programming technique as follows. Take n integrators in cascade and denote their inputs, respectively, by $v^{(n)}(t), v^{(n-1)}(t), ..., v^{(1)}(t), v(t)$. Use formula (3.18) to construct $y(t)$, i.e. multiply the corresponding inputs $v^{(i)}(t)$ to integrators by the corresponding coefficients b_i and add them using an adder (see the top half of Figure 3.2, where $1/s$ represents the integrator block). From (3.17a) we have that

$$v^{(n)}(t) = u(t) - a_{n-1}v^{(n-1)}(t) - \cdots - a_1v^{(1)}(t) - a_0v(t)$$

which can be physically realized by using the corresponding feedback loops in the simulation diagram and adding them as shown in the bottom half of Figure 3.2.

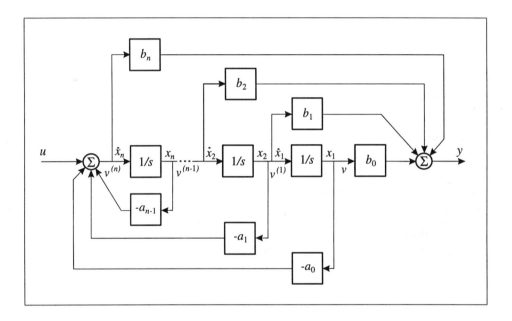

Figure 3.2: Simulation diagram for the direct
programming technique (controller canonical form)

A systematic procedure to obtain the state space form from a simulation diagram is *to choose the outputs of integrators as state variables*. Using this convention, the state space model for the simulation diagram presented in Figure 3.2 is obtained in a straightforward way by reading and recording information from the simulation diagram, which leads to

$$\dot{\mathbf{x}}(t) = \begin{bmatrix} 0 & 1 & 0 & \cdots & \cdots & 0 \\ 0 & 0 & 1 & 0 & \cdots & 0 \\ \vdots & \vdots & \ddots & \ddots & \ddots & \vdots \\ \vdots & \vdots & \vdots & \ddots & \ddots & 0 \\ 0 & 0 & 0 & \cdots & 0 & 1 \\ -a_0 & -a_1 & -a_2 & \cdots & \cdots & -a_{n-1} \end{bmatrix} \mathbf{x}(t) + \begin{bmatrix} 0 \\ 0 \\ \vdots \\ \vdots \\ 0 \\ 1 \end{bmatrix} u(t) \qquad (3.19)$$

and

$$y(t) = [(b_0 - a_0 b_n) \quad (b_1 - a_1 b_n) \quad \cdots \quad (b_{n-1} - a_{n-1} b_n)]\mathbf{x}(t) + b_n u(t)$$
$$(3.20)$$

This form of the system model is called the *controller canonical form*. It is identical to the one obtained in the previous section—equations (3.9) and (3.14). Controller canonical form plays an important role in control theory since it represents the so-called controllable system. System controllability is one of the main concepts of modern control theory. It will be studied in detail in Chapter 5.

It is important to point out that there are many state space forms for a given dynamical system, and that all of them are related by linear transformations. More about this fact, together with the development of other important state space canonical forms, can be found in Kailath (1980; see also similarity transformation in Section 3.4).

Note that the MATLAB function tf2ss produces the state space form for a given transfer function, in fact, it produces the controller canonical form.

Example 3.2: The transfer function of the flexible beam from Section 2.6 is given by

$$G(s) = \frac{1.65s^4 - 0.331s^3 - 576s^2 + 90.6s + 19080}{s^6 + 0.996s^5 + 463s^4 + 97.8s^3 + 12131s^2 + 8.11s}$$

Using the direct programming technique with formulas (3.19) and (3.20), the state space controller canonical form is given by

$$\dot{\mathbf{x}} = \begin{bmatrix} 0 & 1 & 0 & 0 & 0 & 0 \\ 0 & 0 & 1 & 0 & 0 & 0 \\ 0 & 0 & 0 & 1 & 0 & 0 \\ 0 & 0 & 0 & 0 & 1 & 0 \\ 0 & 0 & 0 & 0 & 0 & 1 \\ 0 & -8.11 & -12131 & -97.8 & -463 & -0.996 \end{bmatrix} \mathbf{x} + \begin{bmatrix} 0 \\ 0 \\ 0 \\ 0 \\ 0 \\ 1 \end{bmatrix} u$$

and

$$y = [19080 \quad 90.6 \quad -576 \quad -0.331 \quad 1.65 \quad 0]x$$

◇

Direct Programming Technique and Observer Canonical Form

In addition to controller canonical form, *observer canonical form* is related to another important concept of modern control theory: system observability. Observer canonical form has a very simple structure and represents an observable system. The concept of linear system observability will be considered thoroughly in Chapter 5.

Observer canonical form can be derived as follows. Equation (3.16) is written in the form

$$
\begin{aligned}
Y(s)\left(s^n + a_{n-1}s^{n-1} + \cdots + a_1 s + a_0\right) \\
= U(s)\left(b_n s^n + b_{n-1}s^{n-1} + \cdots + b_1 s + b_0\right)
\end{aligned}
\tag{3.21}
$$

and expressed as

$$
\begin{aligned}
Y(s) = &-\frac{1}{s^n}\left(a_{n-1}s^{n-1} + \cdots + a_1 s + a_0\right)Y(s) \\
&+\frac{1}{s^n}U(s)\left(b_n s^n + b_{n-1}s^{n-1} + \cdots + b_1 s + b_0\right)
\end{aligned}
\tag{3.22}
$$

leading to

$$
Y(s) = -\frac{1}{s}a_{n-1}Y(s) - \frac{1}{s^2}a_{n-2}Y(s) - \cdots - \frac{1}{s^{n-1}}a_1 Y(s) - \frac{1}{s^n}a_0 Y(s)
$$

$$
+ b_n U(s) + \frac{1}{s}b_{n-1}U(s) + \frac{1}{s^2}b_{n-2}U(s) + \cdots + \frac{1}{s^{n-1}}b_1 U(s) + \frac{1}{s^n}b_0 U(s)
\tag{3.23}
$$

This relationship can be implemented by using a simulation diagram composed of n integrators in a cascade, and letting the corresponding signals to pass through the specified number of integrators. For example, terms containing $1/s$ should pass through only one integrator, signals $a_{n-2}y(t)$ and $b_{n-2}u(t)$ should pass through two integrators, and so on. Finally, signals $a_0 y(t)$ and $b_0 u(t)$ should be integrated n-times, i.e. they must pass through all n integrators. The corresponding simulation diagram is given in Figure 3.3.

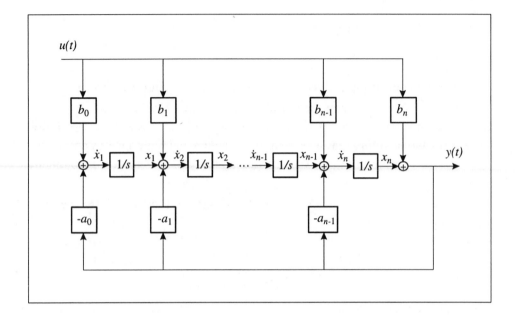

Figure 3.3: Simulation diagram for observer canonical form

Defining the state variables as the outputs of integrators, and recording relationships among state variables and the system output, we get from the above figure

$$y(t) = x_n(t) + b_n u(t) \tag{3.24}$$

$$\dot{x}_1(t) = -a_0 y(t) + b_0 u(t) = -a_0 x_n(t) + (b_0 - a_0 b_n) u(t)$$
$$\dot{x}_2(t) = -a_1 y(t) + b_1 u(t) + x_1(t) = x_1(t) - a_1 x_n(t) + (b_1 - a_1 b_n) u(t)$$
$$\dot{x}_3(t) = -a_2 y(t) + b_2 u(t) + x_2(t) = x_2(t) - a_2 x_n(t) + (b_2 - a_2 b_n) u(t)$$

$$\cdots$$

$$\dot{x}_n(t) = -a_{n-1} y(t) + b_{n-1} u(t) + x_{n-1}(t)$$
$$= x_{n-1}(t) - a_{n-1} x_n(t) + (b_{n-1} - a_{n-1} b_n) u(t) \tag{3.25}$$

The matrix form of observer canonical form is easily obtained from (3.24) and (3.25) as

$$\dot{\mathbf{x}}(t) = \begin{bmatrix} 0 & 0 & \cdots & \cdots & 0 & -a_0 \\ 1 & 0 & \cdots & \cdots & 0 & -a_1 \\ 0 & 1 & \ddots & \cdots & \vdots & -a_2 \\ \vdots & 0 & 1 & \ddots & \vdots & \vdots \\ \vdots & \vdots & \ddots & \ddots & 0 & -a_{n-2} \\ 0 & 0 & \cdots & 0 & 1 & -a_{n-1} \end{bmatrix} \mathbf{x}(t) + \begin{bmatrix} b_0 - a_0 b_n \\ b_1 - a_1 b_n \\ b_2 - a_2 b_n \\ \vdots \\ \vdots \\ b_{n-1} - a_{n-1} b_n \end{bmatrix} u(t) \quad (3.26)$$

and

$$y(t) = \begin{bmatrix} 0 & \cdots & 0 & 1 \end{bmatrix} \mathbf{x}(t) + b_n u(t) \quad (3.27)$$

Example 3.3: The observer canonical form for the flexible beam from Example 3.2 is given by

$$\dot{\mathbf{x}} = \begin{bmatrix} 0 & 0 & 0 & 0 & 0 & 0 \\ 1 & 0 & 0 & 0 & 0 & -8.11 \\ 0 & 1 & 0 & 0 & 0 & -12131 \\ 0 & 0 & 1 & 0 & 0 & -97.8 \\ 0 & 0 & 0 & 1 & 0 & -463 \\ 0 & 0 & 0 & 0 & 1 & -0.996 \end{bmatrix} \mathbf{x} + \begin{bmatrix} 19080 \\ 90.6 \\ -576 \\ -0.331 \\ 1.65 \\ 0 \end{bmatrix} u$$

and

$$y = \begin{bmatrix} 0 & 0 & 0 & 0 & 0 & 1 \end{bmatrix} \mathbf{x}$$

◇

Observer canonical form is very useful for computer simulation of linear dynamical systems since it allows the effect of the system initial conditions to be taken into account. Thus, this form represents an observable system, in the sense to be defined in Chapter 5, which means that all state variables have an impact on the system output, and vice versa, that from the system output and state equations one is able to reconstruct the state variables at any time instant, and of course at zero, and thus, determine $x_1(0), x_2(0), ..., x_n(0)$ in terms of the original initial conditions $y(0^-), dy(0^-)/dt, ..., d^{n-1}y(0^-)/dt^{n-1}$. For more details see Section 5.5 for a subtopic on the observability role in analog computer simulation.

Parallel Programming Technique

For this technique we distinguish two cases: distinct real roots and multiple real roots of the system transfer function denominator.

Distinct Real Roots

This state space form is very convenient for applications. Derivation of this type of the model starts with the transfer function in the partial fraction expansion form. Let us assume, without loss of generality, that the polynomial in the numerator has degree of $m < n$, then

$$
\begin{aligned}
\frac{Y(s)}{U(s)} &= \frac{P_m(s)}{(s + p_1)(s + p_2) \cdots (s + p_n)} \\
&= \frac{k_1}{s + p_1} + \frac{k_2}{s + p_2} + \cdots + \frac{k_n}{s + p_n}
\end{aligned}
\tag{3.28}
$$

Here $p_1, p_2, ..., p_n$ are *distinct real* roots (poles) of the transfer function denominator.

The simulation diagram of such a form is shown in Figure 3.4.

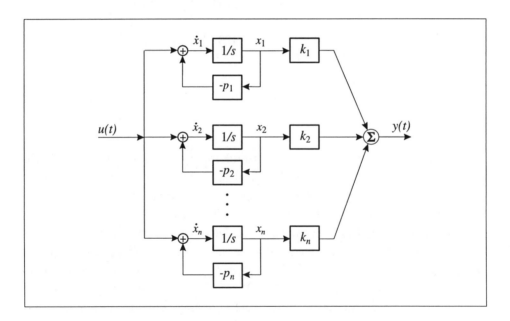

Figure 3.4: The simulation diagram for the parallel programming technique (modal canonical form)

The state space model derived from this simulation diagram is given by

$$\dot{x}(t) = \begin{bmatrix} -p_1 & 0 & \cdots & \cdots & & 0 \\ 0 & -p_2 & \ddots & \cdots & & 0 \\ \vdots & & \ddots & \ddots & \ddots & \vdots \\ \vdots & \vdots & & \ddots & \ddots & 0 \\ 0 & 0 & \cdots & & 0 & -p_n \end{bmatrix} x(t) + \begin{bmatrix} 1 \\ 1 \\ \vdots \\ \vdots \\ 1 \end{bmatrix} u(t) \qquad (3.29)$$

$$y(t) = [k_1 \quad k_2 \quad \cdots \quad \cdots \quad k_n] x(t)$$

This form is known in the literature as the *modal canonical form* (also known as uncoupled form).

Example 3.4: Find the state space model of a system described by the transfer function

$$\frac{Y(s)}{U(s)} = \frac{(s+5)(s+4)}{(s+1)(s+2)(s+3)}$$

using both direct and parallel programming techniques.

The nonfactorized transfer function is

$$\frac{Y(s)}{U(s)} = \frac{s^2 + 9s + 20}{s^3 + 6s^2 + 11s + 6}$$

and the state space form obtained by using (3.19) and (3.20) of the direct programming technique is

$$\dot{x} = \begin{bmatrix} 0 & 1 & 0 \\ 0 & 0 & 1 \\ -6 & -11 & -6 \end{bmatrix} x + \begin{bmatrix} 0 \\ 0 \\ 1 \end{bmatrix} u$$

$$y = [20 \quad 9 \quad 1] x$$

Note that the MATLAB function tf2ss produces

$$\underline{\dot{x}} = \begin{bmatrix} -6 & -11 & -6 \\ 1 & 0 & 0 \\ 0 & 1 & 0 \end{bmatrix} \underline{x} + \begin{bmatrix} 1 \\ 0 \\ 0 \end{bmatrix} u$$

$$y = [1 \quad 9 \quad 20] \underline{x}$$

which only indicates a permutation in the state space variables, that is

$$\mathbf{x} = \begin{bmatrix} 0 & 0 & 1 \\ 0 & 1 & 0 \\ 1 & 0 & 0 \end{bmatrix} \underline{\mathbf{x}}$$

Employing the partial fraction expansion (which can be obtained by the MATLAB function `residue`), the transfer function is written as

$$\frac{Y(s)}{U(s)} = \frac{(s+5)(s+4)}{(s+1)(s+2)(s+3)} = \frac{6}{s+1} - \frac{6}{s+2} + \frac{1}{s+3}$$

The state space model, directly written using (3.29), is

$$\dot{\mathbf{x}} = \begin{bmatrix} -1 & 0 & 0 \\ 0 & -2 & 0 \\ 0 & 0 & -3 \end{bmatrix} \mathbf{x} + \begin{bmatrix} 1 \\ 1 \\ 1 \end{bmatrix} u$$

$$y = \begin{bmatrix} 6 & -6 & 1 \end{bmatrix} \mathbf{x}$$

$\diamond$

Note that the parallel programming technique presented is valid only for the case of real distinct roots. If complex conjugate roots appear they should be combined in pairs corresponding to the second-order transfer functions, which can be independently implemented as demonstrated in the next example.

Example 3.5: Let a transfer function containing a pair of complex conjugate roots be given by

$$G(s) = \frac{4}{s+1-j} + \frac{4}{s+1+j} + \frac{2}{s+5} + \frac{3}{s+10}$$

We first group the complex conjugate poles in a second-order transfer function, that is

$$G(s) = \frac{8s+8}{s^2+2s+2} + \frac{2}{s+5} + \frac{3}{s+10}$$

Then, distinct real poles are implemented like in the case of parallel programming. A second-order transfer function, corresponding to the pair of complex conjugate poles, is implemented using direct programming, and added in parallel to the first-order transfer functions corresponding to the real poles. The simulation diagram is given in Figure 3.5, where the controller canonical form is used to represent a second-order transfer function corresponding to the complex conjugate poles. From this simulation diagram we have

$$\dot{x}_1 = -5x_1 + u$$
$$\dot{x}_2 = -10x_2 + u$$
$$\dot{x}_3 = x_4$$
$$\dot{x}_4 = -2x_3 - 2x_4 + u$$
$$y = 2x_1 + 3x_2 + 8x_3 + 8x_4$$

so that the required state space form is

$$\dot{\mathbf{x}} = \begin{bmatrix} -5 & 0 & 0 & 0 \\ 0 & -10 & 0 & 0 \\ 0 & 0 & 0 & 1 \\ 0 & 0 & -2 & -2 \end{bmatrix} \mathbf{x} + \begin{bmatrix} 1 \\ 1 \\ 0 \\ 1 \end{bmatrix} u$$

$$y = \begin{bmatrix} 2 & 3 & 8 & 8 \end{bmatrix} \mathbf{x}$$

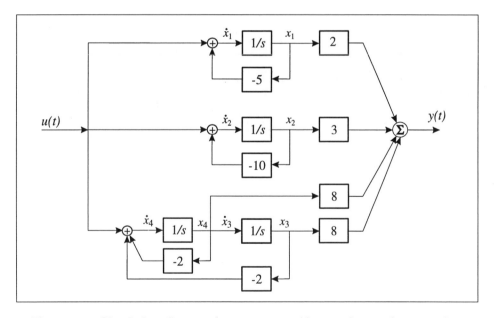

Figure 3.5: Simulation diagram for a system with complex conjugate poles

◇

Multiple Real Roots

When the transfer function has multiple real poles, it is not possible to represent the system in uncoupled form. Assume that a real pole p_1 of the transfer function has multiplicity r and that the other poles are real and distinct, that is

$$\frac{Y(s)}{U(s)} = \frac{N(s)}{(s+p_1)^r(s+p_{r+1})\cdots(s+p_n)}$$

The partial fraction form of the above expression is

$$\frac{Y(s)}{U(s)} = \frac{k_{11}}{s+p_1} + \frac{k_{12}}{(s+p_1)^2} + \cdots + \frac{k_{1r}}{(s+p_1)^r} + \frac{k_{r+1}}{s+p_{r+1}} + \cdots + \frac{k_n}{s+p_n}$$

The simulation diagram for such a system is shown in Figure 3.6.

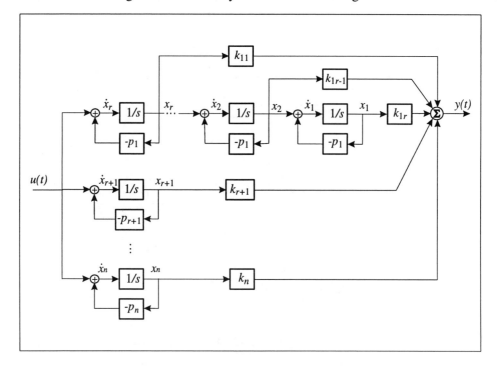

Figure 3.6: The simulation diagram for the Jordan canonical form

Taking for the state variables the outputs of integrators, the state space model is obtained as follows

$$
\mathbf{A} = \begin{bmatrix}
-p_1 & 1 & 0 & \cdots & \cdots & 0 & 0 & \cdots & \cdots & 0 \\
0 & -p_1 & 1 & 0 & \cdots & 0 & 0 & \cdots & \cdots & 0 \\
\vdots & \ddots & \ddots & \ddots & \ddots & \vdots & \vdots & \vdots & \vdots & \vdots \\
0 & \cdots & 0 & -p_1 & 1 & 0 & \vdots & \vdots & \vdots & \vdots \\
0 & \cdots & \cdots & 0 & -p_1 & 1 & 0 & \cdots & \cdots & 0 \\
0 & \cdots & \cdots & \cdots & 0 & -p_1 & 0 & \cdots & \cdots & 0 \\
0 & 0 & \cdots & \cdots & \cdots & 0 & -p_{r+1} & 0 & \cdots & 0 \\
0 & 0 & \cdots & \cdots & \cdots & 0 & \ddots & -p_{r+2} & \ddots & \vdots \\
\vdots & \vdots & \vdots & \vdots & \vdots & \vdots & \vdots & \ddots & \ddots & 0 \\
0 & 0 & \cdots & \cdots & \cdots & 0 & 0 & \cdots & 0 & -p_n
\end{bmatrix}
$$
$$(3.30)$$

$$\mathbf{B} = \begin{bmatrix} 0 & 0 & \cdots & \cdots & 0 & 1 & 1 & \cdots & \cdots & 1 \end{bmatrix}^T$$

$$\mathbf{C} = \begin{bmatrix} k_{1r} & k_{1r-1} & \cdots & \cdots & k_{12} & k_{11} & k_{r+1} & k_{r+2} & \cdots & k_n \end{bmatrix}, \quad \mathbf{D} = 0$$

This form of the system model is known as the *Jordan canonical form*. The complete analysis of the Jordan canonical form requires a lot of space and time. However, understanding the Jordan form is crucial for correct interpretation of system stability, hence in the following chapter, the Jordan form will be completely explained.

Example 3.6: Find the state space model from the transfer function using the Jordan canonical form

$$G(s) = \frac{s^2 + 6s + 8}{(s+1)^2(s+3)}$$

This transfer function can be expanded as

$$G(s) = \frac{1.25}{s+1} + \frac{1.5}{(s+1)^2} - \frac{0.25}{s+3}$$

so that the required state space model is

$$\mathbf{x} = \begin{bmatrix} -1 & 1 & 0 \\ 0 & -1 & 0 \\ 0 & 0 & -3 \end{bmatrix} \mathbf{x} + \begin{bmatrix} 0 \\ 1 \\ 1 \end{bmatrix} u$$

$$y = \begin{bmatrix} 1.5 & 1.25 & -0.25 \end{bmatrix} \mathbf{x}$$

◇

3.2 Time Response from the State Equation

The solution of the state space equations (3.1) and (3.2) can be obtained either in the time domain by solving the corresponding matrix differential equation directly or in the frequency domain by exploiting the power of the Laplace transform. Both methods will be presented in this section.

3.2.1 Time Domain Solution

For the purpose of solving the state equation (3.1), let us first suppose that the system is in the scalar form

$$\dot{x} = ax + bu \tag{3.31}$$

with a known initial condition $x(0) = x_0$. It is very well known from the elementary theory of differential equations that the solution of (3.31) is

$$x(t) = e^{at}x_0 + \int_0^t e^{a(t-\tau)}bu(\tau)d\tau \tag{3.32}$$

The exponential term e^{at} can be expressed using the Taylor series expansion about $t_0 = 0$ as

$$e^{at} = 1 + at + \frac{1}{2!}a^2t^2 + \frac{1}{3!}a^3t^3 + \cdots = \sum_{i=0}^{\infty} \frac{1}{i!}(at)^i \tag{3.33}$$

Analogously, in the following we prove that the solution of a general nth-order matrix state space differential equation (3.1) is given by

$$\mathbf{x}(t) = e^{\mathbf{A}t}\mathbf{x}(0) + \int_0^t e^{\mathbf{A}(t-\tau)}\mathbf{B}\mathbf{u}(\tau)d\tau \tag{3.34}$$

For simplicity, we first consider the homogeneous system without control input, that is

$$\dot{\mathbf{x}} = \mathbf{A}\mathbf{x}, \qquad \mathbf{x}(0) = \mathbf{x}_o \tag{3.35}$$

By analogy with the scalar case, we expect the solution of this differential equation to be

$$\mathbf{x}(t) = e^{\mathbf{A}t}\mathbf{x}(0) \tag{3.36}$$

We shall prove that this is indeed a solution if (3.36) satisfies differential equation (3.35), where *the matrix exponential is defined by using the Taylor series expansion as*

$$e^{\mathbf{A}t} = \mathbf{I} + \mathbf{A}t + \frac{1}{2!}\mathbf{A}^2 t^2 + \frac{1}{3!}\mathbf{A}^3 t^3 + \cdots = \sum_{i=0}^{\infty} \frac{1}{i!}\mathbf{A}^i t^i \tag{3.37}$$

The proof is simple and is obtained by taking the derivative of the right-hand side of (3.37), that is

$$\frac{de^{\mathbf{A}t}}{dt} = \left(\frac{d}{dt}\right)\left(\mathbf{I} + \mathbf{A}t + \frac{1}{2!}\mathbf{A}^2 t^2 + \cdots\right)$$

$$= \mathbf{A} + \frac{2}{2!}\mathbf{A}^2 t + \frac{3}{3!}\mathbf{A}^3 t^2 + \cdots = \mathbf{A}\left(\mathbf{I} + \frac{1}{1!}\mathbf{A}t + \frac{1}{2!}\mathbf{A}^2 t^2 + \cdots\right)$$

$$= \mathbf{A}e^{\mathbf{A}t} = e^{\mathbf{A}t}\mathbf{A}$$

Now, substitution of (3.36) into differential equation (3.35) yields

$$\dot{\mathbf{x}} = \frac{d}{dt}\mathbf{x} = \frac{d}{dt}e^{\mathbf{A}t}\mathbf{x}(0) = \mathbf{A}e^{\mathbf{A}t}\mathbf{x}(0) = \mathbf{A}\mathbf{x}(t)$$

so that matrix differential equation (3.35) is satisfied, and hence $\mathbf{x}(t) = e^{\mathbf{A}t}\mathbf{x}(0)$ is its solution.

The matrix $e^{\mathbf{A}t}$ is known as the *state transition matrix* because it relates the system state at time t to that at time zero, and is denoted by

$$\Phi(t) = e^{\mathbf{A}t} \tag{3.38}$$

The state transition matrix as a time function depends only on the matrix $\mathbf{A}$. Therefore $\Phi(t)$ completely describes the internal behavior of the system, when the external influence (control input $\mathbf{u}(t)$) is absent. The system transition matrix plays a fundamental role in the theory of linear dynamical systems. In the following, we state and verify the main properties of this matrix, which is represented in the symbolic form by $e^{\mathbf{A}t}$ and so far defined only by (3.37).

Properties of the State Transition Matrix

It can be easily verified, by taking the derivative of

$$x(t) = \Phi(t)x(0)$$

that the state transition matrix satisfies the linear homogeneous state equation (3.1) with the initial condition equal to an identity matrix, that is

$$\frac{d\Phi(t)}{dt} = A\Phi(t), \qquad \Phi(0) = I \tag{3.39}$$

The main properties of the matrix $\Phi(t)$, which follow from (3.37) and (3.38), are as follows:

(a) $\Phi(0) = I$

(b) $\Phi^{-1}(t) = \Phi(-t) \Rightarrow \Phi(t)$ is nonsingular for every t

(c) $\Phi(t_2 - t_0) = \Phi(t_2 - t_1)\Phi(t_1 - t_0)$

(d) $\Phi(t)^i = \Phi(it)$, for $i \in N$

The proofs are straightforward. Property (a) is obtained when $t = 0$ is substituted into the series expansion of $e^{At} = I + At + \cdots$.

Property (b) holds, since

$$\left(e^{At}\right)^{-1}e^{At} = I$$

which after multiplication from the right by e^{-At} implies

$$\left(e^{At}\right)^{-1}e^0 = e^{-At} \Rightarrow \Phi^{-1}(t) = \Phi(-t)$$

and (c) follows from

$$\Phi(t_2 - t_0) = e^{A(t_2-t_0)} = e^{A(t_2-t_1+t_1-t_0)}$$
$$= e^{A(t_2-t_0)}e^{A(t_1-t_0)} = \Phi(t_2 - t_1)\Phi(t_1 - t_0)$$

Property (d) is proved by using the fact that

$$\Phi(t)^i = \left(e^{At}\right)^i = e^{A(it)} = \Phi(it)$$

In addition to properties (a), (b), (c), and (d), we have already established one additional property, namely the derivative property, as

(e)

$$\frac{d}{dt}\Phi(t) = \mathbf{A}\Phi(t) \;\Leftrightarrow\; \frac{d}{dt}e^{\mathbf{A}t} = e^{\mathbf{A}t}\mathbf{A} = \mathbf{A}e^{\mathbf{A}t}$$

The state transition matrix $\Phi(t)$ can be found by using several methods. Two of them are given in this chapter—formulas (3.37) and (3.49). The third one, very popular in linear algebra, is based on the Cayley–Hamilton theorem and is given in Appendix C.

In the case when the control vector $\mathbf{u}(t)$ is present in the system (forced response)

$$\dot{\mathbf{x}} = \mathbf{A}\mathbf{x} + \mathbf{B}\mathbf{u}, \quad \mathbf{x}(0) = \mathbf{x}_0$$

we look for the solution of the state space equation in the form

$$\mathbf{x}(t) = e^{\mathbf{A}t}\mathbf{f}(t) \tag{3.40}$$

Then

$$\dot{\mathbf{x}}(t) = \mathbf{A}e^{\mathbf{A}t}\mathbf{f}(t) + e^{\mathbf{A}t}\dot{\mathbf{f}}(t) = \mathbf{A}\mathbf{x} + e^{\mathbf{A}t}\dot{\mathbf{f}} \tag{3.41}$$

It follows from (3.1) and (3.41) that

$$e^{\mathbf{A}t}\dot{\mathbf{f}}(t) = \mathbf{B}\mathbf{u} \tag{3.42}$$

From (3.42) we have

$$\dot{\mathbf{f}}(t) = \left(e^{\mathbf{A}t}\right)^{-1}\mathbf{B}\mathbf{u} = e^{-\mathbf{A}t}\mathbf{B}\mathbf{u} \tag{3.43}$$

Integrating this equation, bearing in mind that $\mathbf{x}(0) = e^{\mathbf{A}\cdot 0}\mathbf{f}(0) = \mathbf{f}(0)$, we get

$$\mathbf{f}(t) - \mathbf{f}(0) = \int_0^t e^{-\mathbf{A}\tau}\mathbf{B}\mathbf{u}(\tau)d\tau \tag{3.44}$$

Substitution of the last expression in (3.40) gives the required solution

$$\mathbf{x}(t) = e^{\mathbf{A}t}\mathbf{x}(0) + \int_0^t e^{\mathbf{A}(t-\tau)}\mathbf{B}\mathbf{u}(\tau)d\tau \tag{3.45}$$

or

$$\mathbf{x}(t) = \Phi(t)\mathbf{x}(0) + \int_0^t \Phi(t-\tau)\mathbf{B}\mathbf{u}(\tau)d\tau \tag{3.46}$$

When the initial state of the system is known at time t_0, rather than at time $t = 0$, the solution of the state equation is similarly obtained as

$$
\mathbf{x}(t) = \Phi(t - t_0)\mathbf{x}(t_0) + \int_{t_0}^{t} \Phi(t - \tau)\mathbf{B}\mathbf{u}(\tau)d\tau
$$

$$
= e^{\mathbf{A}(t-t_0)}\mathbf{x}(t_0) + \int_{t_0}^{t} e^{\mathbf{A}(t-\tau)}\mathbf{B}\mathbf{u}(\tau)d\tau
$$

(3.47)

This can be easily verified by repeating steps (3.40)–(3.45) with $\mathbf{x}(t_0) = \mathbf{x}_o$ and $\mathbf{x}(t_0) = e^{\mathbf{A}t_0}\mathbf{f}(t_0)$.

Example 3.7: For the system given in Example 3.4 find the state transition matrix $\Phi(t)$. Evaluate $\Phi(1)$ using the MATLAB function expm. Assuming that the initial state of the system is zero, find the state response to a unit step. Check the solution obtained by using the MATLAB function step.

At the present time we are able to find the state transition matrix (matrix exponential) by using formula (3.37), which deals with an infinite series, and hence is not very convenient for calculations. Better ways to find $\Phi(t)$ are the method based on the Cayley–Hamilton theorem (see Appendix C) and the formula based on the Laplace transform, see formula (3.49). However, in this problem, if we start with the uncoupled (modal) state space form of the system considered in Example 3.4, we can avoid using any of the above methods in order to find the state transition matrix. Namely, *for diagonal matrices only*, it is easy to show that

$$
\Phi(t) = e^{\begin{bmatrix} -1 & 0 & 0 \\ 0 & -2 & 0 \\ 0 & 0 & -3 \end{bmatrix}t} = \begin{bmatrix} e^{-t} & 0 & 0 \\ 0 & e^{-2t} & 0 \\ 0 & 0 & e^{-3t} \end{bmatrix}
$$

Using the MATLAB function for evaluating the matrix exponential as expm(A*1), we get

$$
\Phi(1) = \begin{bmatrix} e^{-1} & 0 & 0 \\ 0 & e^{-2} & 0 \\ 0 & 0 & e^{-3} \end{bmatrix} = \begin{bmatrix} 0.3679 & 0 & 0 \\ 0 & 0.1353 & 0 \\ 0 & 0 & 0.0498 \end{bmatrix}
$$

The state response to a unit step is computed from (3.46) as

$$\mathbf{x}(t) = \int_0^t \Phi(t-\tau)\mathbf{B}u(\tau)d\tau = \int_0^t \begin{bmatrix} e^{-(t-\tau)} & 0 & 0 \\ 0 & e^{-2(t-\tau)} & 0 \\ 0 & 0 & e^{-3(t-\tau)} \end{bmatrix} \begin{bmatrix} 1 \\ 1 \\ 1 \end{bmatrix} \cdot 1 d\tau$$

$$= \int_0^t \begin{bmatrix} e^{-(t-\tau)} \\ e^{-2(t-\tau)} \\ e^{-3(t-\tau)} \end{bmatrix} d\tau = \begin{bmatrix} 1 - e^{-t} \\ 0.5\left(1 - e^{-2t}\right) \\ 0.333\left(1 - e^{-3t}\right) \end{bmatrix}$$

The step responses of system states, obtained by using MATLAB statements
`[y,x]=step(A,B,C,D)` and `plot(x)`, with

$$\mathbf{A} = \begin{bmatrix} -1 & 0 & 0 \\ 0 & -2 & 0 \\ 0 & 0 & -3 \end{bmatrix}, \ \mathbf{B} = \begin{bmatrix} 1 \\ 1 \\ 1 \end{bmatrix}, \ \mathbf{C} = \begin{bmatrix} 6 & -6 & 1 \end{bmatrix}, \ \mathbf{D} = 0$$

are shown in Figure 3.7.

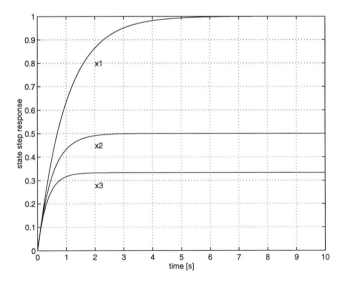

Figure 3.7: The state responses for Example 3.7

◇

3.2.2 Solution Using the Laplace Transform

The time trajectory of the state vector $\mathbf{x}(t)$ can be also found using the Laplace transform method. The main properties of the Laplace transform and common transform pairs are given in Appendix A.

The Laplace transform applied to the state equation (3.1) gives

$$s\mathbf{X}(s) - \mathbf{x}(0^-) = \mathbf{A}\mathbf{X}(s) + \mathbf{B}\mathbf{U}(s)$$

or

$$\mathbf{X}(s) = (s\mathbf{I} - \mathbf{A})^{-1}\mathbf{x}(0^-) + (s\mathbf{I} - \mathbf{A})^{-1}\mathbf{B}\mathbf{U}(s) \qquad (3.48)$$

Let us assume that $\mathbf{x}(0) = \mathbf{x}(0^-)$. Comparing equations (3.46) and (3.48), it is easy to see that the term $(s\mathbf{I} - \mathbf{A})^{-1}$ is the Laplace transform of the state transition matrix, that is

$$\Phi(s) = (s\mathbf{I} - \mathbf{A})^{-1} = \frac{1}{\det(s\mathbf{I} - \mathbf{A})}\mathrm{adj}(s\mathbf{I} - \mathbf{A}) = \mathcal{L}\{\Phi(t)\} \qquad (3.49)$$

The time form of the state vector $\mathbf{x}(t)$ is obtained by applying the inverse Laplace transform to

$$\mathbf{X}(s) = \Phi(s)\mathbf{x}(0) + \Phi(s)\mathbf{B}\mathbf{U}(s) \qquad (3.50)$$

Note that the second term on the right-hand side corresponds in the time domain to the convolution integral, so that we have

$$\mathbf{x}(t) = e^{\mathbf{A}t}\mathbf{x}(0) + \int_0^t e^{\mathbf{A}(t-\tau)}\mathbf{B}\mathbf{u}(\tau)d\tau \qquad (3.51)$$

Once the state vector $\mathbf{x}(t)$ is determined, the output vector $\mathbf{y}(t)$ of the system is simply obtained by substitution of $\mathbf{x}(t)$ into equation (3.2), that is

$$\mathbf{y}(t) = \mathbf{C}\Phi(t)\mathbf{x}(0) + \mathbf{C}\int_0^t \Phi(t-\tau)\mathbf{B}\mathbf{u}(\tau)d\tau + \mathbf{D}\mathbf{u}(t) \qquad (3.52)$$

or, in the complex domain

$$\mathbf{Y}(s) = \mathbf{C}\Phi(s)\mathbf{x}(0) + [\mathbf{C}\Phi(s)\mathbf{B} + \mathbf{D}]\mathbf{U}(s) \qquad (3.53)$$

3.2.3 State Space Model and Transfer Function

The matrix that establishes a relationship between the output vector $\mathbf{Y}(s)$ and the input vector $\mathbf{U}(s)$, for the zero initial conditions, $\mathbf{x}(0) = 0$, is called the *system matrix transfer function*. From (3.53) it is given by

$$\mathbf{G}(s) = \mathbf{C}(s\mathbf{I} - \mathbf{A})^{-1}\mathbf{B} + \mathbf{D} \tag{3.54}$$

Note that (3.54) represents the open-loop system matrix transfer function.

Example 3.8: Find the transfer function for the system given in Example 3.4.

It is the easiest to use modal (uncoupled) canonical form, which leads to

$$G(s) = \begin{bmatrix} 6 & -6 & 1 \end{bmatrix} \begin{bmatrix} s+1 & 0 & 0 \\ 0 & s+2 & 0 \\ 0 & 0 & s+3 \end{bmatrix}^{-1} \begin{bmatrix} 1 \\ 1 \\ 1 \end{bmatrix}$$

$$= \begin{bmatrix} 6 & -6 & 1 \end{bmatrix} \begin{bmatrix} \frac{1}{s+1} & 0 & 0 \\ 0 & \frac{1}{s+2} & 0 \\ 0 & 0 & \frac{1}{s+3} \end{bmatrix} \begin{bmatrix} 1 \\ 1 \\ 1 \end{bmatrix}$$

$$= \frac{6}{s+1} - \frac{6}{s+2} + \frac{1}{s+3} = \frac{(s+5)(s+4)}{(s+1)(s+2)(s+3)}$$

If we start with controller canonical form we will get, after some algebra,

$$G(s) = \begin{bmatrix} 20 & 9 & 1 \end{bmatrix} \begin{bmatrix} s+6 & -1 & 0 \\ 11 & s & -1 \\ 6 & 0 & s \end{bmatrix}^{-1} \begin{bmatrix} 0 \\ 0 \\ 1 \end{bmatrix}$$

$$= \begin{bmatrix} 20 & 9 & 1 \end{bmatrix} \frac{1}{s^3 + 6s^2 + 11s + 6} \begin{bmatrix} s^2 + 6s + 11 & s+6 & 1 \\ -6 & s(s+6) & s \\ -6s & -11s - 6 & s^2 \end{bmatrix} \begin{bmatrix} 0 \\ 0 \\ 1 \end{bmatrix}$$

$$= \frac{s^2 + 9s + 20}{s^3 + 6s^2 + 11s + 6}$$

Note that the MATLAB function `ss2tf` can be used to solve the above problem.

◇

3.3 Discrete-Time Models

Discrete-time systems are either inherently discrete (e.g. models of bank accounts, national economy growth models, population growth models, digital words) or they are obtained as a result of sampling (discretization) of continuous-time systems. In such kinds of systems, inputs, state space variables, and outputs have the discrete form and the system models can be represented in the form of transition tables.

The mathematical model of a discrete-time system can be written in terms of a recursive formula by using linear matrix difference equations as

$$\mathbf{x}[(k+1)T] = \mathbf{A}_d\mathbf{x}(kT) + \mathbf{B}_d\mathbf{u}(kT)$$
$$\mathbf{y}(kT) = \mathbf{C}_d\mathbf{x}(kT) + \mathbf{D}_d\mathbf{u}(kT) \tag{3.55}$$

Here T represents the sampling interval, which may be omitted for brevity. Even more, in the case of inherent discrete systems, there is no need to introduce the notion of the sampling interval T so that these systems are described by (3.55) with $T = 1$.

Similarly to continuous-time systems, discrete state space equations can be derived either from difference equations (Subsection 3.3.1) or from discrete transfer functions using simulation diagrams (Subsection 3.3.2). In Subsection 3.3.3 we show how to discretize continuous-time linear systems and obtain discrete-time ones. In the rest of the section we parallel most of the results obtained in previous sections for continuous-time systems.

3.3.1 Difference Equations and State Space Form

An nth-order difference equation is given by

$$y(k+n) + a_{n-1}y(k+n-1) + \cdots + a_1y(k+1) + a_0y(k)$$
$$= b_nu(k+n) + b_{n-1}u(k+n-1) + \cdots + b_1u(k+1) + b_0u(k) \tag{3.56}$$

This equation expresses all values in terms of discrete-time k.

The corresponding state space equation can be derived by using the same techniques as in the continuous-time case. For example, for phase variable canonical form (controller canonical form) in discrete-time, we have

$$
\begin{bmatrix} x_1(k+1) \\ x_2(k+1) \\ \vdots \\ x_{n-1}(k+1) \\ x_n(k+1) \end{bmatrix} = \begin{bmatrix} 0 & 1 & 0 & \cdots & 0 \\ 0 & 0 & 1 & \cdots & 0 \\ \vdots & \vdots & \vdots & \ddots & \vdots \\ 0 & 0 & 0 & \cdots & 1 \\ -a_0 & -a_1 & -a_2 & \cdots & -a_{n-1} \end{bmatrix} \begin{bmatrix} x_1(k) \\ x_2(k) \\ \vdots \\ x_{n-1}(k) \\ x_n(k) \end{bmatrix} + \begin{bmatrix} 0 \\ 0 \\ \vdots \\ 0 \\ 1 \end{bmatrix} u(k)
$$

$$
y(k) = [(b_0 - a_0 b_n) \quad (b_1 - a_1 b_n) \quad \cdots \quad (b_{n-1} - a_{n-1} b_n)] \begin{bmatrix} x_1(k) \\ x_2(k) \\ \vdots \\ x_n(k) \end{bmatrix}
$$

$$
+ b_n u(k)
$$

(3.57)

Note that the transformation equations, dual to the continuous-time ones (3.11)–(3.13), are given in the discrete-time domain by

$$
\xi(k+n) + a_{n-1}\xi(k+n-1) + \cdots + a_1\xi(k+1) + a_0\xi(k) = u(k) \quad (3.58)
$$

$$
\begin{aligned}
x_1(k) &= \xi(k) \\
x_2(k) &= \xi(k+1) \\
x_3(k) &= \xi(k+2) \\
&\vdots \\
x_n(k) &= \xi(k+n-1)
\end{aligned}
$$

(3.59)

$$
y(k) = b_0\xi(k) + b_1\xi(k+1) + b_2\xi(k+2) + \cdots + b_n\xi(k+n) \quad (3.60)
$$

Eliminating $\xi(k+n)$ from (3.60) by using (3.58) and (3.59), the output equation given in (3.57) is obtained.

3.3.2 Discrete Transfer Function and State Space Model

The derivation of state space equations from discrete transfer functions, based on simulation diagrams, is very similar to the continuous-time case. The only difference is that in simulation diagrams the integration block $1/s$ is replaced by the unit delay element z^{-1}. *The state variables are selected as outputs of these delay elements.* We shall illustrate this method by an example.

Example 3.9: Find two state space forms for the transfer function

$$\frac{Y(z)}{U(z)} = \frac{z + 1.1}{(z - 0.9)(z + 0.7)(z - 0.7)}$$

We solve this problem by using both direct (a) and parallel (b) programming techniques.

(a) The transfer function can be rewritten as

$$\frac{Y(z)}{U(z)} = \frac{z + 1.1}{z^3 - 0.9z^2 - 0.49z + 0.441}$$

The simulation diagram for this transfer function is shown in Figure 3.8.

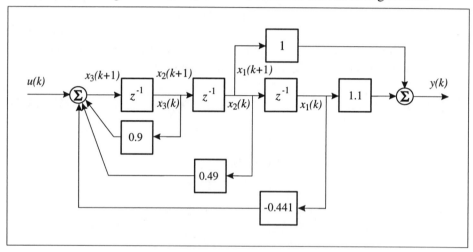

Figure 3.8: Simulation diagram for direct programming
in discrete-time domain (controller canonical form)

The state space model is obtained from this simulation diagram by using the outputs of delay elements as state variables. It is given by

$$\mathbf{x}(k + 1) = \begin{bmatrix} 0 & 1 & 0 \\ 0 & 0 & 1 \\ -0.441 & 0.49 & 0.9 \end{bmatrix} \mathbf{x}(k) + \begin{bmatrix} 0 \\ 0 \\ 1 \end{bmatrix} u(k)$$

$$y(k) = [\,1.1 \quad 1 \quad 0\,]\mathbf{x}(k)$$

Note that controller canonical form could have been obtained without drawing the simulation diagram. We know that this form is identical to phase variable canonical form, which is represented by (3.57). Identifying the corresponding coefficients in the original transfer function, the desired state space form is obtained directly from (3.57). We have used the above method in order to demonstrate at the same time the procedure of drawing simulation diagrams in the discrete-time domain.

(b) Employing the partial fraction expansion (with help of the MATLAB function `residue`), we get

$$\frac{Y(z)}{U(z)} = \frac{6.25}{z - 0.9} + \frac{0.1786}{z + 0.7} + \frac{-6.4286}{z - 0.7}$$

Since the poles of the transfer function are real and distinct we get the modal canonical form as

$$\mathbf{x}(k+1) = \begin{bmatrix} 0.9 & 0 & 0 \\ 0 & -0.7 & 0 \\ 0 & 0 & 0.7 \end{bmatrix} \mathbf{x}(k) + \begin{bmatrix} 1 \\ 1 \\ 1 \end{bmatrix} u(k)$$

$$y(k) = [6.25 \quad 0.1786 \quad -6.4286]\mathbf{x}(k)$$

$\diamond$

3.3.3 Discretization of Continuous-Time Systems

Real physical dynamic systems are continuous in nature. In this section, we show how to obtain discrete-time state space models from continuous-time system models. Assume that the plant is linear, continuous, and time invariant with r-inputs and p-outputs (see Figure 3.9). Inputs are sampled by using the zero-order hold (ZOH) device. This device samples inputs at discrete-time instants kT (see Figure 3.10b) and the values obtained for vector $\mathbf{u}(kT)$ are held until $(k+1)T$. The corresponding signal is given in Figure 3.10c.

The state space model of such a plant is

$$\begin{aligned} \dot{\mathbf{x}}(t) &= \mathbf{A}\mathbf{x}(t) + \mathbf{B}\mathbf{m}(t) \\ \mathbf{y}(t) &= \mathbf{C}\mathbf{x}(t) + \mathbf{D}\mathbf{m}(t) \end{aligned} \tag{3.61}$$

These equations define states and outputs during the sampling interval $kT \leq t < (k+1)T$. Input signals $m_i(t), i = 1, ..., r$, are defined by

$$m_i(t) = m_i(kT) = u_i(kT), \quad kT \leq t < (k+1)T, \quad k = 0, 1, 2, ... \tag{3.62}$$

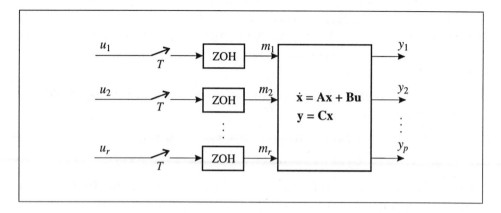

Figure 3.9: Sampling in a multivariable controlled plant

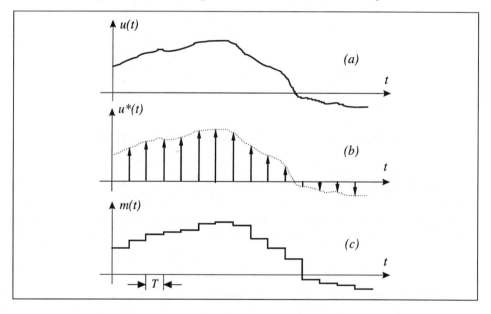

Figure 3.10: Transformation of a continuous-time
input signal by the zero-order hold element

In the following, we show how to perform discretization of a continuous-time state space model (3.61) and obtain a discrete-time state space model having the form of (3.55) together with the corresponding expressions for matrices $\mathbf{A}_d, \mathbf{B}_d, \mathbf{C}_d,$ and $\mathbf{D}_d$.

Consider equation (3.45) with $t = T$

$$\mathbf{x}(T) = e^{\mathbf{A}T}\mathbf{x}(0) + \int_0^T e^{\mathbf{A}(T-\tau)}\mathbf{B}\mathbf{u}(0)d\tau$$

$$= e^{\mathbf{A}T}\mathbf{x}(0) + e^{\mathbf{A}T}\int_0^T e^{-\mathbf{A}\tau}d\tau\mathbf{B}\mathbf{u}(0) = \Phi(T)x(0) + \int_0^T \Phi(T-\tau)d\tau\mathbf{B}\mathbf{u}(0)$$

$$(3.63)$$

which can be written in the form

$$\mathbf{x}(T) = \mathbf{A}_d\mathbf{x}(0) + \mathbf{B}_d\mathbf{u}(0) \qquad (3.64)$$

Comparing (3.63) and (3.64) we can find expressions for $\mathbf{A}_d$ and $\mathbf{B}_d$. They are given by

$$\mathbf{A}_d = e^{\mathbf{A}T} = \Phi(T)$$

$$(3.65)$$

$$\mathbf{B}_d = e^{\mathbf{A}T}\int_0^T e^{-\mathbf{A}\tau}d\tau\mathbf{B} = \int_0^T e^{\mathbf{A}(T-\tau)}d\tau \cdot \mathbf{B} = \int_0^T e^{\mathbf{A}\sigma}d\sigma \cdot \mathbf{B}$$

Note that $\mathbf{A}_d$ and $\mathbf{B}_d$ are obtained for the time interval from 0 to T. However, it can easily be shown that due to system time invariance the same expressions for $\mathbf{A}_d$ and $\mathbf{B}_d$ are obtained for any time interval. Namely, steps (3.63)–(3.65) can be repeated for succeeding time intervals $2T, 3T, ..., (k+1)T$ with initial conditions taken, respectively, as $\mathbf{x}(T), \mathbf{x}(2T), ..., \mathbf{x}(kT)$. Therefore, for the time instant $t = (k+1)T$ and for $t_0 = kT$, we have from (3.47)

$$\mathbf{x}[(k+1)T] = \Phi[(k+1)T - kT]\mathbf{x}(kT) + \int_{kT}^{(k+1)T} \Phi[(k+1)T - \tau]d\tau\mathbf{B}\mathbf{u}(kT)$$

$$= \mathbf{A}_d\mathbf{x}(kT) + \mathbf{B}_d\mathbf{u}(kT)$$

$$(3.66)$$

From the above equation we see that the matrices $\mathbf{A}_d$ and $\mathbf{B}_d$ are given by

$$\mathbf{A}_d = \Phi[(k+1)T - kT] = \Phi(T) = e^{\mathbf{A}T}$$

$$(3.67)$$

$$\mathbf{B}_d = \int_{kT}^{(k+1)T} \Phi[(k+1)T - \tau]d\tau\mathbf{B} = \int_0^T \Phi(\sigma)d\sigma\mathbf{B} = \int_0^T e^{\mathbf{A}\sigma}d\sigma\mathbf{B}$$

The last equality is obtained by using change of variables as $\sigma = (k+1)T - \tau$. Since (3.65) and (3.67) are identical, we conclude that for a time invariant continuous-time linear system, the discretization procedure yields a time invariant discrete-time linear system whose matrices $\mathbf{A}_d$ and $\mathbf{B}_d$ depend only on $\mathbf{A}$, $\mathbf{B}$, and the sampling interval T.

In a similar manner the output equation (3.61) at $t = kT$ is given by

$$\mathbf{y}(kT) = \mathbf{C}\mathbf{x}(kT) + \mathbf{D}\mathbf{u}(kT) \tag{3.68}$$

Comparing this equation with the general output equation of linear discrete-time systems (3.55), we conclude that

$$\mathbf{C}_d = \mathbf{C}, \quad \mathbf{D}_d = \mathbf{D} \tag{3.69}$$

In the literature one can find several methods for discretization of continuous-time linear systems. The discretization technique presented in this section is known as the *integral approximation method.*

In the case of discrete-time linear systems obtained by sampling continuous-time linear systems, the matrix $\mathbf{A}_d$, given by (3.65), can be determined from the infinite series

$$\mathbf{A}_d = e^{\mathbf{A}T} = \mathbf{I} + \mathbf{A}T + \frac{1}{2!}\mathbf{A}^2 T^2 + \cdots = \sum_{i=0}^{\infty} \frac{1}{i!}\mathbf{A}^i T^i \tag{3.70}$$

It can be also obtained either by using formula (3.49) or the method based on the Cayley–Hamilton theorem and setting $t = T$ in $\Phi(t) = e^{\mathbf{A}t}$. Also, in order to evaluate $e^{\mathbf{A}T}$ we can use MATLAB function expm(A*T).

To find $\mathbf{B}_d$, the second expression in (3.65) is integrated to give (see Appendix C—matrix integrals)

$$\mathbf{B}_d = e^{\mathbf{A}T}\left(-e^{-\mathbf{A}T}\mathbf{A}^{-1} + \mathbf{A}^{-1}\right)\mathbf{B} = (\mathbf{A}_d - \mathbf{I})\mathbf{A}^{-1}\mathbf{B} \tag{3.71}$$

which is valid under the assumption that $\mathbf{A}$ is invertible. If $\mathbf{A}$ is singular, $\mathbf{B}_d$ can be determined from

$$\mathbf{B}_d = \left(\sum_{i=1}^{\infty} \frac{T^i}{i!}\mathbf{A}^{i-1}\right)\mathbf{B} = \left(\sum_{i=0}^{\infty} \frac{T^{i+1}}{(i+1)!}\mathbf{A}^i\right)\mathbf{B} \tag{3.72}$$

which is obtained by using (3.37) in (3.67) and performing the corresponding integration. Note that the above sum converges quite quickly so that only a few terms give quite an accurate expression for $\mathbf{B}_d$.

Example 3.10: Find the discrete-time state space model of a continuous-time system

$$\dot{\mathbf{x}} = \begin{bmatrix} 0 & 1 \\ -2 & -3 \end{bmatrix} \mathbf{x} + \begin{bmatrix} 0 \\ 1 \end{bmatrix}$$

$$y = \begin{bmatrix} 1 & 0 \end{bmatrix} \mathbf{x}$$

The sampling period T is equal to 0.1.

According to (3.65) and (3.69), we have from (3.49)

$$\mathbf{A}_d = \Phi(T) = \begin{bmatrix} 2e^{-T} - e^{-2T} & e^{-T} - e^{-2T} \\ 2e^{-2T} - 2e^{-T} & 2e^{-2T} - e^{-T} \end{bmatrix} = \begin{bmatrix} 0.9909 & 0.0861 \\ -0.1722 & 0.7326 \end{bmatrix}$$

$$\mathbf{B}_d = (\mathbf{A}_d - \mathbf{I})\mathbf{A}^{-1}\mathbf{B} = \begin{bmatrix} \frac{1}{2}(1 + e^{-2T}) - e^{-T} \\ e^{-T} - e^{-2T} \end{bmatrix} = \begin{bmatrix} 0.0045 \\ 0.0861 \end{bmatrix}$$

$$\mathbf{C}_d = \begin{bmatrix} 1 & 0 \end{bmatrix}, \quad \mathbf{D}_d = 0$$

The same result is obtained by using the MATLAB function for discretization of a continuous state space model as [Ad,Bd]=c2d(A,B,T).

◇

3.3.4 Solution of the Discrete-Time State Equation

The objective of this section is to find the solution of the difference state equation (3.55) for the given initial state $\mathbf{x}(0)$ and the control signal $\mathbf{u}(k)$ at the sampling instants $T, 2T, ..., kT$. For simplicity we assume $T = 1$.

From the state equation $\mathbf{x}(k+1) = \mathbf{A}_d x(k) + \mathbf{B}_d \mathbf{u}(k)$, for $k = 0, 1, ...,$ $N - 1$, it follows

$$\mathbf{x}(1) = \mathbf{A}_d \mathbf{x}(0) + \mathbf{B}_d \mathbf{u}(0)$$

$$\mathbf{x}(2) = \mathbf{A}_d \mathbf{x}(1) + \mathbf{B}_d \mathbf{u}(1) = \mathbf{A}_d^2 \mathbf{x}(0) + \mathbf{A}_d \mathbf{B}_d \mathbf{u}(0) + \mathbf{B}_d \mathbf{u}(1)$$

$$\mathbf{x}(3) = \mathbf{A}_d \mathbf{x}(2) + \mathbf{B}_d \mathbf{u}(2) = \mathbf{A}_d^3 \mathbf{x}(0) + \mathbf{A}_d^2 \mathbf{B}_d \mathbf{u}(0) + \mathbf{A}_d \mathbf{B}_d \mathbf{u}(1) + \mathbf{B}_d \mathbf{u}(2)$$

$\vdots$

$$x(N) = \mathbf{A}_d x(N-1) + \mathbf{B}_d \mathbf{u}(N-1) = \mathbf{A}_d^N x(0) + \sum_{i=0}^{N-1} \mathbf{A}_d^{N-i-1} \mathbf{B}_d \mathbf{u}(i) \qquad (3.73)$$

Using the notion of the *discrete-time state transition matrix* defined by

$$\Phi_d(k) = \mathbf{A}_d^k \qquad (3.74)$$

we get

$$x(N) = \Phi_d(N)x(0) + \sum_{i=0}^{N-1} \Phi_d(N-i-1)\mathbf{B}_d \mathbf{u}(i) \qquad (3.75)$$

Note that the discrete-time state transition matrix relates the state of an input-free system at initial time $(k = 0)$ to the state of the system at any other time $k > 0$, that is

$$x(k) = \Phi_d(k)x(0) = \mathbf{A}_d^k x(0) \qquad (3.76)$$

It is easy to verify that the discrete-time state transition matrix has the following properties

(a) $\Phi_d(0) = \mathbf{A}_d^0 = \mathbf{I} \;\Leftarrow\; x(0) = \Phi_d(0)x(0)$

(b) $\Phi_d(k_2 - k_1) = \Phi_d(k_2 - k_1)\Phi_d(k_1 - k_0) = \mathbf{A}_d^{k_2-k_1}\mathbf{A}_d^{k_1-k_o} = \mathbf{A}_d^{k_2-k_o}$

(c) $\Phi_d^i(k) = \Phi_d(ik) \;\Leftarrow\; \left(\mathbf{A}_d^k\right)^i = \mathbf{A}_d^{ik}$

(d) $\Phi_d(k+1) = \mathbf{A}_d\Phi_d(k), \quad \Phi_d(0) = \mathbf{I}$

The last property follows from

$$x(k+1) = \mathbf{A}_d x(k) \;\Rightarrow\; \Phi_d(k+1)x(0) = \mathbf{A}_d\Phi_d(k)x(0)$$

It is important to point out that the discrete-time state transition matrix may be singular, which follows from the fact that $\mathbf{A}_d^k$ is nonsingular if and only if the matrix $\mathbf{A}_d$ is nonsingular. In the case of inherent discrete-time systems, the matrix $\mathbf{A}_d$ may be singular in general. However, if $\mathbf{A}_d$ is obtained through the discretization procedure of a continuous-time linear system, like in (3.65), then

$$(\mathbf{A}_d)^{-1} = \left(e^{\mathbf{A}T}\right)^{-1} = e^{-\mathbf{A}T}$$

so that the discrete-time state transition matrix is nonsingular in this case.

Remark 1: If the initial value of the state vector is not $\mathbf{x}(0)$ but $\mathbf{x}(k_0)$, then the solution (3.75) is

$$\mathbf{x}(k_0 + N) = \Phi_d(N)\mathbf{x}(k_0) + \sum_{i=0}^{N-1} \Phi_d(N - i + 1)\mathbf{B}_d\mathbf{u}(k_0 + i) \qquad (3.77)$$

The output of the system at sampling instant $k = N$ is obtained by substituting $\mathbf{x}(k)$ from (3.75) into the output equation, producing

$$\mathbf{y}(N) = \mathbf{C}_d\Phi_d(N)\mathbf{x}(0) + \mathbf{C}_d\sum_{i=0}^{N-1} \Phi(N - i + 1)\mathbf{B}_d\mathbf{u}(i) + \mathbf{D}_d\mathbf{u}(N) \qquad (3.78)$$

Note that for $T \neq 1$, equations (3.75) and (3.78) are modified as

$$\mathbf{x}(NT) = \Phi_d(NT)\mathbf{x}(0) + \sum_{i=0}^{N-1} \Phi_d[(N - i - 1)T]\mathbf{B}_d\mathbf{u}(iT) \qquad (3.79)$$

$$\mathbf{y}(NT) = \mathbf{C}_d\Phi_d(NT)\mathbf{x}(0) + \mathbf{C}_d\sum_{i=0}^{N-1} \Phi[(N - i + 1)T]\mathbf{B}_d\mathbf{u}(iT) + \mathbf{D}_d\mathbf{u}(NT)$$

$$(3.80)$$

Remark 2: The discrete-time state transition matrix defined by $\mathbf{A}_d^k$ can be evaluated efficiently for large values of k by using a method based on the Cayley–Hamilton theorem and described in Appendix C. It can be also evaluated by using the $\mathcal{Z}$-transform method as given in formula (3.85), see Subsection 3.3.5.

Example 3.11: For the system given in Example 3.10, use MATLAB to find the unit step and impulse responses assuming that the initial condition is $\mathbf{x}(0) = [0 \ \ 0]^T$.

The required time responses can be obtained directly by using MAT-LAB statements [y,x]=dstep(Ad,Bd,Cd,Dd) (for step response) and [y,x]=dimpulse(Ad,Bd,Cd,Dd) (for impulse response) with the discrete-time model matrices obtained in the last example. The corresponding state and output responses are presented in Figure 3.11.

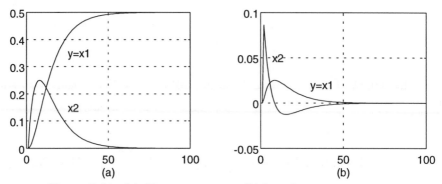

Figure 3.11: (a) Step responses; (b) impulse responses

The same problem could have been solved analytically as follows. Since the initial condition is zero and $u(k) = 1$ for $k \geq 0$, we get from (3.73) the state response as

$$\mathbf{x}(N) = \sum_{i=0}^{N-1} \mathbf{A}_d^{N-i-1} \mathbf{B}_d, \quad N = 1, 2, \ldots$$

The output response, obtained from (3.78), is given by

$$y(N) = \mathbf{C}\mathbf{x}(N), \quad N = 1, 2, \ldots$$

However, at this point, for large N one is faced with the problem of efficiently calculating the powers of matrix $\mathbf{A}_d$. This can be facilitated analytically by using either the Cayley–Hamilton theorem (see Appendix C) or the $\mathcal{Z}$-transform method to be presented in the next subsection.

By the Cayley–Hamilton method, we have for a 2×2 matrix that

$$\mathbf{A}_d^k = \alpha_0 \mathbf{I} + \alpha_1 \mathbf{A}_d, \quad k = 2, 3, 4, \ldots$$

with α_0 and α_1 satisfying

$$\lambda_1^k = \alpha_0 + \alpha_1 \lambda_1$$
$$\lambda_2^k = \alpha_0 + \alpha_1 \lambda_2$$

where λ_1 and λ_2 are distinct eigenvalues of $\mathbf{A}_d$. System eigenvalues will be considered in Section 3.4.

◇

3.3.5 Solution of the Discrete State Equation by the $\mathcal{Z}$-transform

Applying the $\mathcal{Z}$-transform (see Appendix B) to the state space equation of a discrete-time system

$$\mathbf{x}(k+1) = \mathbf{A}_d \mathbf{x}(k) + \mathbf{B}_d \mathbf{u}(k) \tag{3.81}$$

we get

$$z\mathbf{X}(z) - z\mathbf{x}(0) = \mathbf{A}_d \mathbf{X}(z) + \mathbf{B}_d \mathbf{U}(z) \tag{3.82}$$

The complex state vector $\mathbf{X}(z)$ can be expressed as

$$\mathbf{X}(z) = (z\mathbf{I} - \mathbf{A}_d)^{-1} z\mathbf{x}(0) + (z\mathbf{I} - \mathbf{A}_d)^{-1} \mathbf{B}_d \mathbf{U}(z) \tag{3.83}$$

The inverse z-transform of the last equation gives $\mathbf{x}(k)$, that is

$$\mathbf{x}(k) = \mathcal{Z}^{-1}\left[(z\mathbf{I} - \mathbf{A}_d)^{-1} z \right] x(0) + \mathcal{Z}^{-1}\left[(z\mathbf{I} - \mathbf{A}_d)^{-1} \mathbf{B}_d \mathbf{U}(z) \right] \tag{3.84}$$

Comparing equations (3.75) and (3.84) we conclude that

$$\Phi_d(k) = \mathcal{Z}^{-1}\left[(z\mathbf{I} - \mathbf{A}_d)^{-1} z \right] = \mathbf{A}_d^k, \quad k = 1, 2, 3, \dots \tag{3.85}$$

Let us repeat and emphasize that the discrete state transition matrix $\Phi_d(k)$ of a general discrete-time invariant linear system can be obtained either by using (3.85) or the Cayley–Hamilton method given in Appendix C.

The inverse transform of the second term on the right-hand side of (3.84) is obtained directly by the application of the discrete convolution theorem (see Appendix B), leading to

$$\mathcal{Z}^{-1}\left\{ (z\mathbf{I} - \mathbf{A}_d)^{-1} \mathbf{B}_d \mathbf{U}(z) \right\} = \sum_{i=0}^{k-1} \Phi_d(k - 1 - i) \mathbf{B}_d \mathbf{u}(i) \tag{3.86}$$

Combining (3.84) and (3.86) we get the required solution of the discrete-time state space equation as

$$x(k) = \Phi_d(k)x(0) + \sum_{i=0}^{k-1} \Phi_d(k-i-1)B_d u(i) \qquad (3.87)$$

The complex form of the output vector $Y(z)$ is obtained if the $\mathscr{Z}$-transform is applied to the output equation

$$y(k) = C_d x(k) + D_d u(k)$$

and $X(z)$ is substituted from (3.83), leading to

$$Y(z) = C_d(zI - A_d)^{-1}zx(0) + \left[C_d(zI - A_d)^{-1}B_d + D_d\right]U(z)$$

From the above expression, for the zero initial condition, i.e. $x(0) = 0$, the *discrete matrix transfer function* is given by

$$G_d(z) = C_d(zI - A_d)^{-1}B_d + D_d \qquad (3.88)$$

3.3.6 Response Between Sampling Instants

An important feature of the state variable method is that it can be modified to determine the output between sampling instants. Let $t_0 = kT$ and $t = (k + \Delta)T$, where $0 \le \Delta < 1$. Equation (3.47) gives

$$x[(k + \Delta)T] = e^{A\Delta T}x(kT) + \int_{kT}^{(k+\Delta)T} e^{A[(k+\Delta)T-\tau]}Bu(\tau)d\tau \qquad (3.89)$$

Replacing $(k + \Delta)T - \tau$ by β and assuming that $u(\tau)$ is constant during $kT \le \tau < (k + \Delta)T$, we get

$$x[(k + \Delta)T] = e^{A\Delta T}x(kT) + \int_0^{\Delta T} e^{A\beta}d\beta Bu(kT) \qquad (3.90)$$

$$= A_d(\Delta T)x(kT) + B_d(\Delta T)u(kT)$$

where

$$A_d(\Delta T) = e^{A\Delta T} \qquad (3.91)$$

and

$$\mathbf{B}_d(\Delta T) = \int\limits_0^{\Delta T} e^{\mathbf{A}\beta} d\beta \mathbf{B} \qquad (3.92)$$

Therefore, the matrix $\mathbf{A}_d(\Delta T)$ is obtained by replacing T by ΔT in $\mathbf{A}_d$. Similarly, $\mathbf{B}_d(\Delta T)$ is obtained by replacing T by ΔT in $\mathbf{B}_d$.

3.3.7 Euler's Approximation

Discretization of a continuous-time linear model, as presented in Subsection 3.3.3, by the integral approximation method, gives a desired discrete-time linear model. However, in the case of high-order systems, computation of the state transition matrix is very involved, so that in those cases the matrices $\mathbf{A}_d$ and $\mathbf{B}_d$ are calculated approximately by using some simpler methods. The simplest such a method, known as Euler's approximation, is just one of several methods used for numerical solution of differential equations.

The objective of numerical integration is to find a discrete-time counterpart to a continuous-time model

$$\dot{\mathbf{x}}(t) = \mathbf{A}\mathbf{x}(t) + \mathbf{B}\mathbf{m}(t)$$

in the form of a difference equation. The equation obtained, given by a recursive formula, is then easily solved by a digital computer. The integration of the above equation gives

$$\mathbf{x}(t) = \int\limits_{-\infty}^{t} [\mathbf{A}\mathbf{x}(\tau) + \mathbf{B}\mathbf{m}(\tau)] d\tau$$

For simplicity, the main idea of the Euler method is explained for a scalar case. Consider the first-order system $\dot{x} = ax + bu$. The integration is analogous to the problem of finding the area, within the imposed integration limits, between the curve defined by $f(t) = ax(t) + bu(t)$ and the time axis. This area is approximately equal to the sum of the rectangles in Figure 3.12.

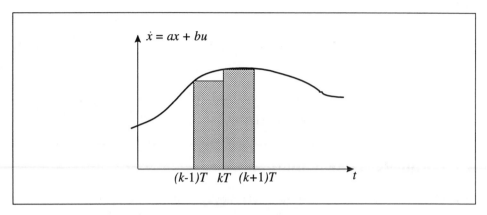

Figure 3.12: Euler's integration method

If the area is calculated according to Figure 3.12, then from the last expression for $t = (k+1)T$ we get

$$\mathbf{x}[(k+1)T] = \int_{-\infty}^{kT} [\mathbf{Ax}(\tau) + \mathbf{Bm}(\tau)]d\tau + \int_{kT}^{(k+1)T} [\mathbf{Ax}(\tau) + \mathbf{Bm}(\tau)]d\tau$$

$$= \mathbf{x}(kT) + T\mathbf{Ax}(kT) + T\mathbf{Bm}(kT)$$

or

$$\mathbf{x}[(k+1)T] = (\mathbf{I} + T\mathbf{A})\mathbf{x}(kT) + T\mathbf{Bm}(kT) \qquad (3.93)$$

From the last equation, we conclude that for the Euler approximation the state and input matrices are given by

$$\mathbf{A}_d = \mathbf{I} + T \cdot \mathbf{A}, \quad \mathbf{B}_d = T \cdot \mathbf{B} \qquad (3.94)$$

It can be observed from (3.70), (3.72), and (3.94) that (3.94) produces only the first two terms of the series expansion given in (3.70) and only the first term of the series expansion given in (3.72). Thus, the Euler approximation is less accurate than the integral approximation considered in Subsection 3.3.3, and for Euler's approximation the sampling interval T must be chosen sufficiently small in order to get satisfactory results.

In general, for more accurate computation of the discrete-time model one can use any known method for numerical solution of differential equations, e.g. the fourth-order Runge–Kutta method or the Adams–Moulton method (Gear, 1971).

3.4 The System Characteristic Equation and Eigenvalues

The characteristic equation is very important in the study of both linear time invariant continuous and discrete systems. No matter what model type is considered (ordinary nth-order differential equation, state space or transfer function), the characteristic equation always has the same form.

If we start with a differential nth-order system model in the operator form

$$\left(p^n + a_{n-1}p^{n-1} + \cdots + a_1 p + a_0\right)y(t)$$
$$= \left(b_m p^m + b_{m-1}p^{m-1} + \cdots + b_1 p + b_0\right)u(t)$$

where the operator p is defined as

$$p^i = \frac{d^i}{dt^i}, \quad i = 1, 2, \ldots, n-1$$

and $m \leq n$, then the *characteristic equation*, according to the mathematical theory of linear differential equations (Boyce and DiPrima, 1992), is defined by

$$s^n + a_{n-1}s^{n-1} + \cdots + a_1 s + a_0 = 0 \tag{3.95}$$

Note that the operator p is replaced by the complex variable s playing the role of a derivative in the Laplace transform context.

In the state space variable approach we have seen from (3.54) that

$$\mathbf{G}(s) = \mathbf{C}(s\mathbf{I} - \mathbf{A})^{-1}\mathbf{B} + \mathbf{D} = \frac{1}{|s\mathbf{I} - \mathbf{A}|}\mathbf{C}[\mathrm{adj}(s\mathbf{I} - \mathbf{A})]\mathbf{B} + \mathbf{D}$$

$$= \frac{1}{|s\mathbf{I} - \mathbf{A}|}\{\mathbf{C}[\mathrm{adj}(s\mathbf{I} - \mathbf{A})]\mathbf{B} + |s\mathbf{I} - \mathbf{A}|\mathbf{D}\}$$

The characteristic equation here is defined by

$$|s\mathbf{I} - \mathbf{A}| = 0 \tag{3.96}$$

A third form of the characteristic equation is obtained in the context of the transfer function approach. The transfer function of a single-input single-output system is

$$G(s) = \frac{b_m s^m + b_{m-1}s^{m-1} + \cdots + b_1 s + b_0}{s^n + a_{n-1}s^{n-1} + \cdots + a_1 s + a_0} \tag{3.97}$$

The characteristic equation in this case is obtained by equating the denominator of this expression to zero. Note that for multivariable systems, the characteristic polynomial (obtained from the corresponding characteristic equation) appears in denominators of all entries of the matrix transfer function.

No matter what form of the system model is considered, the characteristic equation is always the same. This is obvious from (3.95) and (3.97), but is not so clear from (3.96). It is left as an exercise to the reader to show that (3.95) and (3.96) are identical (Problem 3.30).

The *eigenvalues* are defined in linear algebra as scalars, λ, satisfying (Fraleigh and Beauregard, 1990)

$$\mathbf{A}\mathbf{v} = \lambda\mathbf{v} \tag{3.98}$$

where the vectors $\mathbf{v} \neq 0$ are called the *eigenvectors*. This system of n linear algebraic equations (λ is fixed) has a solution $\mathbf{v} \neq 0$ if and only if

$$|\lambda\mathbf{I} - \mathbf{A}| = 0 \tag{3.99}$$

Obviously, (3.96) and (3.99) have the same form. Since (3.96) = (3.95), it follows that the last equation is the characteristic equation, and hence the eigenvalues are the zeros of the characteristic equation. For the characteristic equation of order n, the number of eigenvalues is equal to n. Thus, the roots of the characteristic equation in the state space context are the eigenvalues of the matrix $\mathbf{A}$. These roots in the transfer function context are called the *system poles*, according to the mathematical tools for analysis of these systems—the complex variable methods.

Similarity Transformation

We have pointed out before that a system modeled by the state space technique may have many state space forms. Here, we establish a relationship among those state space forms by using a linear transformation known as the similarity transformation.

For a given system

$$\dot{\mathbf{x}} = \mathbf{A}\mathbf{x} + \mathbf{B}\mathbf{u}, \quad \mathbf{x}(0) = \mathbf{x}_o$$
$$\mathbf{y} = \mathbf{C}\mathbf{x} + \mathbf{D}\mathbf{u}$$

we can introduce a new state vector $\hat{\mathbf{x}}$ by a linear coordinate transformation as follows

$$\mathbf{x} = \mathbf{P}\hat{\mathbf{x}}$$

where $\mathbf{P}$ is some nonsingular $n \times n$ matrix. A new state space model is obtained as

$$\dot{\hat{\mathbf{x}}} = \hat{\mathbf{A}}\hat{\mathbf{x}} + \hat{\mathbf{B}}\mathbf{u}, \quad \hat{\mathbf{x}}(0) = \hat{\mathbf{x}}_o$$

$$\mathbf{y} = \hat{\mathbf{C}}\hat{\mathbf{x}} + \hat{\mathbf{D}}\mathbf{u} \tag{3.100}$$

where

$$\hat{\mathbf{A}} = \mathbf{P}^{-1}\mathbf{A}\mathbf{P}, \ \hat{\mathbf{B}} = \mathbf{P}^{-1}\mathbf{B}, \ \hat{\mathbf{C}} = \mathbf{C}\mathbf{P}, \ \hat{\mathbf{D}} = \mathbf{D}, \ \hat{\mathbf{x}}(0) = \mathbf{P}^{-1}\mathbf{x}(0) \tag{3.101}$$

This transformation is known in the literature as the *similarity transformation*. It plays an important role in linear control system theory and practice.

Very important features of this transformation are that under similarity transformation both the system eigenvalues and the system transfer function are invariant.

Eigenvalue Invariance

A new state space model obtained by the similarity transformation does not change internal structure of the model, that is, the eigenvalues of the system remain the same. This can be shown as follows

$$\left| s\mathbf{I} - \hat{\mathbf{A}} \right| = \left| s\mathbf{I} - \mathbf{P}^{-1}\mathbf{A}\mathbf{P} \right| = \left| \mathbf{P}^{-1}(s\mathbf{I} - \mathbf{A})\mathbf{P} \right|$$

$$= \left| \mathbf{P}^{-1} \right| \left| s\mathbf{I} - \mathbf{A} \right| \left| \mathbf{P} \right| = \left| s\mathbf{I} - \mathbf{A} \right| \tag{3.102}$$

Note that in this proof the following properties of the matrix determinant have been used

$$\det(\mathbf{M}_1 \mathbf{M}_2 \mathbf{M}_3) = \det\mathbf{M}_1 \times \det\mathbf{M}_2 \times \det\mathbf{M}_3$$

$$\det\mathbf{M}^{-1} = \frac{1}{\det\mathbf{M}}$$

see Appendix C.

Transfer Function Invariance

Another important feature of the similarity transformation is that the transfer function remains the same for both models, which can be shown as follows

$$\hat{\mathbf{G}}(s) = \hat{\mathbf{C}}\left(s\mathbf{I} - \hat{\mathbf{A}}\right)^{-1}\hat{\mathbf{B}} + \hat{\mathbf{D}} = \mathbf{C}\mathbf{P}\left(s\mathbf{I} - \mathbf{P}^{-1}\mathbf{A}\mathbf{P}\right)^{-1}\mathbf{P}^{-1}\mathbf{B} + \mathbf{D}$$

$$= \mathbf{C}\mathbf{P}\left[\mathbf{P}^{-1}(s\mathbf{I} - \mathbf{A})\mathbf{P}\right]^{-1}\mathbf{P}^{-1}\mathbf{B} + \mathbf{D} \tag{3.103}$$

$$= \mathbf{C}\mathbf{P}\mathbf{P}^{-1}(s\mathbf{I} - \mathbf{A})^{-1}\mathbf{P}\mathbf{P}^{-1}\mathbf{B} + \mathbf{D}$$

$$= \mathbf{C}(s\mathbf{I} - \mathbf{A})^{-1}\mathbf{B} + \mathbf{D} = \mathbf{G}(s)$$

Note that we have used in (3.103) the matrix inversion property (Appendix C)

$$(\mathbf{M}_1\mathbf{M}_2\mathbf{M}_3)^{-1} = \mathbf{M}_3^{-1}\mathbf{M}_2^{-1}\mathbf{M}_1^{-1}$$

The above result is quite logical—the system preserves its input–output behavior no matter how it is mathematically described.

Modal Transformation

One of the most interesting similarity transformations is the one that puts matrix $\mathbf{A}$ into diagonal form. Assume that $\mathbf{P} = \mathbf{V} = [\mathbf{v}_1, \mathbf{v}_2, \ldots, \mathbf{v}_n]$, where $\mathbf{v}_i$ are the eigenvectors. We then have

$$\mathbf{V}^{-1}\mathbf{A}\mathbf{V} = \hat{\mathbf{A}} = \Lambda = diag(\lambda_1, \lambda_2, \ldots, \lambda_n) \qquad (3.104)$$

It is easy to show that the elements λ_i, $i = 1, \ldots, n$, on the matrix diagonal of Λ are the roots of the characteristic equation $|s\mathbf{I} - \Lambda| = |s\mathbf{I} - \mathbf{A}| = 0$, i.e. they are the eigenvalues. This can be shown in a straightforward way

$$|s\mathbf{I} - \Lambda| = \det\{diag(s - \lambda_1, s - \lambda_2, \ldots, s - \lambda_n)\}$$
$$= (s - \lambda_1)(s - \lambda_2) \cdots (s - \lambda_n)$$

The state transformation (3.104) is known as the *modal transformation*.

Note that the pure diagonal state space form defined in (3.104) can be obtained only in the following three cases.

1. The system matrix has distinct eigenvalues, namely $\lambda_1 \neq \lambda_2 \neq \cdots \neq \lambda_n$.
2. The system matrix is symmetric (see Appendix C).
3. The system minimal polynomial does not contain multiple eigenvalues. For the definition of the minimal polynomial and the corresponding pure diagonal Jordan form, see Subsection 4.2.4.

In the above three cases we say that the system matrix is diagonalizable.

Remark: Relation (3.104) may be represented in another form, that is

$$\mathbf{V}^{-1}\mathbf{A} = \Lambda\mathbf{V}^{-1}$$

or

$$\mathbf{W}^T\mathbf{A} = \Lambda\mathbf{W}^T$$

where

$$\mathbf{W}^T = \mathbf{V}^{-1} \Rightarrow \mathbf{W}^T\mathbf{V} = \mathbf{I}$$

In this case the *left eigenvectors* $\mathbf{w}_i$, $i = 1, 2, ..., n$, can be computed from

$$\mathbf{w}_i^T \mathbf{A} = \lambda_i \mathbf{w}_i^T \quad \Rightarrow \quad \mathbf{A}^T \mathbf{w}_i = \lambda_i \mathbf{w}_i$$

where $\mathbf{W} = [\mathbf{w}_1, \mathbf{w}_2, \ldots, \mathbf{w}_n]$. Since $|\lambda \mathbf{I} - \mathbf{A}| = |\lambda \mathbf{I} - \mathbf{A}^T|$, then λ_i is also an eigenvalue of $\mathbf{A}^T$.

There are numerous program packages available to compute both the eigenvalues and eigenvectors of a matrix. In MATLAB this is done by using the function eig.

3.4.1 Multiple Eigenvalues

If the matrix $\mathbf{A}$ has multiple eigenvalues, it is possible to transform it into a block diagonal form by using the transformation

$$\mathbf{J} = \mathbf{V}^{-1} \mathbf{A} \mathbf{V} \tag{3.105}$$

where the matrix $\mathbf{V}$ is composed of n linearly independent, so-called *generalized eigenvectors* and $\mathbf{J}$ is known as the Jordan canonical form. This block diagonal form contains simple Jordan blocks on the diagonal. Simple Jordan blocks have the given eigenvalue on the main diagonal, ones above the main diagonal with all other elements equal to zero. For example, a simple Jordan block of order four is given by

$$\mathbf{J}_i(\lambda_i) = \begin{bmatrix} \lambda_i & 1 & 0 & 0 \\ 0 & \lambda_i & 1 & 0 \\ 0 & 0 & \lambda_i & 1 \\ 0 & 0 & 0 & \lambda_i \end{bmatrix}$$

Let the eigenvalue λ_1 have multiplicity of order 3 in addition to two real and distinct eigenvalues, $\lambda_2 \neq \lambda_3$; then a $\mathbf{J}$ matrix of order 5×5 may contain the following three simple Jordan blocks

$$\mathbf{J} = \begin{bmatrix} \lambda_1 & 1 & 0 & 0 & 0 \\ 0 & \lambda_1 & 1 & 0 & 0 \\ 0 & 0 & \lambda_1 & 0 & 0 \\ 0 & 0 & 0 & \lambda_2 & 0 \\ 0 & 0 & 0 & 0 & \lambda_3 \end{bmatrix}$$

However, other choices are also possible. For example, we may have the following distribution of simple Jordan blocks

$$\mathbf{J} = \begin{bmatrix} \lambda_1 & 1 & 0 & 0 & 0 \\ 0 & \lambda_1 & 0 & 0 & 0 \\ 0 & 0 & \lambda_1 & 0 & 0 \\ 0 & 0 & 0 & \lambda_2 & 0 \\ 0 & 0 & 0 & 0 & \lambda_3 \end{bmatrix} \quad \text{or} \quad \mathbf{J} = \begin{bmatrix} \lambda_1 & 0 & 0 & 0 & 0 \\ 0 & \lambda_1 & 0 & 0 & 0 \\ 0 & 0 & \lambda_1 & 0 & 0 \\ 0 & 0 & 0 & \lambda_2 & 0 \\ 0 & 0 & 0 & 0 & \lambda_3 \end{bmatrix}$$

The study of the Jordan form is quite complex. Much more about the Jordan form will be presented in Chapter 4, where we study system stability.

3.4.2 Modal Decomposition

Diagonalization of matrix $\mathbf{A}$ using transformation $\mathbf{x} = \mathbf{V}\hat{\mathbf{x}}$ makes the system $\dot{\mathbf{x}} = \mathbf{A}\mathbf{x} + \mathbf{B}\mathbf{u}$ diagonal, that is

$$\dot{\hat{\mathbf{x}}} = \Lambda\hat{\mathbf{x}} + \left(\mathbf{V}^{-1}\mathbf{B}\right)\mathbf{u} = \Lambda\hat{\mathbf{x}} + \left(\mathbf{W}^T\mathbf{B}\right)\mathbf{u}, \quad \hat{\mathbf{x}}(0) = \hat{\mathbf{x}}_0$$

In such a case the homogeneous equation $\dot{\mathbf{x}} = \mathbf{A}\mathbf{x}$, $\mathbf{x}(0) = \mathbf{x}_0$, becomes

$$\dot{\hat{\mathbf{x}}} = \Lambda\hat{\mathbf{x}}, \quad \hat{\mathbf{x}}(0) = \mathbf{V}^{-1}\mathbf{x}(0) = \mathbf{V}^{-1}\mathbf{x}_0$$

or

$$\dot{\hat{x}}_i = \lambda_i\hat{x}_i, \quad i = 1,\dots,n$$

This system is represented by n independent differential equations. The modal response to the initial condition is

$$\hat{\mathbf{x}}(t) = e^{\Lambda t}\hat{\mathbf{x}}_0 = e^{\Lambda t}\mathbf{V}^{-1}\mathbf{x}_0 = e^{\Lambda t}\mathbf{W}^T\mathbf{x}_0$$

or

$$\hat{x}_i(t) = \hat{x}_i(0)e^{\lambda_i t} = \left(\mathbf{w}_i^T\mathbf{x}_0\right)e^{\lambda_i t}$$

The response $\mathbf{x}(t)$ is a combination of the modal components

$$\begin{aligned} \mathbf{x}(t) = \mathbf{V}\hat{\mathbf{x}}(t) &= \mathbf{V}e^{\Lambda t}\mathbf{V}^{-1}\mathbf{x}_0 = \mathbf{V}e^{\Lambda t}\mathbf{W}^T\mathbf{x}_0 \\ &= \left(\mathbf{w}_1^T\mathbf{x}_0\right)e^{\lambda_1 t}\mathbf{v}_1 + \left(\mathbf{w}_2^T\mathbf{x}_0\right)e^{\lambda_2 t}\mathbf{v}_2 + \cdots + \left(\mathbf{w}_n^T\mathbf{x}_0\right)e^{\lambda_n t}\mathbf{v}_n \end{aligned} \quad (3.106)$$

This equation represents the modal decomposition of $\mathbf{x}(t)$ and it shows that the total response consists of a sum of responses of all individual modes. Note that $\mathbf{w}_i^T \mathbf{x}_0$ are scalars.

It is customary to call the reciprocals of λ_i the *system time constants* and denote them by τ_i, that is

$$\tau_i = \frac{1}{\lambda_i}, \quad i = 1, 2, ..., n$$

This has physical meaning since the system dynamics is determined by its time constants and these do appear in the system response in the form e^{-t/τ_i}.

The transient response of the system may be influenced differently by different modes, depending of the eigenvalues λ_i. Some modes may decay faster than the others. Some modes might be dominant in the system response. These cases will be illustrated in Chapter 6.

Remark: A similarity transformation $\Lambda = \mathbf{V}^{-1}\mathbf{A}\mathbf{V}$ can be used for the state transition matrix calculation. Recall

$$\hat{\mathbf{x}}(t) = e^{\Lambda t}\hat{\mathbf{x}}(0), \quad \hat{\mathbf{x}}(t) = \mathbf{V}\mathbf{x}(t), \quad \hat{\mathbf{x}}(0) = \mathbf{V}\mathbf{x}(0)$$

and

$$\mathbf{x}(t) = \mathbf{V}^{-1}e^{\Lambda t}\mathbf{V}\mathbf{x}(0) = \Phi(t)\mathbf{x}(0)$$

Hence,

$$\Phi(t) = e^{\Lambda t} = \mathbf{V}^{-1}e^{\Lambda t}\mathbf{V} = \mathbf{W}^T e^{\Lambda t}\mathbf{V}$$

or, in the complex domain

$$\Phi(s) = \mathbf{V}^{-1}(s\mathbf{I} - \Lambda)^{-1}\mathbf{V}$$

$$= \mathbf{V}^{-1}diag\{s - \lambda_1, s - \lambda_2, ..., s - \lambda_n\}^{-1}\mathbf{V}$$

$$= \mathbf{V}^{-1}diag\left\{\frac{1}{s - \lambda_1}, \frac{1}{s - \lambda_2}, ..., \frac{1}{s - \lambda_n}\right\}\mathbf{V}$$

Remark: The presented theory about the system characteristic equation, eigenvalues, eigenvectors, similarity and modal transformations can be applied directly to discrete-time linear systems with $\mathbf{A}_d$ replacing $\mathbf{A}$.

3.5 State Space MATLAB Laboratory Experiments

In this section we present three MATLAB laboratory experiments on the state space method in control systems. These experiments can be used either as supplements for lectures or independently in the corresponding control system laboratory. Most of the required MATLAB functions have been already introduced in the examples done in this chapter. Students should also consult Appendix D, where a shortened MATLAB manual is given. It is advisable that before using any MATLAB function, the students check all its options by typing `help function name`.

3.5.1 Experiment 1—The Inverted Pendulum

Part 1. The linearized equations of the inverted pendulum, obtained by assuming that the pendulum mass is concentrated at its center of gravity (Kwakernaak and Sivan, 1972; Kamen, 1990) are given by

$$(J + mL^2)\ddot{\theta}(t) - mgL\theta(t) + mL\ddot{d}(t) = 0$$

$$(M + m)\ddot{d}(t) + mL\ddot{\theta}(t) = u(t)$$

(3.107)

where $\theta(t)$ is the angle of the pendulum from the vertical position, $d(t)$ is the position of the cart, $u(t)$ is the force applied to the cart, M is the mass of the cart, m is the mass of the pendulum, g is the gravitational constant, and J is the moment of inertia about the center of mass. Assuming that normalized values are given by $J = 1$, $L = 1$, $g = 9.81$, $M = 1$, $m = 0.1$, derive the state space form

$$\dot{\mathbf{x}}(t) = \mathbf{A}\mathbf{x}(t) + \mathbf{B}u(t)$$

where

$$\mathbf{x}(t) = [\theta(t) \quad \dot{\theta}(t) \quad d(t) \quad \dot{d}(t)]^T$$

and $\mathbf{A}^{4\times4}$ and $\mathbf{B}^{4\times1}$ are the corresponding matrices.

Part 2. Using MATLAB determine the following:

(a) The eigenvalues, eigenvectors, and characteristic polynomial of matrix $\mathbf{A}$.
(b) The state transition matrix at the time instant $t = 1$.
(c) The unit impulse response (take $\theta(t)$ and $d(t)$ as the output variables) for $0 \le t \le 1$ with the step size $\Delta t = 0.1$ and draw the system response using the MATLAB function `plot`.

(d) The unit step response for $0 \leq t \leq 1$ and $\Delta t = 0.1$. Draw the system response.

(e) The unit ramp response for $0 \leq t \leq 1$ and $\Delta t = 0.1$ and draw the system response. Compare the response diagrams obtained in (c), (d), and (e).

(f) The state response resulting from the initial state $\mathbf{x}(0) = \begin{bmatrix} -1 & 1 & 1 & 1 \end{bmatrix}^T$ and the input $\mathbf{u}(t) = \sin(t)$ for $0 \leq t \leq 5$ and $\Delta t = 0.1$.

(g) The inverse of the state transition matrix $\left(e^{\mathbf{A}t} \right)^{-1}$ for $t = 5$.

(h) The state $\mathbf{x}(t)$ at time $t = 5$ assuming that $\mathbf{x}(10) = \begin{bmatrix} 10 & 0 & 5 & 2 \end{bmatrix}^T$ and $u(t) = 0$ by using the result from (g).

(i) Find the system transfer function.

Part 3. Discretize the continuous-time system given in (3.107) with $T = 0.02$ and find the discrete space model

$$\mathbf{x}(k+1) = \mathbf{A}_d\mathbf{x}(k) + \mathbf{B}_d u(k)$$

Assuming that the output equation of the discrete system is given by

$$\mathbf{y}(k) = \begin{bmatrix} 1 & 0 & 0 & 0 \\ 0 & 0 & 1 & 0 \end{bmatrix} \mathbf{x}(k) = \mathbf{C}_d\mathbf{x}(k)$$

find the system (output) response for $0 \leq k \leq 50$ due to initial conditions $\mathbf{x}_0 = \begin{bmatrix} -1 & 1 & -1 & 1 \end{bmatrix}^T$ and unit step input (note that $u(k)$ should be generated as a column vector of 50 elements equal to 1).

Part 4. Consider the continuous-time system given by

$$\frac{d^2 y(t)}{dt^2} + 0.1\frac{dy(t)}{dt} = u(t) \tag{3.108}$$

(a) Discretize this system with $T = 1$ by using the Euler approximation.

(b) Find the response of the obtained discrete system for $k = 1, 2, 3, ..., 20$, when $u(t) = \sin(0.1\pi t)$ and $y(0) = \dot{y}(0) = 0$.

(c) Find discrete transfer function, characteristic equation, eigenvalues, and eigenvectors.

Part 5. Discretize the state space form of (3.108) obtained by using MATLAB function c2d with $T = 1$. Find the discrete system response for the initial condition and the input function defined in Part 4b. Compare the results obtained in Parts 4 and 5. Comment on the results obtained.

3.5.2 Experiment 2—Response of Continuous Systems

Part 1. Consider a continuous-time linear system represented by its transfer function

$$G(s) = \frac{(s+5)}{s^2 + 5s + 6}$$

(a) Find the impulse response by using the MATLAB function `impulse`. In this case you have to use `[y,x]=impulse(num,den)`, where `num` and `den` are row vectors that contain the polynomial coefficients in descending powers of s. Plot both state space variable and output responses (use function `plot`).

(b) Find the step response by using the function `step` and plot both the state response and the output response.

(c) Find the zero-state response due to an input given by $f(t) = e^{-3t}, \ t \geq 0$. Note that you have to use the function `lsim` and specify input at every time instant of interest. That can be obtained by `t=0:0.1:5` (defines t at $0, 0.1, 0.2, ..., 4.9, 5$) and by `f(t)=exp(-3*t)`. Check that the results obtained in (c) agree with analytical results at $t = 1$.

(d) Obtain the state space form for this system by using the function `tf2ss`. Repeat parts (a), (b), and (c) for the corresponding state space representation. Use the following MATLAB instructions

```
[y,x]=impulse(A,B,C,D,1);
[y,x]=step(A,B,C,D,1);
[y,x]=lsim(A,B,C,D,f,t);
```

respectively, with f and t defined in (c). Compare the results obtained.

Part 2. Consider the continuous-time linear system represented by

$$\frac{d^2y(t)}{dt^2} + 4\frac{dy(t)}{dt} + 4y(t) = \frac{df(t)}{dt} + f(t)$$
$$f(t) = e^{-4t}, \quad t \geq 0, \qquad y(0^-) = 2, \quad \dot{y}(0) = 1$$

(a) Find the complete system response by using the MATLAB function `lsim`. Compare the simulation results obtained with analytical results. Hint: Use

```
[y,x]=lsim(A,B,C,D,f,t,X0);
```

with $t = 0:0.1:5$. Note that the initial condition for the state vector, `X0`, has to be found. This can be obtained by playing algebra with the state and output equations and setting $t = 0$.

(b) Find the zeros and poles of this system by using the function `tf2zp`.

(c) Find the system response due to initial conditions specified in Part 2a and the impulse delta function as an input. Since you are not able to specify the system input in time (the delta function has no time structure), you cannot use the `lsim` function. Instead use the `initial` function (zero-input response). The required response is obtained analytically as follows

$$\mathbf{x}(t) = e^{\mathbf{A}t}(\mathbf{x}(0) + \mathbf{B})$$

where $\mathbf{A}$ and $\mathbf{B}$ stand for the system and input matrices in the state space. Thus, the new initial condition is given by $\mathbf{x}(0) + \mathbf{B}$.

(d) Justify the answer obtained in (c). Solve the same problem analytically by using the Laplace transform. Plot results from (c) and compare with results obtained in (d). Can you draw any conclusion for this "nonstandard" problem from the point of view of the system initial conditions at $t = 0^+$. (The standard problem requires that for the impulse response all initial conditions are set to zero.)

Part 3. Given the following dynamical system represented in the state space form by (Gajić and Shen, 1993)

$$\mathbf{A} = \begin{bmatrix} -0.01357 & -32.2 & -46.3 & 0 \\ 0.00012 & 0 & 1.214 & 0 \\ -0.0001212 & 0 & -1.214 & 1 \\ 0.00057 & 0 & -9.1 & -0.6696 \end{bmatrix}, \quad \mathbf{B} = \begin{bmatrix} -0.433 \\ 0.1394 \\ -0.1394 \\ -0.1577 \end{bmatrix}$$

$$\mathbf{C} = \begin{bmatrix} 0 & 0 & 0 & 1 \\ 1 & 0 & 0 & 0 \end{bmatrix}, \quad \mathbf{D} = \mathbf{0}^{2\times1}$$

This is a real mathematical model of an F-8 aircraft (Teneketzis and Sandell, 1977). Using MATLAB, determine the following quantities.

(a) The eigenvalues, eigenvectors, and characteristic polynomial. Take `p=poly(A)` and verify that `roots(p)` produces also the eigenvalues of matrix $\mathbf{A}$.

(b) The state transition matrix at the time instant $t = 1$. Use the `expm` function.

(c) The unit impulse response and plot output variables. Hint: Use

```
impulse(A,B,C,D);
```

(d) The unit step response and plot the corresponding output variables.

(e) Let the initial system state be $\mathbf{x}(0) = \begin{bmatrix} -1 & 1 & 0.5 & 1 \end{bmatrix}^T$. Find the response due to an input given by $f(t) = \sin t$, $0 < t < 1000$. Hint: Take $t=0:10:1000$ and find the corresponding values for $f(t)$ by using the function sin in the form $f = \sin(t)$. Then use the lsim function.

(f) Find the system transfer functions. Note that you have one input and two outputs which implies two transfer functions. Hint: Use the function ss2tf.

(g) Find the inverse of the state transition matrix $\left(e^{\mathbf{A}t} \right)^{-1} = e^{-\mathbf{A}t}$ at $t = 2$.

Part 4. Consider a linear continuous-time dynamical system represented by its transfer function

$$G(s) = \frac{(s+1)(s+3)(s+5)(s+7)}{s(s+2)(s+4)(s+6)(s+8)(s+10)}$$

(a) Input the system zeros and poles as column vectors. Note that in this case the static gain $k = 1$. Use the function zp2ss(z,p,k) in order to get the state space matrices.

(b) Find the eigenvalues and eigenvectors of matrix $\mathbf{A}$.

(c) Verify that the transformation $\mathbf{x} = \mathbf{P}\tilde{\mathbf{x}}$, where $\mathbf{P}$ is the matrix whose columns are the eigenvectors of matrix $\mathbf{A}$, produces in the new coordinates the diagonal system matrix $\mathbf{\Lambda} = \mathbf{P}^{-1}\mathbf{AP}$ with diagonal elements equal to the eigenvalues of matrix $\mathbf{A}$.

(d) Find the remaining state space matrices in the new coordinates. Find the transfer function in the new coordinates and compare it with the original one.

(e) Compare the unit step responses of the original and transformed systems.

3.5.3 Experiment 3—Response of Discrete Systems

Part 1. Consider a discrete-time linear system represented by its transfer function

$$G(z) = \frac{z}{4z^2 + 4z + 1}$$

(a) Find the impulse response by using the MATLAB function dimpulse. In this case you have to use [y,x]=dimpulse(num,den), where num and den are row vectors which contain the polynomial coefficients in descending powers of s. Plot both state and output responses (use function plot).

(b) Find the step response by using the function dstep and plot both the state and output responses.

(c) Find the system (output) response due to a unit step function, $f(k) = h(k)$, and initial conditions specified by $y(-1) = 0$, $y(-2) = 1$. Note that you have to use the function dlsim and to specify input at every time instant of interest. That can be obtained by k=0:1:20 (defines k at $0, 1, 2, ..., 19, 20$) and by f(k)=1. Check analytically that the results obtained in (c) agree with the analytical results for $k = 10$.

(d) Obtain the state space form for this system by using the function tf2ss. Repeat parts (a), (b), and (c). Use the following MATLAB statements

```
[y,x]=dimpulse(A,B,C,D,1);
[y,x]=dstep(A,B,C,D,1);
[y,x]=dlsim(A,B,C,D,f,k);
```

respectively, with f and k defined in (c). Compare the results obtained.

Part 2. Consider the discrete-time linear system represented by

$$y(k + 2) + \frac{5}{6}y(k + 1) + \frac{1}{6}y(k) = f(k + 1)$$

$$f(k) = (0.8)^k u(k), \quad y(-1) = 2, \quad y(-2) = 3$$

(a) Find the system response by using the MATLAB function dlsim. Hint: Use

```
[y,x]=dlsim(A,B,C,D,f,X0); with k=0:1:10.
```

Note that the initial condition has to be found. This can be obtained by playing algebra with the state space and output equations. Compare the simulation results obtained with analytical results.

(b) Find the zeros and poles of this system by using the function tf2zp.

(c) Find the system response due to initial conditions specified in Part 2a and with the impulse delta function as an input. Use the dlsim function.

(d) Solve the same problem analytically by using the $\mathcal{Z}$-transform. Plot results from (c) and compare with results obtained in (d).

Part 3. Given a dynamical system represented in the continuous-time state space form in Section 3.5.2, Experiment 2, Part 3.

(a) Discretize the continuous-time system by using the MATLAB function c2d. Assume that the sampling period is $T = 1$.

(b) Find the eigenvalues, eigenvectors, and characteristic polynomial of the obtained discrete-time system.

(c) Find the state transition matrix at time instant $k = 5$.

(d) Find the unit impulse response and plot output variables. Hint: Use

```
dimpulse(A,B,C,D);
```

(e) Find the unit step response and plot the corresponding output variables.

(f) Assume that the initial system state is $\mathbf{x}(0) = \begin{bmatrix} -1 & 0 & 1 & -0.5 \end{bmatrix}^T$. Find the response due to an input given by $f(t) = sink, \ 0 < k < 1000$. Hint: Take $k=0:10:1000$ and find the corresponding values for $f(k)$ by using the function $\sin$ in the form $f = \sin(k)$. Then use the $dlsim$ function. Compare the obtained discrete-time results with the continuous-time results for the same system studied in Section 3.5.2, Experiment 2.

(g) Find the system transfer functions. Note that you have one input and two outputs which implies two transfer functions. The matrices $\mathbf{C}$ and $\mathbf{D}$ are not changed due to discretization procedure. Hint: Use the function $ss2tf$.

3.6 References

Arkun, Y. and S. Ramakrishnan, "Bounds of the optimum quadratic cost of structure constrained regulators," *IEEE Transactions on Automatic Control*, vol. AC-28, 924–927, 1983.

Boyce, W. and R. DiPrima, *Elementary Differential Equations*, Wiley, New York, 1992.

Chen, C., *Linear System Theory Design*, CBS College Publishing, New York, 1984.

Dressler, R. and D. Tabak, "Satellite tracking by combined optimal estimation and control techniques," *IEEE Transactions on Automatic Control*, vol. AC-16, 833–840, 1971.

Fraleigh, J. and R. Beauregard, *Linear Algebra*, Addison-Wesley, Reading, Massachusetts, 1990.

Gajić, Z. and X. Shen, "Study of the discrete singularly perturbed linear quadratic control problem by a bilinear transformation," *Automatica*, vol. 27, 1025–1028, 1991.

Gajić, Z. and X. Shen, *Parallel Algorithms for Optimal Control of Large Scale Linear Systems*, Springer-Verlag, London, 1993.

Gear, W., *Numerical Initial Value Problems in Ordinary Differential Equations*, Prentice Hall, Englewood Cliffs, New Jersey, 1971.

Grodt, T. and Z. Gajić, "The recursive reduced order numerical solution of the singularly perturbed differential Riccati equation," *IEEE Transactions on Automatic Control*, vol. AC-33, 751–754, 1988.

Kailath, T., *Linear Systems*, Prentice Hall, Englewood Cliffs, New Jersey, 1980.

Kamen, E., *Introduction to Signals and Systems*, Macmillan, New York, 1990.

Kokotović, P., J. Allemong, J. Winkelman, and J. Chow, "Singular perturbation and iterative separation of the time scales," *Automatica*, vol. 16, 23–33, 1980.

Kwakernaak, H. and R. Sivan, *Linear Optimal Control Systems*, Wiley, New York, 1972.

Litkouhi, B., *Sampled-Data Control Systems with Slow and Fast Modes*, Ph.D. Dissertation, Michigan State University, 1983.

Petkovski, D., N. Harkara, and Z. Gajić, "Fast suboptimal solution to the static output control problem of linear singularly perturbed systems," *Automatica*, vol. 27, 721–724, 1991.

Teneketzis, D. and N. Sandell, "Linear regulator design for stochastic systems by multiple time-scale method," *IEEE Transactions on Automatic Control*, vol. AC-22, 615–621, 1977.

Spong, M., "The swing up control problem for the acrobot," *IEEE Control Systems Magazine*, vol. 15, 49–55, 1995.

3.7 Problems

3.1 An antenna control problem (Dressler and Tabak, 1971) is represented by the open-loop transfer function

$$G(s) = \frac{K(s+1)}{s^2(s+6)(s+11.5)(s^2+8s+256)}$$

Find state space matrices for the following forms:

(a) Controller canonical form.

(b) Observer canonical form.

3.2 A robotic manipulator called the acrobot has the following linearized model (Spong, 1995)

$$\mathbf{A} = \begin{bmatrix} 0 & 0 & 1 & 0 \\ 0 & 0 & 0 & 1 \\ 12.49 & -12.54 & 0 & 0 \\ -14.49 & 29.36 & 0 & 0 \end{bmatrix}, \quad \mathbf{B} = \begin{bmatrix} 0 \\ 0 \\ -2.98 \\ 5.98 \end{bmatrix}$$

Assume that the output matrices are given by

$$\mathbf{C} = \begin{bmatrix} 1 & 0 & 1 & 0 \end{bmatrix}, \quad \mathbf{D} = 0$$

Use MATLAB in order to find the following quantities:

(a) Eigenvalues and characteristic polynomial.
(b) Modal canonical form.
(c) Open-loop transfer function.
(d) Controller and observer canonical forms.

3.3 Consider the harmonic oscillator in the state space form

$$\begin{bmatrix} \dot{x}_1 \\ \dot{x}_2 \end{bmatrix} = \begin{bmatrix} 0 & 1 \\ -1 & 0 \end{bmatrix} \begin{bmatrix} x_1 \\ x_2 \end{bmatrix} + \begin{bmatrix} 0 \\ 1 \end{bmatrix} u, \quad \begin{bmatrix} x_1(0) \\ x_2(0) \end{bmatrix} = \begin{bmatrix} 0 \\ 1 \end{bmatrix}$$

$$y = \begin{bmatrix} 0 & 1 \end{bmatrix} \begin{bmatrix} x_1 \\ x_2 \end{bmatrix}$$

(a) Find the state transition matrix.
(b) Find the system response due to a unit step input.
(c) Verify the answer obtained by using the MATLAB functions `ss2zp` and `lsim`.

3.4 Given the matrix

$$\mathbf{A} = \begin{bmatrix} 0 & 1 \\ -1 & -2 \end{bmatrix}$$

Find $e^{\mathbf{A}t}$ by the Cayley–Hamilton and the Laplace transform methods.

3.5 For the system

$$\ddot{y}(t) + 2\dot{y}(t) + y(t) = 6\dot{u}(t) + u(t)$$

(a) Draw the simulation block diagram. Use any method.

(b) Find the system impulse response by the MATLAB function im-pulse.

3.6 Given a discrete system $x(k + 1) = \mathbf{A}x(k)$, where

$$\mathbf{A} = \begin{bmatrix} 1 & 2 \\ 0 & 3 \end{bmatrix}$$

Find its response due to the initial condition given by $x(0) = \begin{bmatrix} 1 & 1 \end{bmatrix}^T$.

3.7 Given a linear time invariant continuous system

$$\dot{x}(t) = \mathbf{A}x(t) + \begin{bmatrix} 0 \\ 1 \end{bmatrix} u(t)$$

with

$$x(1) = \begin{bmatrix} x_1(1) \\ x_2(1) \end{bmatrix} = \begin{bmatrix} 5 \\ 0 \end{bmatrix}, \quad u(t) = \begin{cases} 1 & t \geq 2 \\ 0 & 0 < t \leq 2 \end{cases}$$

Assuming that the state transition matrix has a known (given) form as

$$\Phi(t - t_0) = \begin{bmatrix} \phi_{11}(t - t_0) & 0 \\ \phi_{21}(t - t_0) & \phi_{22}(t - t_0) \end{bmatrix}$$

find the system response for any $t \geq 0$.

3.8 Find the impulse response of the system

$$\ddot{y} + 2\dot{y} + 10y = \ddot{u} - 3\dot{u} + 5u$$

Find the system transfer function and the state space form.

3.9 Find the system response for $t > 1$ due to its initial condition at $t = 1$

$$\frac{dy(t)}{dt} + 4y(t) + 3 \int y(\tau)d\tau = 0, \quad y(1) = 2$$

3.10 Find the response of the discrete system

$$y(k + 2) + y(k) = (-1)^k, \quad y(0) = 1, \ y(1) = 0$$

Verify the answer by the MATLAB function dlsim.

3.11 A continuous-time system is represented by

$$\ddot{y} + 4\dot{y} + 3y = u$$

(a) Find the transfer function and the impulse response.

(b) Compute the response $y(t)$ for $y(0^-) = -2$, $\dot{y}(0^-) = 1$ and $u(t)$ equal to a unit step function.

3.12 Given a linear continuous-time system with

$$A = \begin{bmatrix} 0 & 1 \\ -2 & -3 \end{bmatrix}, \ B = \begin{bmatrix} 0 \\ -1 \end{bmatrix}, \ x(0) = \begin{bmatrix} -1 \\ 0 \end{bmatrix}, \ C = I, \ D = 0$$

(a) Find the state transition matrix.
(b) Find the system transfer function.
(c) Find the system response due to a unit step input.
(d) Verify the answers obtained by using MATLAB.

3.13 A discrete system is given by

$$y(k+1) - 0.5y(k) = 2u(k+1) + u(k)$$

Compute the impulse response. Verify the result by the MATLAB function `dimpulse`.

3.14 Discretize the following system by using the Euler approximation

$$\ddot{y} + 2\dot{y} + 3\sin(y(t)) = u(t), \quad y(0) = 1, \ \dot{y}(0) = 2$$

3.15 Given a time invariant linear system with the impulse response equal to e^{-t}. Find the response of this system due to an input $2\delta(t-1) + 3h(t-2)$, where $\delta(t)$ is the impulse delta function and $h(t)$ is a unit step function. What MATLAB functions can be used to solve this problem?

3.16 A linear discrete system is represented by

$$A = \begin{bmatrix} 0 & 1 \\ -2 & -3 \end{bmatrix}, \ B = \begin{bmatrix} 1 \\ 1 \end{bmatrix}, \ x(0) = \begin{bmatrix} -1 \\ 0 \end{bmatrix}, \ C = [0 \ \ 1], \ D = 0$$

(a) Find its state transition matrix.
(b) Find the transfer function.
(c) Find the system response due to $u(k) = k$ assuming that $x_1(0) = 1$ and $x_2(0) = 3$.

3.17 Find the response of the system given by

$$\ddot{y} + 2\dot{y} + y = \dot{u} + u, \quad u(t) = 2e^{-t}, \quad y(0) = 1, \ \dot{y}(0) = 1$$

3.18 Given a second-order linear system at rest (initial conditions are zero)

$$\ddot{y} + 2\xi\omega_n\dot{y} + \omega_n^2 y = \omega_n^2 u(t)$$

Find its unit step response for $\xi < 1$.

3.19 Find the response of the discrete system

$$y(k+2) - 6y(k+1) + 8y(k) = 3k + 2, \quad y(0) = 1, \ y(1) = 1$$

3.20 Find the response of the continuous system

$$\ddot{y} + 3\dot{y} - 10y = 2\dot{u} + 5u, \quad y(0) = 1, \ \dot{y}(0) = -1, \quad u(t) = t$$

3.21 Discretize the system $\dot{y} = u$ by using both the Euler and the integral approximations. Compare the discrete systems obtained.

3.22 Given a linear continuous system

$$\dot{x}(t) = Ax(t)$$

with

$$A = \begin{bmatrix} 2 & 1 \\ 2 & 3 \end{bmatrix}$$

Find the similarity transformation such that this system has the diagonal form in the new coordinates.

3.23 Find the state transition matrix of a continuous system with

$$A = \begin{bmatrix} 0 & -1 \\ 1 & 0 \end{bmatrix}$$

Use the Taylor series expansion method.

3.24 Find the response of a discrete system represented by

$$y(k+2) + 2y(k) + 1 = (-1)^k, \quad y(0) = y(1) = 1$$

3.25 Find the transition matrix in the complex domain for the system represented by

$$A = \begin{bmatrix} 1 & 0 & 0 \\ 0 & 0 & 0 \\ 1 & 0 & 0 \end{bmatrix}$$

3.26 Consider a fifth-order industrial reactor (Arkun and Ramakrishnan, 1983; Petkovski *et al.*, 1991) represented by

$$A = \begin{bmatrix} -16.11 & -0.39 & 27.2 & 0 & 0 \\ 0.01 & -16.99 & 0 & 0 & 12.47 \\ 15.11 & 0 & -53.6 & -16.57 & 71.78 \\ -53.36 & 0 & 0 & -107.2 & 232.11 \\ 2.27 & 60.1 & 0 & 2.273 & -102.99 \end{bmatrix}$$

$$B = \begin{bmatrix} 11.12 & -2.61 & -21.91 & -53.5 & 69.1 \\ -12.6 & 3.36 & 0 & 0 & 0 \end{bmatrix}^T$$

$$C = \begin{bmatrix} 0 & 0 & 0 & 0 & 1 \\ 0 & 1 & 1 & 0 & 0 \end{bmatrix}$$

Using MATLAB, find the following:

(a) The system transfer function.
(b) The impulse response.
(c) The response due to inputs $u_1(t) = e^{-t} + \sin(t)$ and $u_2(t) = 0$.

3.27 Discretize the system given in Problem 3.26 by the MATLAB function c2d with $T = 0.1$ and repeat the steps (a), (b), and (c) from Problem 3.26.

3.28 The model of a synchronous machine connected to an infinite bus (Koko-tović *et al.*, 1980; Grodt and Gajić, 1988) has the system matrix

$$A = \begin{bmatrix} -0.58 & 0 & 0 & -0.27 & 0 & 0.2 & 0 \\ 0 & -1 & 0 & 0 & 0 & 1 & 0 \\ 0 & 0 & -5 & 2.1 & 0 & 0 & 0 \\ 0 & 0 & 0 & 0 & 337 & 0 & 0 \\ -0.14 & 0 & 0.14 & -0.2 & -0.28 & 0 & 0 \\ 0 & 0 & 0 & 0 & 0 & 0.08 & 2 \\ -17.2 & 66.7 & -11.6 & 40.9 & 0 & -66.7 & -16.7 \end{bmatrix}$$

(a) Find the eigenvalues, eigenvectors, and similarity transformation that puts this system into a diagonal form.
(b) Discretize this system with $T = 1$.
(c) Find the response of the discrete system obtained in part (b) due to the initial condition $x(0) = \begin{bmatrix} 1 & 1 & 1 & 1 & 1 & 1 & 1 \end{bmatrix}^T$ and draw the corresponding response for the time interval $0 \le k \le 10$. Use MATLAB.

3.29 A linearized mathematical model of an aircraft considered in Litkouhi (1983) and Gajić and Shen (1991) has the form

$$
A = \begin{bmatrix}
-0.015 & -0.0805 & -0.0011666 & 0 \\
0 & 0 & 0 & 0.03333 \\
-2.28 & 0 & -0.84 & 1 \\
0.6 & 0 & -4.8 & -0.49
\end{bmatrix}
$$

$$
B = \begin{bmatrix}
-0.0000916 & 0.0007416 \\
0 & 0 \\
-0.11 & 0 \\
-8.7 & 0
\end{bmatrix}
$$

Obtain the following (using MATLAB):

(a) Discretize this model with $T = 1$.

(b) Find its response due to a unit ramp input.

(c) Find the system transfer function and the system poles.

3.30 Show by induction that the characteristic equation (3.96) of a system in the phase variable canonical form is indeed given by (3.95).

3.31 Write a MATLAB program to obtain the system's modal form for Example 3.4. Using that program, check the corresponding results in Examples 3.4 and 3.9.

Chapter Four

Linear System Stability

Stability is the most important concept for control systems. Unstable systems left alone (without stabilizing control inputs) can eventually destroy themselves; hence the main task of external inputs (controls) is to stabilize systems. In this chapter we introduce the general concept of system stability through several definitions. These definitions are specialized for time invariant linear systems. Methods for examining system stability are presented for both continuous- and discrete-time, time invariant, linear systems.

In Section 4.1 we introduce two basic definitions for stability of linear time invariant systems, namely the concepts of *stable* and *asymptotically stable* systems are defined.

System stability is related to system eigenvalues in Section 4.2. Conditions for both stable and asymptotically stable systems having either or both distinct and multiple eigenvalues are given. In order to completely clarify the stability of multiple eigenvalues on the imaginary axis, the Jordan canonical form is presented in sufficient detail and the notion of the system minimal polynomial is introduced.

In Section 4.3 system stability in the sense of Lyapunov is defined, and the corresponding Lyapunov algebraic equation test is formulated. This method, in the case of linear time invariant systems, is primarily important from a theoretical point of view, and its main result will be used in Section 4.4 to provide a simple and elegant proof for the Routh–Hurwitz stability criterion.

Since it is not easy to find eigenvalues for high-order dynamic systems, two algebraic tests for examining system stability without actually finding the system eigenvalues are given. The Routh–Hurwitz test, applicable to continuous-time systems, is presented in Section 4.4. Complete derivations and ideas of the proof are given in terms of known techniques so that both the problem of forming the Routh table and its final stability conclusion become meaningful to students. The dual method for examining the stability of discrete-time systems, known as Jury's test, is given in Section 4.5.

The frequency domain stability technique based on the Nyquist plot and important notions of gain and phase stability margins are discussed in Section 4.6. Only the main results and definitions are given and several examples are provided. The reason for that is the fact that the Nyquist method is extremely impractical for testing stability even for low-order systems since it is based on the tedious chore of plotting the open-loop system transfer function over all frequencies $\omega \in (-\infty, +\infty)$. However, two notions derived from the Nyquist method—stability *phase margin* and *gain margin*—are very useful for controller design methods. Stability phase and gain margins will be used in Chapter 9. At the end of this chapter a MATLAB laboratory experiment is given.

Chapter Objectives and Study Guide

The main objective of this chapter is to introduce the reader to the notions of stable, unstable, and asymptotically stable linear time invariant systems. This chapter provides several methods for testing whether or not given systems are either stable, asymptotically stable, or unstable. Students should master completely both the Routh–Hurwitz and Jury tests as the working tools for actually examining the stability of continuous- and discrete-time, time invariant, linear systems.

Instructors may skip Subsections 4.2.4–4.2.6 on the Jordan canonical form and stability of multiple eigenvalues on the imaginary axis since the study of Jordan forms may be characterized as an advanced linear algebra topic. However, in that case, instructors must point out to the students that the multiple eigenvalues on the imaginary axis are in most cases unstable, but the definite answer cannot be given before the system minimal polynomial is constructed (from the Jordan canonical form). *There are stable systems with multiple eigenvalues on the imaginary axis.* Unfortunately, some standard undergraduate control and systems textbooks *wrongly state that all systems with multiple eigenvalues on the imaginary axis are unstable.* In addition, while studying the frequency domain stability based on the Nyquist plot, Section 4.6, the main emphasis should be put on two important notions: gain and phase stability margins (they are important for controller design techniques)—the Nyquist plot method by itself is very inefficient for actually examining system stability.

4.1 Stability Concept and Main Definitions

For linear time invariant systems, system stability is strictly related to the system matrix $\mathbf{A}$. It can be roughly said that stability of these systems is a property of the system matrix $\mathbf{A}$, the property that the solution of an input-free continuous-time (or discrete-time) linear system

$$\dot{\mathbf{x}}(t) = \mathbf{A}\mathbf{x}(t), \qquad \mathbf{x}(t_0) = \mathbf{x}_0 \tag{4.1}$$

$$\mathbf{x}(k+1) = \mathbf{A}\mathbf{x}(k), \qquad \mathbf{x}(k_0) = \mathbf{x}_0 \tag{4.2}$$

remains bounded for all times. Solutions of (4.1) and (4.2) are given by

$$\mathbf{x}(t) = e^{\mathbf{A}(t-t_0)}\mathbf{x}_0, \qquad \mathbf{x}(k+1) = \mathbf{A}^{k-k_0}\mathbf{x}_0 \tag{4.3}$$

so that the system stability requirement is defined by

$$\begin{aligned}
\|\mathbf{x}(t)\| \leq const < \infty, & \qquad \forall t \\
\|\mathbf{x}(k)\| \leq const < \infty, & \qquad \forall k
\end{aligned} \tag{4.4}$$

where $\|.\|$ is any norm, e.g. the Euclidean norm

$$\|\mathbf{x}\| = \left(\sum_{i=1}^{n} x_i^2 \right)^{1/2} \tag{4.5}$$

Note that when an input is applied to a system, then formulas (4.3) represent the system *transient response*, and hence the stability requirement (4.4) can be reformulated as the boundness of the system transient response.

The main stability definitions important for studying the stability of time invariant linear systems and sufficient for the purpose of this course are given below.

Definition 4.1 A system is *stable* if the system motion is bounded, in other words, if the norm of the state vector is bounded by a constant.

Definition 4.2 A system is *asymptotically stable* if in addition to being stable it satisfies $\mathbf{x}(t) \to 0$ when $t \to \infty$, ($\mathbf{x}(k) \to 0$ when $k \to \infty$ for discrete-time systems).

For linear time invariant systems, system stability can be related to the system eigenvalues. This important observation is fully explained in the next section.

4.2 System Eigenvalues and Stability

Looking at expressions (4.3) we have to find out what conditions matrix $\mathbf{A}$ has to satisfy in order that inequalities (4.4) hold. In order to clarify this issue we start with a scalar version of (4.1), whose solution is given by

$$x(t) = e^{a(t-t_0)}x_0, \qquad t > t_0 \tag{4.6}$$

It is obvious that if $a < 0$ then $x(t) \to 0$ as $t \to \infty$. Also, if $a \leq 0$ then $|x(t)| \leq |x_0|$, i.e. the system motion is bounded. Note that for a scalar case, a represents the system eigenvalue. This indicates that system stability can be connected to system eigenvalues.

In the following we consider a general nth-order system and distinguish two cases: systems with distinct and multiple eigenvalues. The simpler case of distinct eigenvalues is considered first.

4.2.1 Stability of Distinct Eigenvalues

We have seen in Section 3.4 that the linear system response, for a general nth-order system, can be represented in terms of the system eigenvalues and eigenvectors. Assuming that *all eigenvalues are distinct*, it has been shown in (3.106) that

$$\mathbf{x}(t) = c_1 e^{\lambda_1 t}\mathbf{v}_1 + c_2 e^{\lambda_2 t}\mathbf{v}_2 + \cdots + c_n e^{\lambda_n t}\mathbf{v}_n \tag{4.7}$$

where $c_i, i = 1, 2, ..., n$, are constants, $\lambda_i, i = 1, 2, ..., n$, are eigenvalues, and $\mathbf{v}_1, \mathbf{v}_2, ..., \mathbf{v}_n$ are the eigenvectors of matrix $\mathbf{A}$. It is obvious from (4.7) that if all eigenvalues are real and distinct and $\lambda_i < 0, \forall i$ then $x(t) \to 0$ as $t \to \infty$, and hence the system is asymptotically stable. If $\lambda_i \leq 0, \forall i$ the system is stable. In addition, if only one of λ_j for any $j = 1, 2, ..., n$ is positive, the state response blows up and the system is unstable.

It is also easy to observe that for complex conjugate eigenvalues (which are distinct eigenvalues), e.g. $\alpha_i \pm j\beta_i$, the term

$$c_i e^{(\alpha_i \pm j\beta_i)t}\mathbf{v}_i = c_i e^{\alpha_i t}e^{\pm j\beta_i t}\mathbf{v}_i \tag{4.8}$$

grows in time if $\alpha_i > 0$ and decays if $\alpha_i < 0$. Note that $\left|e^{\pm j\beta_i t}\right| = 1$. If all eigenvalues in (4.7) satisfy $R_e\{\lambda_i\} = \alpha_i < 0$, we have that $\mathbf{x}(t) \to 0$ as

$t \to \infty$ and the system is *asymptotically stable*. Observe that if just one of the eigenvalues has a positive real part, i.e. $R_e\{\lambda_j\} = \alpha_j > 0$ for some j, then $\mathbf{x}(t) \to \infty$ and the system is *unstable*. In the case when $R_e\{\lambda_i\} = \alpha_i \leq 0$, the terms corresponding to $\alpha_i = 0$ in (4.7) will be bounded by constants since

$$\left\| c_i e^{\pm j\beta_i t} \mathbf{v}_i \right\| \leq \left\| e^{\pm j\beta_i t} \right\| \|c_i \mathbf{v}_i\| = \|c_i \mathbf{v}_i\| = const \tag{4.9}$$

so that

$$\|\mathbf{x}(t)\| \leq const \tag{4.10}$$

and the system is *stable*.

From the previous analysis, we can summarize the result for the case of *distinct eigenvalues* in the form of the following theorem.

Theorem 4.1 *A continuous-time linear, time invariant system with* distinct eigenvalues *is asymptotically stable if* $R_e(\lambda_i) = \alpha_i < 0, \forall i$. *It is stable for* $R_e(\lambda_i) = \alpha_i \leq 0, \forall i$, *and unstable if there exists any eigenvalue* λ_j *such that* $R_e(\lambda_j) = \alpha_j > 0$.

Similar stability analysis can be repeated for time invariant linear discrete systems. A scalar discrete-time system

$$x(k+1) = ax(k), \qquad x(k_0) = x_0 \tag{4.11}$$

has the solution

$$x(k) = a^{k-k_0} x_0, \qquad k - k_0 \geq 0 \tag{4.12}$$

which is bounded in time if $|a| \leq 1$. If $|a| < 1$ then $x(k) \to 0$ as $k \to \infty$ and the system is *asymptotically stable*. For $|a| > 1$ as $k \to \infty$ then $x(k) \to \infty$, which causes instability. Note that in the scalar case the parameter a represents the system's eigenvalue. For a general discrete-time system with distinct eigenvalues the solution can be represented in a form corresponding to (4.7), that is

$$\mathbf{x}(k) = c_1(\lambda_1)^k \mathbf{v}_1 + c_2(\lambda_2)^k \mathbf{v}_2 + \cdots + c_n(\lambda_n)^k \mathbf{v}_n \tag{4.13}$$

It is obvious that if $|\lambda_i| < 1$ for $\forall i$ then $\mathbf{x}(k) \to 0$ as $k \to \infty$ and the system is *asymptotically stable*. Conversely, if for some λ_j we have $|\lambda_j| > 1$, the motion of the system is unbounded, i.e. the system is *unstable*. Also, in another interesting case, i.e. when $|\lambda_i| \leq 1$, terms corresponding to $\lambda_i = 1$ will produce constants in formula (4.13) so that the motion of the system will be bounded by a sum of constants and the system will be *stable*. A dual theorem to Theorem 4.1 can be now formulated as follows.

Theorem 4.2 *A time invariant discrete system with* distinct eigenvalues *is asymptotically stable if* $|\lambda_i| < 1$ *for* $\forall i$*; it is stable for* $|\lambda_i| \leq 1, \forall i$*; and it is unstable if for some* λ_j*,* $|\lambda_j| > 1$.

4.2.2 Stability of Multiple Eigenvalues

In this subsection we show that both multiple real and multiple complex conjugate eigenvalues located in the open left half plane (which does not include the imaginary axis) are asymptotically stable. The same procedure indicates that the corresponding eigenvalues (system poles) located in the open right half plane are unstable. In order to be able to determine stability of multiple eigenvalues on the imaginary axis, we have to present the Jordan canonical form in sufficient detail and introduce the notion of the system minimal polynomial, which will be done in subsequent subsections.

Multiple Real Poles in the Open Left Half Complex Plane

Multiple real poles in the open left half of the complex plane are asymptotically stable, which will be demonstrated as follows. The system characteristic polynomial for a multiple real pole of multiplicity n_1 located in the open left half complex plane is given by

$$\Delta(\lambda) = (\lambda + a)^{n_1} \Delta_1(\lambda), \quad a > 0, \quad n > n_1 > 1$$

so that the corresponding system transfer function can be written as

$$H(s) = \frac{1}{(s+a)^{n_1}} H_1(s)$$

Using partial fraction expansion and the inverse Laplace transform (see Appendix A), we obtain the time domain contribution of the multiple real poles at $s = -a$ as

$$\mathcal{L}^{-1} \left\{ \sum_{i=0}^{n_1-1} \frac{k_{n_1-i}}{(s+a)^{n_1}} \right\} = \sum_{i=0}^{n_1-1} \frac{k_{n_1-i}}{(n_1 - 1 - i)!} t^{n_1-1-i} e^{-at}$$

Since

$$\lim_{t \to \infty} \left\{ t^{n_1-1-i} e^{-at} \right\} \to 0, \quad \forall i = 0, 1, 2, ..., n_1 - 1$$

we conclude that multiple real poles (eigenvalues) located in the open left half complex plane are asymptotically stable. Note that for $a > 0$ the above limit is equal to infinity, and we can therefore conclude that multiple real poles located in the open right half complex plane are unstable.

Multiple Complex Conjugate Poles in the Open Left Complex Plane

Consider the general case of complex conjugate poles located in the open left half of the complex plane. The system transfer function in this case, assuming that the poles' multiplicity is equal to $n_2 < n$, has the form

$$H(s) = \frac{1}{(s + \alpha - j\beta)^{n_2}} \frac{1}{(s + \alpha + j\beta)^{n_2}} H_2(s), \quad \alpha > 0$$

which can be expanded as

$$H(s) = \frac{k_{n_2}}{(s + \alpha - j\beta)^{n_2}} + \frac{k_{n_2-1}}{(s + \alpha - j\beta)^{n_2-1}} + \cdots + \frac{k_1}{(s + \alpha - j\beta)}$$

$$+ \frac{k_{n_2}^*}{(s + \alpha + j\beta)^{n_2}} + \frac{k_{n_2-1}^*}{(s + \alpha + j\beta)^{n_2-1}} + \cdots + \frac{k_1^*}{(s + \alpha + j\beta)} + H_3(s)$$

The contribution of this n_2th-order complex conjugate pole in the time domain is obtained by taking the Laplace inverse, which produces the following terms

$$\mathcal{L}^{-1}\left\{ \frac{1}{(s + \alpha \mp j\beta)^i} \right\} = \frac{t^{i-1}}{(i-1)!} e^{-\alpha t} e^{\pm j\beta t}, \quad i = 1, 2, ..., n_2 \qquad (4.14)$$

Since the limits of all these components as $t \to \infty$ are zeros, we conclude that the system multiple eigenvalues located in the open left half plane are asymptotically stable. Also, if $\alpha < 0$, i.e. in the case of multiple complex conjugate poles in the open right half complex plane, it is obvious that the above terms blow up in time, which causes system instability (makes system unstable).

The above discussion, together with the results obtained in Subsection 4.2.1, is summarized in the form of the following theorem.

Theorem 4.3 *The linear time invariant system (4.1) having either or both distinct and multiple eigenvalues is asymptotically stable if all eigenvalues of the matrix **A** are in the open left half of the complex plane. It is unstable if only one of the eigenvalues is in the open right half plane. Distinct eigenvalues on the imaginary axis are stable. The stability of multiple eigenvalues on the imaginary* *axis should be independently examined using the notion of the system minimal polynomial.*

The dual theorem is valid for discrete-time, time invariant, linear systems (4.2) with the imaginary axis replaced by the unit circle.

In Figures 4.1 and 4.2 we present stability regions in the complex plane of eigenvalues for continuous- and discrete-time, time invariant, linear systems.

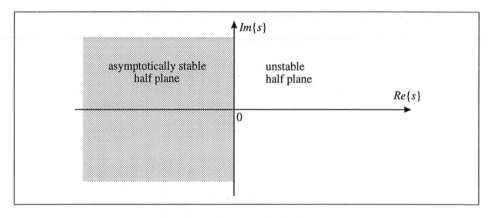

Figure 4.1: Stability region for linear continuous-time systems

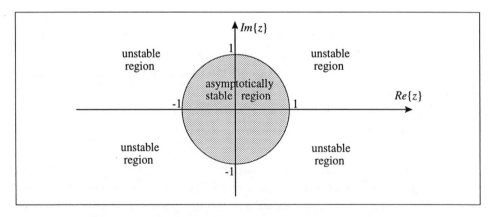

Figure 4.2: Stability region for linear discrete-time systems

4.2.3 Case Studies

The stability of three real physical systems—an inverted pendulum, a flexible beam, and an F-15 aircraft—is studied in view of the results stated in Theorem 4.3. These systems represent, respectively, unstable, stable, and asymptotically stable systems.

Inverted Pendulum

Consider the system matrix $\mathbf{A}$ of a linearized model of the inverted pendulum studied in Section 1.6. For the following data $m_2 = 10m_1$, $l = 0.5$, $g = 9.8$, we get from (1.80)

$$\mathbf{A} = \begin{bmatrix} 0 & 1 & 0 & 0 \\ 0 & 0 & -0.98 & 0 \\ 0 & 0 & 0 & 1 \\ 0 & 0 & 21.56 & 0 \end{bmatrix}$$

The eigenvalues of $\mathbf{A}$ are given by

$$\lambda_1 = -4.6433, \quad \lambda_2 = 0, \quad \lambda_3 = 0, \quad \lambda_4 = 4.6433$$

According to Theorem 4.3 this system is unstable due to the fact that the eigenvalue λ_4 is located in the right half of the complex plane. This conclusion is expected since the linearized model is obtained for the operating point corresponding to the inverted pendulum in the upright position. Apparently, without an external forcing function the inverted pendulum cannot by itself maintain a given upright position. It is unstable, and it must fall down.

Flexible Beam

The characteristic equation of a sixth-order model of the flexible beam system from Section 2.6 produces the following set of eigenvalues (obtained by using the MATLAB function `roots`)

$$\lambda_{1,2} = -0.394 \pm j21.4487, \quad \lambda_{3,4} = -0.1006 \pm j1.62$$
$$\lambda_5 = -0.0067, \quad \lambda_6 = 0$$

This represents a stable system since all eigenvalues, except for λ_6, are in the open left half of the complex plane and $\lambda_6 = 0$ is a simple (distinct) pole on the imaginary axis.

F-15 Aircraft

The F-15 aircraft under supersonic flight conditions, whose linearized mathematical model is given in Example 1.4, has the following eigenvalues

$$\lambda_{1,2} = -2.6616 \pm j1.7279, \quad \lambda_{3,4} = -0.0064 \pm j0.06782$$

It can be seen that this aircraft is asymptotically stable. However, it has a pair of complex conjugate poles very close to the imaginary axis and a small perturbation in system parameters could make this aircraft unstable.

It should be pointed out that all the above conclusions hold for *open-loop system stability*. Due to the presence of forcing functions in linear control systems one can use linear state feedback to stabilize systems under consideration. That is, for

$$\dot{x} = Ax + Bu$$

linear state feedback of the form $u = -Fx$ produces a closed-loop system

$$\dot{x} = (A - BF)x = A_c x$$

A feedback gain matrix F has to be chosen such that all eigenvalues of the closed-loop system matrix A_c are asymptotically stable. Conditions under which *closed-loop system stabilization* by linear state feedback can be achieved and the corresponding procedure are presented in Section 8.2.

4.2.4 Jordan Canonical Form[1]

Several standard undergraduate controls and systems textbooks *wrongly state that all multiple eigenvalues on the imaginary axis are unstable.* It will be easily seen, by using the Jordan canonical form, that the above conclusion is at least superficial and sometimes wrong. This misinterpretation is the consequence of using the characteristic polynomial, instead of the minimal polynomial, in interpreting system stability. Since the minimal polynomial is obtained from the Jordan form, in order to be able to define correctly and precisely the stability of

---

[1]   This section may be skipped with a reference given to Theorem 4.3. This subsection is an integral part of Subsection 4.2.5. However, Examples 4.4 and 4.6–4.7 can be independently studied in order to sustain the conclusions of Theorem 4.3.

multiple eigenvalues on the imaginary axis, the presentation of the Jordan form
is mandatory.

We have already seen in Subsection 3.4.1 that systems with multiple eigenvalues can be transferred into the Jordan canonical form by using a similarity
transformation. That is, there exists a transformation $\mathbf{V}$, obtained by using system generalized eigenvectors, such that

$$\mathbf{x} = \mathbf{V}\hat{\mathbf{x}} \tag{4.15}$$

transforms (4.1) into

$$\dot{\hat{\mathbf{x}}} = \mathbf{V}^{-1}\mathbf{A}\mathbf{V}\hat{\mathbf{x}} = \mathbf{J}\hat{\mathbf{x}} \tag{4.16}$$

where $\mathbf{J}$ has the Jordan form. Here, we present more details about the procedure
for obtaining the Jordan canonical form.

Assume that for a system of order n there are l distinct eigenvalues with
multiplicities l_i. The Jordan form of this system is given by the following block
diagonal matrix

$$\mathbf{J} = \begin{bmatrix} \mathbf{J}_1(\lambda_1) & 0 & \cdots & 0 \\ 0 & \mathbf{J}_2(\lambda_2) & \cdots & \vdots \\ \vdots & \cdots & \ddots & 0 \\ 0 & \cdots & 0 & \mathbf{J}_l(\lambda_l) \end{bmatrix} \tag{4.17}$$

where each square block $\mathbf{J}_i(\lambda_i)$ is constructed according to the following simple
rules:

1. Dimension of $\mathbf{J}_i(\lambda_i)$ is equal to l_i.
2. If $l_i = 1$ then $\mathbf{J}_i(\lambda_i) = \lambda_i$.
3. For $l_i > 1$ the diagonal blocks are composed of *simple Jordan blocks*, that is

$$\mathbf{J}_i(\lambda_i) = \begin{bmatrix} \mathbf{J}_i^{(1)}(\lambda_i) & 0 & \cdots & 0 \\ 0 & \mathbf{J}_i^{(2)}(\lambda_i) & \cdots & \vdots \\ \vdots & \cdots & \ddots & 0 \\ 0 & \cdots & 0 & \mathbf{J}_i^{(n_i)}(\lambda_i) \end{bmatrix} \tag{4.18}$$

where the simple Jordan block is defined by

$$\mathbf{J}_i^{(j)}(\lambda_i) = \begin{bmatrix} \lambda_i & 1 & 0 & \cdots & 0 \\ 0 & \lambda_i & 1 & \cdots & \vdots \\ \vdots & 0 & \ddots & \ddots & 0 \\ \vdots & \cdots & \cdots & \ddots & 1 \\ 0 & \cdots & \cdots & 0 & \lambda_i \end{bmatrix} \tag{4.19}$$

4. *The number of simple Jordan blocks* in $\mathbf{J}_i(\lambda_i)$ is determined by the formula

$$n_i = \dim\{\mathcal{N}(\mathbf{A} - \lambda_i\mathbf{I})\} \tag{4.20}$$

where $\mathcal{N}$ stands for the null space (see Appendix C). An useful formula for determining the dimensions of $\mathbf{J}_i^{(j)}(\lambda_i)$ blocks can be found in Lancaster and Tismenetsky (1985, page 235). It says that the number of simple Jordan blocks of dimension k, $(k = 1, 2, ..., l_i)$, in the block $\mathbf{J}_i(\lambda_i)$ is given by

$$2v_k - v_{k-1} - v_{k+1} \tag{4.21}$$

where

$$v_k = \dim\left\{\mathcal{N}(\mathbf{A} - \lambda_i\mathbf{I})^k\right\} \tag{4.22}$$

Rules (1)–(3) are easy to remember. As far as rule (4) is concerned, it is very often sufficient to use only formula (4.20) in order to obtain complete information about the Jordan form. Two examples below demonstrate the use of both formula (4.20) and formulas (4.21) and (4.22). Of course, finding the dimension of the null space of a matrix sometimes requires a lot of computation, but this can be easily accomplished, even for high-order systems, using MATLAB and its functions `null` and `rank`.

In order to complete the presentation of the Jordan canonical form, it remains to show the procedure for constructing the matrix $\mathbf{V}$ whose columns are the generalized eigenvectors of the matrix $\mathbf{A}$. This can be achieved by searching for linearly independent eigenvectors from the standard eigenvalue–eigenvector problem

$$\mathbf{A}\mathbf{v}_i = \lambda_i\mathbf{v}_i, \quad i = 1, 2, ..., l \tag{4.23}$$

which produces, in general, $\nu \leq n$ linearly independent vectors, and completing the n-dimensional basis by using the so-called vectors of Jordan chains defined by

$$\mathbf{A}\mathbf{v}_j = \lambda_i\mathbf{v}_j + \mathbf{v}_{j-1}; \quad \mathbf{v}_{j-1} = \mathbf{v}_i \tag{4.24}$$

It should be pointed out that the construction of the matrix $\mathbf{V}$ is not crucial for the stability study. The reader interested in this subject is referred to Chen (1984), Lancaster and Tismenetsky (1985), and Strang (1988) for the complete procedure for obtaining the generalized eigenvectors of matrices that have multiple eigenvalues.

Note that MATLAB does not have a function for finding the Jordan form directly so that one has to use four steps outlined in this section. The main reason for the omission of the corresponding MATLAB function is the fact that the general procedure for finding the generalized eigenvectors and the Jordan form by using (4.16) is numerically unstable (Kailath, 1980).

Example 4.1: Consider the following matrix

$$A = \begin{bmatrix} 1 & 0 & 1 & 0 & 1 & 0 & 1 \\ 0 & 1 & 0 & 0 & 0 & 0 & 0 \\ 0 & 0 & 1 & 0 & 0 & 0 & 1 \\ 0 & 0 & 0 & 1 & 0 & 0 & 1 \\ 0 & 0 & 0 & 0 & -2 & 1 & 0 \\ 0 & 0 & 0 & 0 & 0 & -2 & 3 \\ 0 & 0 & 0 & 0 & 0 & 0 & -2 \end{bmatrix}$$

It is easy to see that the eigenvalues are $\lambda_1 = 1$ with multiplicity $l_1 = 4$, and $\lambda_2 = -2$ with multiplicity $l_2 = 3$. Using formula (4.20) and MATLAB functions null (to find null space) and rank (to find the dimension of the corresponding null space), we get $n_1 = 3$ and $n_2 = 1$. Thus, the number of simple Jordan blocks associated with the eigenvalue $\lambda_1 = 1$ is 3 and since the total dimension of the block is 4, we conclude that there are two simple Jordan blocks of dimension 1 and one simple Jordan block of dimension 2. The fact that $n_2 = 1$ indicates that there exists only one Jordan block corresponding to the eigenvalue $\lambda_2 = -2$. Since the multiplicity of this eigenvalue is 3, the dimension of the simple Jordan block must be equal to 3. Thus, we have the complete information and we are able to write the Jordan form, which is given by

$$J = \begin{bmatrix} 1 & 1 & 0 & 0 & 0 & 0 & 0 \\ 0 & 1 & 0 & 0 & 0 & 0 & 0 \\ 0 & 0 & 1 & 0 & 0 & 0 & 0 \\ 0 & 0 & 0 & 1 & 0 & 0 & 0 \\ 0 & 0 & 0 & 0 & -2 & 1 & 0 \\ 0 & 0 & 0 & 0 & 0 & -2 & 1 \\ 0 & 0 & 0 & 0 & 0 & 0 & -2 \end{bmatrix}$$

◇

Example 4.2: In this example the information obtained from (4.20) is not sufficient to construct the Jordan form so that we have also to use formulas (4.21) and (4.22). For the following matrix

$$A = \begin{bmatrix} -1 & 0 & 1 & 0 & -1 & 0 & 0 & 1 \\ 0 & -1 & 2 & 0 & 1 & 2 & -2 & -1 \\ 0 & 0 & -1 & 2 & 1 & 0 & 1 & 2 \\ 0 & 0 & 0 & -1 & 2 & 1 & 2 & 1 \\ 0 & 0 & 0 & 0 & -2 & 0 & 1 & 2 \\ 0 & 0 & 0 & 0 & 0 & -2 & 2 & 2 \\ 0 & 0 & 0 & 0 & 0 & 0 & -2 & -2 \\ 0 & 0 & 0 & 0 & 0 & 0 & 0 & -2 \end{bmatrix}$$

the eigenvalues and their multiplicities are given by $\lambda_1 = -1$, $l_1 = 4$, and $\lambda_2 = -2$, $l_2 = 4$. From formula (4.20) and MATLAB functions `null` and `rank`, we get $n_1 = n_2 = 2$, i.e. for each eigenvalue there are two simple Jordan blocks. Since these blocks can be of dimensions 1, 2, or 3, further analysis is needed. Using formulas (4.21) and (4.22) and MATLAB (of course), we get for the eigenvalue $\lambda_1 = -1$

$$v_0 = 0, \ v_1 = 2, \ v_2 = 3$$
$$k = 1 \ \Rightarrow \ 2v_1 - v_0 - v_2 = 1$$
$$k = 2 \ \Rightarrow \ 2v_2 - v_1 - v_3 = 0$$

i.e. there is one simple Jordan block of dimension 1, there are no simple Jordan blocks of dimension 2, and since the total dimension of the diagonal block matrix corresponding to the eigenvalue $\lambda_1 = -1$ is equal to 4, we must also have one simple Jordan block of dimension 3. Using the same strategy, we get that for the second eigenvalue $\lambda_1 = -2$ there is one simple Jordan block of dimension 1 and one simple Jordan block of dimension 3, so that the Jordan canonical form looks like

$$A = \begin{bmatrix} -1 & 1 & 0 & 0 & 0 & 0 & 0 & 0 \\ 0 & -1 & 1 & 0 & 0 & 0 & 0 & 0 \\ 0 & 0 & -1 & 0 & 0 & 0 & 0 & 0 \\ 0 & 0 & 0 & -1 & 0 & 0 & 0 & 0 \\ 0 & 0 & 0 & 0 & -2 & 0 & 0 & 0 \\ 0 & 0 & 0 & 0 & 0 & -2 & 1 & 0 \\ 0 & 0 & 0 & 0 & 0 & 0 & -2 & 1 \\ 0 & 0 & 0 & 0 & 0 & 0 & 0 & -2 \end{bmatrix}$$

◇

Rules (1)–(4) are sufficient to obtain the Jordan form. Another useful fact about the Jordan canonical form is that *the Jordan canonical form is unique up to ordering of simple Jordan blocks* (Lancaster and Tismenetsky, 1985). Now we turn to the definition of an important notion in system stability theory, the notion of the *system minimal polynomial.*

Definition 4.3 For an nth-order time invariant dynamic system with $l < n$ distinct eigenvalues the *minimal polynomial* is defined by

$$m_A(\lambda) = (\lambda - \lambda_1)^{m_1}(\lambda - \lambda_2)^{m_2} \cdots (\lambda - \lambda_l)^{m_l} \tag{4.25}$$

where m_i stands for the *maximal dimensions of simple Jordan blocks corresponding to λ_i in the Jordan canonical form.*

The following examples easily show how to get minimal polynomials from the Jordan canonical forms. In addition, characteristic polynomials are also obtained and compared with the corresponding minimal polynomials.

Example 4.3: Given a matrix A of order 8 with eigenvalues $\lambda_1 = -1$, $\lambda_2 = -2, \lambda_3 = -3$ whose multiplicities are 3, 1, and 4, respectively. Assume that the corresponding Jordan canonical form is given by

$$J = \begin{bmatrix}
-1 & 1 & 0 & 0 & 0 & 0 & 0 & 0 \\
0 & -1 & 0 & 0 & 0 & 0 & 0 & 0 \\
0 & 0 & -1 & 0 & 0 & 0 & 0 & 0 \\
0 & 0 & 0 & -2 & 0 & 0 & 0 & 0 \\
0 & 0 & 0 & 0 & -3 & 1 & 0 & 0 \\
0 & 0 & 0 & 0 & 0 & -3 & 1 & 0 \\
0 & 0 & 0 & 0 & 0 & 0 & -3 & 0 \\
0 & 0 & 0 & 0 & 0 & 0 & 0 & -3
\end{bmatrix}$$

The maximal dimensions of the corresponding simple Jordan blocks are $m_1 = 2$, $m_2 = 1, m_3 = 3$; hence the minimal polynomial is given by

$$m_A(\lambda) = (\lambda + 1)^2(\lambda + 2)(\lambda + 3)^3$$

Note that the characteristic polynomial for this matrix has the form

$$\Delta_A(\lambda) = (\lambda + 1)^3(\lambda + 2)(\lambda + 3)^4$$

so that we can notice different multiplicities of the eigenvalues appearing in the minimal and characteristic polynomials. It is obvious that the order of the minimal polynomial is always less than or equal to the order of the characteristic polynomial, which justifies its name.

◇

Example 4.4: Consider the linear dynamic system of order 5 given by

$$
\begin{bmatrix} \dot{x}_1 \\ \dot{x}_2 \\ \dot{x}_3 \\ \dot{x}_4 \\ \dot{x}_5 \end{bmatrix} = \begin{bmatrix} -1 & 0 & 0 & 0 & 0 \\ 0 & -2 & 0 & 0 & 0 \\ 0 & 0 & 0 & 0 & 0 \\ 0 & 0 & 0 & 0 & 0 \\ 0 & 0 & 0 & 0 & 0 \end{bmatrix} \begin{bmatrix} x_1 \\ x_2 \\ x_3 \\ x_4 \\ x_5 \end{bmatrix}, \quad \begin{bmatrix} x_1(t_0) \\ x_2(t_0) \\ x_3(t_0) \\ x_4(t_0) \\ x_5(t_0) \end{bmatrix} = \begin{bmatrix} x_{10} \\ x_{20} \\ x_{30} \\ x_{40} \\ x_{50} \end{bmatrix}
$$

It is easy to see that for this system $m_1 = 1, m_2 = 1, m_3 = 1$ and $l_1 = 1, l_2 = 1,$ $l_3 = 3$; hence the minimal and characteristic polynomials are given by

$$
m_{\mathbf{A}}(\lambda) = (\lambda + 1)(\lambda + 2)\lambda, \quad \Delta_{\mathbf{A}}(\lambda) = (\lambda + 1)(\lambda + 2)\lambda^3
$$

It is obvious from the system equation that the state trajectories satisfy

$$
\begin{aligned}
x_1(t) &= e^{-t} x_{10} \\
x_2(t) &= e^{-2t} x_{20} \\
x_3(t) &= x_{30} = const \\
x_4(t) &= x_{40} = const \\
x_5(t) &= x_{50} = const
\end{aligned}
$$

so that this system is stable despite having a triple eigenvalue on the imaginary axis. Note that in this particular example the minimal polynomial contains only simple (linear) factors, whereas the characteristic polynomial contains multiple (nonlinear) factors.

◇

In the next subsection we utilize knowledge about the Jordan canonical form and the notion of the system minimal polynomial in order to clarify completely and rigorously the stability nature of multiple eigenvalues (poles) located on the imaginary axis.

4.2.5 Multiple Eigenvalues on the Imaginary Axis[2]

From the previous discussion about the Jordan canonical form and the definition of the minimal polynomial we can conclude that the Jordan form is diagonal if and only if the minimal polynomial does not contain multiple eigenvalues. The following theorem has been established.

Theorem 4.4 *An arbitrary square matrix with real entries is diagonalizable if and only if its minimal polynomial is composed of linear factors.*

Another theorem, which follows directly from Theorem 4.4, is the stability theorem for linear systems that have multiple eigenvalues on the imaginary axis. It can be stated as follows.

Theorem 4.5 *The linear system (4.1) is stable if all eigenvalues of the matrix* **A** *are in the open left half of the complex plane and those on the imaginary axis are distinct roots of the minimal polynomial of* **A**.

In the following we solve several examples in order to demonstrate the results of Theorem 4.5.

Example 4.5: Consider a third-order system

$$\dot{x}(t) = \mathbf{A}x(t), \quad x(0) = \begin{bmatrix} x_{10} \\ x_{20} \\ x_{30} \end{bmatrix}$$

with the system matrix having one of the following three forms

$$\text{(a)} \ \ \mathbf{A} = \begin{bmatrix} 0 & 0 & 0 \\ 0 & 0 & 0 \\ 0 & 0 & 0 \end{bmatrix}, \quad \text{(b)} \ \ \mathbf{A} = \begin{bmatrix} 0 & 1 & 0 \\ 0 & 0 & 0 \\ 0 & 0 & 0 \end{bmatrix}, \quad \text{(c)} \ \ \mathbf{A} = \begin{bmatrix} 0 & 1 & 0 \\ 0 & 0 & 1 \\ 0 & 0 & 0 \end{bmatrix}$$

It is easy to conclude that the characteristic polynomials are the same in all three cases, that is

$$\Delta_a(\lambda) = \Delta_b(\lambda) = \Delta_c(\lambda) = \lambda^3$$

However, the minimal polynomials have different forms

$$m_a(\lambda) = \lambda, \quad m_b(\lambda) = \lambda^2, \quad m_c(\lambda) = \lambda^3$$

In all three cases the matrix **A** has a triple eigenvalue at $\lambda = 0$. However, by Theorem 4.5 only the case (a) is stable; cases (b) and (c) are unstable.

---

[2]   This section may be skipped with a reference given to Theorem 4.3.

The above conclusion can be easily justified by finding the actual system trajectories in all cases. It is easy to show that (see Problem 4.1[3])

$$\mathbf{x}_a(t) = \begin{bmatrix} x_{10} \\ x_{20} \\ x_{30} \end{bmatrix}, \quad \mathbf{x}_b(t) = \begin{bmatrix} x_{10} - x_{20}t \\ x_{20} \\ x_{30} \end{bmatrix}, \quad \mathbf{x}_c(t) = \begin{bmatrix} x_{10} - x_{20}t + \frac{1}{2}x_{30}t^2 \\ x_{20} - x_{30}t \\ x_{30} \end{bmatrix}$$

Thus, the motion of the system (a) is bounded by a constant, whereas the norms of the system trajectories in cases (b) and (c) are unbounded in time, i.e. $\|\mathbf{x}_b(t)\| \to \infty, \quad \|\mathbf{x}_c(t)\| \to \infty$ as $t \to \infty$.

◇

It can be concluded from the previous example (and Theorem 4.5) that the *roots of the minimal polynomial*, but not the roots of the characteristic polynomial, *shape the time behavior* of linear time invariant systems. In addition, the order of the system transfer function is determined by the order of system's minimal polynomial (see Problem 4.2).

Example 4.6: *Simple Eigenvalues (Poles) on the Imaginary Axis*

The *minimal* polynomial in this case is given by

$$m(\lambda) = (\lambda + j\omega)(\lambda - j\omega)m_1(\lambda)$$

After performing the partial fraction expansion of the corresponding transfer function we get

$$H(s) = \frac{N(s)}{D(s)} = \frac{k_1}{s + j\omega} + \frac{k_1^*}{s - j\omega} + H_1(s), \quad k_1 = \alpha + j\beta$$

The time domain contribution from the simple imaginary axis poles at $s = \pm j\omega$ can be obtained by using the inverse Laplace transform. It is given by

$$\mathcal{L}^{-1}\left\{ \frac{\alpha + j\beta}{s + j\omega} + \frac{\alpha - j\beta}{s - j\omega} \right\} = \mathcal{L}^{-1}\left\{ \frac{2\alpha s}{s^2 + \omega^2} + \frac{2\beta\omega}{s^2 + \omega^2} \right\}$$

$$= 2(\alpha\cos\omega t + \beta\sin\omega t)$$

Since

$$|\alpha\cos\omega t + \beta\sin\omega t| \le |\alpha\cos\omega t| + |\beta\sin\omega t| \le |\alpha| + |\beta| = const$$

---

[3] This problem, together with Problem 4.2, should be assigned to students as mandatory homework problems.

the system motion arising from the simple pure imaginary poles is bounded.

We conclude from Example 4.6 that the simple poles on the imaginary axis produce the system's oscillatory behavior. Of course, systems that have eigenvalues on the imaginary axis cannot be asymptotically stable.

◇

Example 4.7: *Multiple Pure Imaginary Eigenvalues*

From Subsection 4.2.2 and formula (4.14) we can also draw conclusions about the instability of the multiple pure imaginary eigenvalues of the *minimal polynomial*. For $\alpha = 0$, we have from the last expression

$$\mathcal{L}^{-1}\left\{\frac{1}{(s \mp j\beta)^i}\right\} = \frac{t^{i-1}}{(i-1)!}e^{\pm j\beta t}, \quad i = 1, 2, ..., n_2$$

which apparently tends to ∞ for $i > 1$.

◇

Similar stability analysis using the notion of the system minimal polynomial can be performed for discrete-time, time invariant, linear systems. In that study the imaginary axis has to be replaced by the unit circle, and therefore, corresponding to Theorem 4.5, we can state the following theorem.

Theorem 4.6 *The linear discrete-time system (4.2) is stable if all eigenvalues of the matrix* **A** *are inside of the unit circle and those on the unit circle are distinct roots of the minimal polynomial of* **A**.

Example 4.8: Let the following matrices represent linear discrete-time systems

$$\mathbf{A}_1 = \begin{bmatrix} -1 & 1 & 0 & 0 \\ 0 & -1 & 0 & 0 \\ 0 & 0 & 0.5 & 0 \\ 0 & 0 & 0 & 0.5 \end{bmatrix}, \quad \mathbf{A}_2 = \begin{bmatrix} -1 & 0 & 0 & 0 \\ 0 & -1 & 0 & 0 \\ 0 & 0 & 0.5 & 1 \\ 0 & 0 & 0 & 0.5 \end{bmatrix}$$

By Theorem 4.6 the matrix $\mathbf{A}_1$ is unstable since its minimal polynomial has a double pole on the unit circle located at −1, but the matrix $\mathbf{A}_2$ is stable since the double pole of its minimal polynomial at 0.5 is inside the unit circle.

◇

4.2.6 Case Study: Flexible Space Structure

A flexible space structure having the following system matrix is considered in Calise and Moerder (1985)

$$\mathbf{A} = \begin{bmatrix} \mathbf{A}_1 & \mathbf{A}_2 \\ \mathbf{A}_3 & \mathbf{A}_4 \end{bmatrix}$$

where

$$\mathbf{A}_1 = \begin{bmatrix} 0 & 1 & 0 & 0 \\ -0.1764 & 0 & 0 & 0 \\ 0 & 0 & 0 & 1 \\ 0 & 0 & -0.1764 & 0 \end{bmatrix}, \quad \mathbf{A}_2 = 0$$

$$\mathbf{A}_4 = \begin{bmatrix} 0 & 1 & 0 & 0 \\ -4.41 & 0 & 0 & 0 \\ 0 & 0 & 0 & 1 \\ 0 & 0 & -4.41 & 0 \end{bmatrix}, \quad \mathbf{A}_3 = 0$$

The matrix $\mathbf{A}$ has four double eigenvalues on the imaginary axis located at

$$\lambda_{1,2} = -j0.42, \quad \lambda_{3,4} = +j0.42, \quad \lambda_{5,6} = -j2.1, \quad \lambda_{7,8} = j2.1$$

Using formula (4.20) in order to find the number of the simple Jordan blocks corresponding to the above multiple eigenvalues, we get that for all of them $n_i = 2$, $i = 1, 2, 3, 4$. This means that the Jordan form has eight simple blocks of dimension one, i.e. the Jordan form is pure diagonal leading to the minimal polynomial with only distinct roots. Thus, by Theorem 4.5 the considered flexible space structure represents a stable system despite the fact that it has multiple eigenvalues on the imaginary axis.

In reaching the above conclusion we have used MATLAB in order to form the matrix $\mathbf{A} - \lambda_i \mathbf{I}$, find its null space, and determine the dimension of the null space obtained. This has been done as follows

```
NA1=null(A+i*0.42*eye(8));
```

```
n1=rank(NA1)
```

Note that MATLAB handles the imaginary numbers in a fairly simple way, and that eye(8) defines an identity matrix of dimension 8. The above statements have to be repeated three times for the remaining three eigenvalues.

4.3 Lyapunov Stability of Linear Systems

In this section we present the Lyapunov stability method specialized for the linear time invariant systems studied in this book. The method has more theoretical importance than practical value and can be used to derive and prove other stability results. Its final statement for linear time invariant systems is elegant and easily tested using MATLAB. However, it is computationally more involved than the other methods for examining the stability of linear systems. Its importance lies in its generality since it can be applied to all nonlinear and linear systems without taking into account whether or not these systems are time invariant or time varying. More about the general study of Lyapunov stability can be found in several books on nonlinear systems (see for example Khalil, 1992). Here, we study the Lyapunov stability theory for time invariant continuous and discrete linear systems only.

In 1892 the Russian mathematician Alexander Mikhailovitch Lyapunov introduced his famous stability theory for nonlinear and linear systems. A complete English translation of Lyapunov's doctoral dissertation was published in the *International Journal of Control* in March 1992. The stability definition given in Section 4.1, Definition 4.1, in fact corresponds to the Lyapunov stability definition, so that "stable" used in this book also means "stable in the sense of Lyapunov". According to Lyapunov, one can check stability of a system by finding some function $V(\mathbf{x})$, called the Lyapunov function, which for time invariant systems satisfies

$$V(\mathbf{x}) > 0, \qquad V(\mathbf{0}) = 0 \qquad\qquad (4.26a)$$

$$\dot{V}(\mathbf{x}) = \frac{dV}{dt} = \frac{\partial V}{\partial \mathbf{x}}\frac{d\mathbf{x}}{dt} \leq 0 \qquad\qquad (4.26b)$$

There is no general procedure for finding the Lyapunov functions for nonlinear systems, but for linear time invariant systems, the procedure comes down to the problem of solving a linear algebraic equation, called the Lyapunov algebraic equation.

In view of (4.26a) and (4.26b), a linear time invariant system is *stable* if one is able to find a scalar function $V(\mathbf{x})$ such that when this function is associated with the system, both conditions given in (4.26) are satisfied. If the condition (4.26b) is a strict inequality, then the result is *asymptotic* stability. It can be shown that for a linear system (4.1) the Lyapunov function can be chosen to be quadratic, that is

$$V(\mathbf{x}) = \mathbf{x}^T \mathbf{P} \mathbf{x}, \qquad \mathbf{P} = \mathbf{P}^T > 0 \tag{4.27}$$

which with the use of (4.1) leads to

$$\dot{V}(\mathbf{x}) = \mathbf{x}^T \left(\mathbf{A}^T \mathbf{P} + \mathbf{P} \mathbf{A} \right) \mathbf{x}$$

i.e. the system is asymptotically stable if the following condition is satisfied

$$\mathbf{A}^T \mathbf{P} + \mathbf{P} \mathbf{A} < 0$$

or, equivalently

$$\mathbf{A}^T \mathbf{P} + \mathbf{P} \mathbf{A} = -\mathbf{Q}, \qquad \mathbf{Q} = \mathbf{Q}^T > 0 \tag{4.28}$$

where $\mathbf{Q}$ is any positive definite matrix. Recall that positive definite matrices have all eigenvalues in the closed right-hand half of the complex plane (see Appendix C). The matrix algebraic equation (4.28) is known as the Lyapunov algebraic equation. More about this important equation and its role in system stability and control can be found in Gajić and Qureshi (1995). Now we are able to formulate the Lyapunov stability theory for linear continuous time invariant systems.

Theorem 4.7 *The linear time invariant system (4.1) is asymptotically stable if and only if for any* $\mathbf{Q} = \mathbf{Q}^T > 0$ *there exists a unique* $\mathbf{P} = \mathbf{P}^T > 0$ *such that (4.28) is satisfied.*

Example 4.9: In this example we demonstrate the necessary steps required in applying the Lyapunov stability test. Consider the following continuous time invariant system represented by

$$\mathbf{A} = \begin{bmatrix} 0 & 1 & 0 \\ 0 & 0 & 1 \\ -1 & -2 & -3 \end{bmatrix}$$

It is easy to check by MATLAB function `eig` that the eigenvalues of this system are $\lambda = -2.3247, -0.3376 \pm j0.5623$, and hence this system is asymptotically stable. In order to apply the Lyapunov method, we first choose a positive definite matrix $\mathbf{Q}$. The standard "initial guess" for $\mathbf{Q}$ is identity, i.e. $\mathbf{Q} = \mathbf{I}_3$. With the help of the MATLAB function `lyap` (used for solving the algebraic Lyapunov equation), we can execute the following statement `P=lyap(A',Q)` and obtain the solution for $\mathbf{P}$ as

$$\mathbf{P} = \begin{bmatrix} 2.3 & 2.1 & 0.5 \\ 2.1 & 4.6 & 1.3 \\ 0.5 & 1.3 & 0.6 \end{bmatrix}$$

Note that we have used a transpose for the system matrix in the MATLAB function `lyap`, i.e. $\left(\mathbf{A}' = \mathbf{A}^T\right)$, since that function solves the equation that represents the transpose of the algebraic Lyapunov equation (4.28). Examining the positive definiteness of the matrix $\mathbf{P}$ (all eigenvalues of $\mathbf{P}$ must be in the closed right half plane), we get that the eigenvalues of this matrix are given by 6.1827, 1.1149, 0.2024; hence $\mathbf{P}$ is positive definite and the Lyapunov test indicates that the system under consideration is stable.

◇

It can be seen from this particular example that the Lyapunov stability test is not numerically very efficient since we have first to solve the linear algebraic Lyapunov equation and then to test the positive definiteness of the matrix $\mathbf{P}$, which requires finding its eigenvalues. Of course, we can find the eigenvalue of the matrix $\mathbf{A}$ immediately and from that information determine the system stability. It is true that the Lyapunov stability test is not the right method to test the stability of linear systems when the system matrix is given by numerical entries. However, it can be used as a *useful concept in theoretical considerations*, e.g. to prove some other stability results. This will be demonstrated in Section 4.4 where we will give a very simple and elegant proof of the very well-known Routh–Hurwitz stability criterion.

Note that Theorem 4.7 can be generalized to include the case when the matrix $\mathbf{Q}$ is positive semidefinite, $\mathbf{Q} = \mathbf{C}^T\mathbf{C} \geq 0$. Recall that positive semidefinite matrices have eigenvalues in the open right half of the complex plane (see Appendix C). Another form of Theorem 4.7 can be formulated as follows (Chen, 1984).

Theorem 4.8 *The time invariant linear system (4.1) is asymptotically stable if and only if the pair* $(\mathbf{A}, \mathbf{C})$ *is observable*[4] *and the algebraic Lyapunov equation (4.28) has a unique positive definite solution.*

---

[4]   For a definition of observability, see Section 5.2.

The observability of the pair $(\mathbf{A}, \mathbf{C})$ can be relaxed to its detectability.[5] This is natural since the detectability implies the observability of the modes which are not asymptotically stable.

Example 4.10: Consider the same system matrix $\mathbf{A}$ as in Example 4.9 with the matrix $\mathbf{Q}$ obtained from

$$\mathbf{Q}_1 = \mathbf{C}^T\mathbf{C} = \begin{bmatrix} 0 \\ 0 \\ 1 \end{bmatrix} \begin{bmatrix} 0 & 0 & 1 \end{bmatrix} = \begin{bmatrix} 0 & 0 & 0 \\ 0 & 0 & 0 \\ 0 & 0 & 1 \end{bmatrix}$$

Note that the pair $(\mathbf{A}, \mathbf{C})$ is observable since $\mathrm{rank}\{\mathcal{O}(\mathbf{A}, \mathbf{C})\} = 3$. The algebraic Lyapunov equation

$$\mathbf{A}^T\mathbf{P}_1 + \mathbf{P}_1\mathbf{A} = \mathbf{Q}_1$$

has the positive definite solution

$$\mathbf{P}_1 = \begin{bmatrix} 0.1 & 0.2 & 0 \\ 0.2 & 0.7 & 0.1 \\ 0 & 0.1 & 0.2 \end{bmatrix} > 0$$

which can be confirmed by finding the eigenvalues of $\mathbf{P}_1$, so that the considered linear system is asymptotically stable.

$\diamond$

Theorems corresponding to Theorems 4.7 and 4.8 can be stated for stability of discrete-time systems. For a linear discrete-time system (4.2) the Lyapunov function has a quadratic form, which, according to the Lyapunov stability theory, must satisfy (Ogata, 1987)

$$\begin{aligned} V(k) &= \mathbf{x}^T(k)\mathbf{P}\mathbf{x}(k) > 0 \\ \Delta V(k) &= V(k+1) - V(k) \leq 0 \end{aligned} \tag{4.29}$$

Since
$$\begin{aligned} V(k+1) - V(k) &= \mathbf{x}^T(k+1)\mathbf{P}\mathbf{x}(k+1) - \mathbf{x}^T(k)\mathbf{P}\mathbf{x}(k) \\ &= \mathbf{x}^T(k)\left(\mathbf{A}^T\mathbf{P}\mathbf{A} - \mathbf{P}\right)\mathbf{x}(k) \leq 0 \end{aligned}$$

---

[5]   For a definition of detectability, see Section 5.5, Definition 5.2.

the stability requirement imposed in (4.29) leads (similarly to the continuous-time argument) to the discrete-time algebraic Lyapunov equation

$$\mathbf{A}^T \mathbf{P} \mathbf{A} - \mathbf{P} = -\mathbf{Q} \qquad (4.30)$$

which for asymptotic stability, according to the Lyapunov stability theory (dual result to Theorem 4.7), must have a unique positive definite solution for some positive definite matrix $\mathbf{Q}$. Thus, we have the following theorem.

Theorem 4.9 *The linear time invariant discrete system (4.2) is asymptotically stable if and only if for any* $\mathbf{Q} = \mathbf{Q}^T > 0$ *there exists a unique* $\mathbf{P} = \mathbf{P}^T > 0$ *such that (4.30) is satisfied.*

The relaxed form of this theorem, valid for a positive semidefinite matrix $\mathbf{Q}$, requires the observability of the pair $(\mathbf{A}, \sqrt{\mathbf{Q}})$. Note that by the very well-known theorem from linear algebra every symmetric positive semidefinite matrix can be written in the form $\mathbf{Q} = \mathbf{C}^T \mathbf{C}$ with matrix $\mathbf{C}$ being known as the square root of $\mathbf{Q}$, i.e. $\mathbf{C} = \sqrt{\mathbf{Q}}$.

A theorem corresponding to Theorem 4.8 is given as follows.

Theorem 4.10 *The time invariant linear discrete system (4.2) is asymptotically stable if and only if the pair* $(\mathbf{A}, \mathbf{C})$ *is observable,* $\mathbf{Q} = \mathbf{Q}^T \geq 0$, *and the algebraic Lyapunov equation (4.30) has a unique positive definite solution.*

Example 4.11: Consider the following discrete-time linear system represented by

$$\mathbf{A} = \begin{bmatrix} 0.1 & 0.2 & -0.1 \\ 0.2 & 0.4 & 0 \\ 0 & -0.1 & 0.7 \end{bmatrix}$$

The eigenvalues of this matrix obtained by using MATLAB function `eig` are 0.0058, 0.4810, 0.7132, which indicates that this system is asymptotically stable in the discrete-time domain. If we want to check the stability of this system by using the Lyapunov theory, we have to choose a positive definite matrix $\mathbf{Q}$, say $\mathbf{Q} = \mathbf{I}_3$, and to solve the discrete-time algebraic Lyapunov equation (4.30). Using the MATLAB function `dlyap` and the statement `P=dlyap(A',Q)`, we get the following solution for $\mathbf{P}$

$$\mathbf{P} = \begin{bmatrix} 1.0692 & 0.1437 & -0.0511 \\ 0.1437 & 1.3180 & -0.2424 \\ -0.0511 & -0.2424 & 1.9958 \end{bmatrix} > 0$$

This is a positive definite matrix since its eigenvalues are given by 1, 1.3008, 2.0822.

◇

4.4 The Routh–Hurwitz Criterion

Finding eigenvalues for high-order systems is not, in general, an easy task. Algebraic tests for examining system stability without actually finding eigenvalues of linear systems were obtained independently by Routh and Hurwitz. Due to the fact that both methods produce similar results, the method to be presented is known in the literature as the Routh–Hurwitz method. It comes down to a simple algebraic game involving only coefficients of the characteristic polynomial.

In this section we present complete derivations and proof of this important method. In most undergraduate control textbooks the Routh–Hurwitz method for examining system stability is presented as a problem of forming a table of numbers by using coefficients of the characteristic polynomial and playing a simple algebraic game; the students have no idea how the game is related to system stability. They only get an *ad hoc answer*: "the number of sign changes in the first column of the Routh table indicates the number of unstable poles". Here, we use simple steps, already known to students, to derive and justify the Routh–Hurwitz test. The presentation is done along the results of Routh (Routh, 1887; Barnett and Siljak, 1977). The proof is based on a paper by Parks (1962).

Before we proceed to the real Routh–Hurwitz test, we introduce the definition of the so-called Hurwitz polynomial and establish simply a nonstability (instability) result, which is sometimes useful to determine quickly whether or not the given system is unstable.

Definition 4.3 A polynomial whose roots are in the closed left half of the complex plane is called the Hurwitz polynomial.

Note that in our terminology Hurwitz polynomials are asymptotically stable.

Consider now the following polynomial, which corresponds to the system characteristic equation

$$s^n + a_{n-1}s^{n-1} + \cdots + a_1 s + a_0$$

Since this polynomial can have either real roots or complex conjugate roots, it can be factored as

$$s^n + a_{n-1}s^{n-1} + \cdots + a_1 s + a_0$$

$$= \prod_{k=1}^{n_1} (s + \alpha_k) \prod_{i=n_1+1}^{n} (s + \alpha_i + j\beta_i)(s + \alpha_i - j\beta_i)$$

$$= \prod_{k=1}^{n_1} (s + \alpha_k) \prod_{i=n_1+1}^{n} \left(s^2 + 2\alpha_i s + \alpha_i^2 + \beta_i^2\right)$$

For a Hurwitz polynomial it is valid that $\alpha_k > 0, \forall k$ and $\alpha_i > 0, \forall i$, hence if we multiply factors in the last expression we see that all coefficients a_i, $i = 0, 1, ..., n-1$, must be positive. Thus, if only one of the coefficients a_i is zero or negative, the given polynomial cannot be Hurwitz. Since the existence of the Hurwitz polynomial implies system asymptotic stability, we have the following instability theorem.

Theorem 4.11 *Instability Theorem*

If only one of the coefficients of the characteristic polynomial is zero or has a different sign to the remaining coefficients, then the given system is not asymptotically stable.

Note that Theorem 4.11 provides the *sufficient condition* for system instability and at the same time gives the *necessary condition* for system asymptotic stability. That is, not all polynomials that have all coefficients strictly positive are asymptotically stable, but it is true that *all asymptotically stable polynomials have all coefficients strictly positive.*

Example 4.12: The following are obvious examples of systems that are not asymptotically stable according to the statement of Theorem 4.11

$$s^3 + 2s + 3$$
$$s^4 + 2s^3 - s^2 + 2s + 1$$
$$s^6 + s^4 + s^3 + s^2 + s - 1$$

$\diamond$

Example 4.13: The following polynomial is unstable despite the fact that all its coefficients are strictly positive

$$s^8 + 2s^7 + 3s^6 + 4s^5 + 5s^4 + 4s^3 + 3s^2 + 2s + 1$$

(see Problem 4.10).

◇

Now, we start with the development of the Routh–Hurwitz criterion. Consider the system characteristic polynomial

$$\Delta(s) = a_n s^n + a_{n-1} s^{n-1} + \cdots + a_1 s + a_0 \tag{4.31}$$

Derivations and proof of the Routh–Hurwitz method are based on the following three steps:

1. Using the characteristic polynomial (4.31), obtain a transfer function of a specific artificial linear system and record the procedure in a table. The table obtained is known as the Routh table.
2. From the transfer function obtained, get a block diagram and the corresponding state space form.
3. Apply the Lyapunov stability test given in Theorem 4.8 to the state space form obtained and draw conditions required for system asymptotic stability. *The conclusion will be that the first column in the Routh table must have all entries positive.*

Step 1. From the given characteristic polynomial (4.31) form the following two polynomials

$$\Delta_0(s) = a_n s^n + a_{n-2} s^{n-2} + a_{n-4} s^{n-4} + \cdots \tag{4.32}$$

and

$$\Delta_1(s) = a_{n-1} s^{n-1} + a_{n-3} s^{n-3} + a_{n-5} s^{n-5} + \cdots \tag{4.33}$$

By performing a series of long divisions, find coefficients $\alpha_i, i = 1, 2, ..., n$, for the following expansion

$$\frac{\Delta_0(s)}{\Delta_1(s)} = \alpha_1 s + \cfrac{1}{\alpha_2 s + \cfrac{1}{\alpha_3 s + \cfrac{1}{\ddots + \cfrac{1}{\alpha_n s}}}} \tag{4.34}$$

The procedure of dividing polynomials (4.32) and (4.33) in a manner specified in (4.34) produces after each long division a new polynomial whose coefficients can be recorded in a table, the well-known Routh table. Here, we give just a simple example, which demonstrates the procedure of obtaining (4.34) by a series of long divisions. The table itself will be constructed in a systematic way in the next subsection.

Example 4.14: Consider the following two polynomials

$$\Delta_0(s) = 2s^4 + 3s^2 + 2, \quad \Delta_1(s) = s^3 + s$$

After a long division, we get

$$\frac{\Delta_0(s)}{\Delta_1(s)} = 2s + \frac{s^2 + 2}{s^3 + s} = 2s + \cfrac{1}{\cfrac{s^3 + s}{s^2 + 2}}$$

which by repeating long division operation produces

$$2s + \cfrac{1}{s + \cfrac{1}{\cfrac{s^2 + 2}{-s}}} = 2s + \cfrac{1}{s + \cfrac{1}{-s + \cfrac{1}{-0.5s}}}$$

so that $\alpha_1 = 2, \alpha_2 = 1, \alpha_3 = -1, \alpha_4 = -0.5$.

◇

From (4.34) we form the following transfer function

$$H(s) = \frac{\Delta_1(s)}{\Delta_0(s) + \Delta_1(s)} = \frac{1}{1 + \frac{\Delta_0(s)}{\Delta_1(s)}} \tag{4.35}$$

Step 2. It is not hard to show that the transfer function (4.35) can be represented by a block diagram given in Figure 4.3. From this block diagram we can easily obtain the corresponding state space form

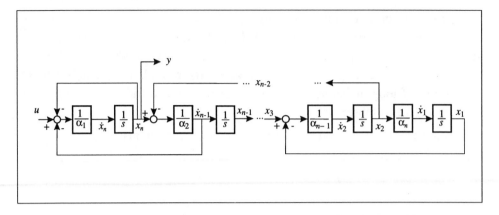

Figure 4.3: Block diagram of the system defined in (4.35)

$$
\begin{bmatrix} \dot{x}_1 \\ \dot{x}_2 \\ \dot{x}_3 \\ \vdots \\ \dot{x}_{n-1} \\ \dot{x}_n \end{bmatrix}
=
\begin{bmatrix}
0 & \frac{1}{\alpha_n} & 0 & \cdots & 0 & 0 & 0 \\
-\frac{1}{\alpha_{n-1}} & 0 & \frac{1}{\alpha_{n-1}} & \cdots & 0 & 0 & 0 \\
0 & -\frac{1}{\alpha_{n-2}} & 0 & \cdots & 0 & 0 & 0 \\
\vdots & \vdots & \vdots & \vdots & \vdots & \vdots & \vdots \\
0 & 0 & 0 & \cdots & -\frac{1}{\alpha_2} & 0 & \frac{1}{\alpha_2} \\
0 & 0 & 0 & \cdots & 0 & -\frac{1}{\alpha_1} & -\frac{1}{\alpha_1}
\end{bmatrix}
\begin{bmatrix} x_1 \\ x_2 \\ x_3 \\ \vdots \\ x_{n-1} \\ x_n \end{bmatrix}
+
\begin{bmatrix} 0 \\ 0 \\ 0 \\ \vdots \\ 0 \\ 1 \end{bmatrix} u
$$

$$\dot{x} = Ax + Bu$$

$$(4.36)$$

and

$$y = \begin{bmatrix} 0 & 0 & 0 & \cdots & 0 & 0 & 1 \end{bmatrix} x = Cx \qquad (4.37)$$

Step 3. In order to complete the proof at this point we evoke Theorem 4.8. That is, if we guess a solution for the matrix P in the form

$$
P =
\begin{bmatrix}
\alpha_n & 0 & \cdots & 0 \\
0 & \alpha_{n-1} & \cdots & \vdots \\
\vdots & \cdots & \ddots & 0 \\
0 & \cdots & 0 & \alpha_1
\end{bmatrix}
\qquad (4.38)
$$

we get from the algebraic Lyapunov equation (4.28)

$$\mathbf{A}^T\mathbf{P} + \mathbf{PA} = -\begin{bmatrix} 0 & \cdots & \cdots & 0 \\ \vdots & \ddots & \cdots & \vdots \\ \vdots & \cdots & 0 & 0 \\ 0 & \cdots & 0 & 2 \end{bmatrix} = -\mathbf{Q} = -\mathbf{C}^T\mathbf{C} \qquad (4.39)$$

with the matrix $\mathbf{Q}$ being positive semidefinite. It can be easily checked that the pair $(\mathbf{A}, \mathbf{C})$ is observable (see Problem 5.17); hence Theorem 4.8 implies that the asymptotic stability of $\mathbf{A}$ is obtained if and only if the matrix $\mathbf{P}$ is positive definite, which is apparently the case for $\alpha_i > 0, \forall i = 1, 2, ..., n$.

This completes the derivations and proof of the Routh–Hurwitz criterion. It remains to show how to obtain the coefficients α_i in a systematic way. This is done in the next subsection where we introduce the Routh table and show how to use it to test system stability.

Note that another simple proof of the Routh–Hurwitz stability test has been obtained very recently (Meinsma, 1995).

4.4.1 The Routh Table

For the given polynomial (4.31) the Routh table is formed as follows. In the first row the coefficients of the polynomial (4.32) are recorded. The second row contains the coefficients of the polynomial (4.33). The third row is formed by using the coefficients of the remainder obtained by performing a long division of $\Delta_0(s)$ and $\Delta_1(s)$, that is

$$\frac{\Delta_0(s)}{\Delta_1(s)} = \frac{a_n}{a_{n-1}}s + \frac{A_1 s^{n-2} + A_2 s^{n-4} + \cdots}{\Delta_1(s)} = \alpha_1 s + \frac{\Delta_2(s)}{\Delta_1(s)} \qquad (4.40)$$

The coefficients of the fourth row are obtained similarly by

$$\frac{\Delta_1(s)}{\Delta_2(s)} = \frac{a_{n-1}}{A_1}s + \frac{B_1 s^{n-3} + B_2 s^{n-5} + \cdots}{\Delta_2(s)} = \alpha_2 s + \frac{\Delta_3(s)}{\Delta_2(s)} \qquad (4.41)$$

The next row is formed from

$$\frac{\Delta_2(s)}{\Delta_3(s)} = \alpha_3 s + \frac{\Delta_4(s)}{\Delta_3(s)} \qquad (4.42)$$

and this procedure is continued until all the rows are filled. The corresponding Routh table is then shown in Table 4.1

s^n	a_n	a_{n-2}	a_{n-4}	$\cdots$	
s^{n-1}	a_{n-1}	a_{n-3}	a_{n-5}	$\cdots$	
s^{n-2}	A_1	A_2	A_3	$\cdots$	
s^{n-3}	B_1	B_2	B_3	$\cdots$	
s^{n-4}	C_1	C_2	C_3	$\cdots$	
$\cdots$	$\cdots$	$\cdots$	$\cdots$	$\cdots$	0
s^1	$\cdots$	$\cdots$	0		
s^0	$\cdots$	0			

Table 4.1: The Routh table

where

$$A_1 = \frac{a_{n-1}a_{n-2} - a_n a_{n-3}}{a_{n-1}}, \quad A_2 = \frac{a_{n-1}a_{n-4} - a_n a_{n-5}}{a_{n-1}}, \quad \cdots$$

$$B_1 = \frac{A_1 a_{n-3} - a_{n-1} A_2}{A_1}, \quad B_2 = \frac{A_1 a_{n-5} - a_{n-1} A_3}{A_1}, \quad \cdots \tag{4.43}$$

$$C_1 = \frac{B_1 A_2 - A_1 B_2}{B_1}, \quad C_2 = \frac{B_1 A_3 - A_1 B_3}{B_1}, \quad \cdots$$

$$\cdots \qquad\qquad\qquad \cdots \qquad\qquad\qquad \cdots$$

It can be observed from (4.40)–(4.42) that

$$\alpha_1 = \frac{a_n}{a_{n-1}}, \quad \alpha_2 = \frac{a_{n-1}}{A_1}, \quad \alpha_3 = \frac{A_1}{B_1}, \quad \alpha_4 = \frac{B_1}{C_1}, \quad \cdots \tag{4.44}$$

Since from the derivations and proof of the Routh–Hurwitz criterion all α_i, $i = 1, 2, ..., n$, must be positive for asymptotic stability, we easily establish the famous Routh criterion.

Routh's Criterion: *For asymptotically stable polynomials (Hurwitz polynomials) all coefficients in the first column of the Routh table must be positive.*

The following two examples demonstrate the procedure for forming the Routh table and drawing conclusions about a system's asymptotic stability.

Example 4.15: A stable system with double poles at -1 and -2 has the characteristic polynomial given by

$$\Delta(s) = (s+1)^2(s+2)^2 = s^4 + 6s^3 + 13s^2 + 12s + 4$$

The Routh table for this polynomial has the form

s^4	1	13	4
s^3	6	12	0
s^2	11	4	0
s^1	108/11	0	
s^0	4	0	

Since all coefficients in the first column of the Routh table are positive, we conclude that this polynomial is asymptotically stable, which agrees with previous information about the location of the system poles.

◇

Example 4.16: The following is an example of a polynomial that is not asymptotically stable. It is given by

$$\Delta(s) = s^6 + s^5 + 3s^4 + 2s^3 + s^2 + 2s + 1$$

Its Routh table is obtained as

s^6	1	3	1	1
s^5	1	2	2	0
s^4	1	−1	1	0
s^3	3	1	0	
s^2	−4/3	1	0	
s^1	13/4	0		
s^0	1	0		

Since one coefficient in the first column of the Routh table is negative, this polynomial is not asymptotically stable. In this example we can use MATLAB and its function `roots` to find the actual location of the poles of this system. This is achieved by the following MATLAB statements

```
c=[1 1 3 2 1 2 1];
p=roots(c)
```

The obtained result shows the following poles:

$$p_{1,2} = -0.2609 \pm j1.6067; \quad p_{3,4} = -0.6742 \pm j0.1154$$
$$p_{5,6} = 0.4352 \pm j0.7857$$

which indicates four asymptotically stable and two unstable poles.

◇

It is interesting to point out that by checking the number of sign changes in the first column of the Routh table, it is possible to determine exactly the number of poles located in the closed right half of the complex plane (unstable poles). As a matter of fact, this result comprises the original Routh theorem (Routh, 1887).

Theorem 4.12 *The number of sign changes in the first column of the Routh table determines the number of unstable poles.*

It can be seen from the previous example that the sign in the first column is changed twice: first from 3 to –4/3, and second from –4/3 to 13/4. Thus, in this example, according to Theorem 4.12, we have two unstable poles, which has been confirmed by MATLAB.

An important advantage of the Routh–Hurwitz method over the direct procedure for finding and examining eigenvalues lies in the fact that in the case of linear systems containing some parameters, the Routh–Hurwitz method gives the range of values for the given parameters such that the system under consideration is asymptotically stable. Of course, in this case the eigenvalue method is not applicable since unless we know exactly all coefficients, the eigenvalue method is useless except in trivial cases of low-order dimensional systems.[6]

The following example illustrates the above feature.

Example 4.17: Consider a polynomial whose coefficients contain a parameter, that is

$$\Delta(s) = s^4 + 2s^3 + Ks^2 + s + 3$$

The Routh table is given by

s^4	1	K	3
s^3	2	1	0
s^2	$\frac{2K-1}{2}$	3	0
s^1	$\frac{2K-13}{2K-1}$	0	
s^0	3	0	

Since all elements in the first column must be positive for asymptotic stability, it follows that

$$2K - 1 > 0 \quad \text{and} \quad 2K - 13 > 0 \quad \Rightarrow \quad K > 6.5$$

preserves polynomial's asymptotic stability.

$\diamond$

While forming the Routh table, singular cases may appear. That is, zero elements may appear in the first column of the Routh table, and/or everywhere else in the table. A zero element in the first column does not allow completion of the procedure for forming the Routh table as outlined so far in this section.

---

6   Polynomial algebraic equations can be analytically solved only for $n \leq 4$.

Singular Cases:

While forming the Routh table, two singular cases may appear.

1. The first column of the Routh table contains a zero element.
2. An all-zero row appears in the table.

The second singular case may happen if the polynomial has either:
(a) pure imaginary axis poles, or
(b) pairs of real poles symmetrically distributed with respect to the origin, or
(c) quadruplets of complex poles symmetrical with respect to the origin.

These cases are presented in Figure 4.4.

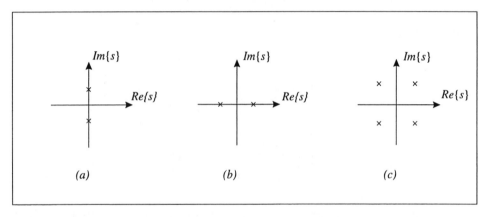

Figure 4.4: Possible locations of system poles for singular case 2

Note that in either singular case 1 or 2, the polynomials under consideration cannot be Hurwitz (asymptotically stable). However, it is interesting to study these cases in order to determine whether or not systems are *stable* (see Example 4.19), and to find out the exact number of unstable poles.

In order to remedy singular case 1, Routh suggested the so-called ϵ-method. In that method a zero in the first column is replaced by a small positive parameter $\epsilon > 0$, and the standard procedure for forming the Routh table is continued. After the table is completed the signs of all elements in the first row (these elements now depend on ϵ) are determined by taking limits as $\epsilon \to 0^+$. The next example demonstrates the ϵ-method.

Example 4.18: Applying the ϵ-method to the polynomial

$$\Delta(s) = s^5 + 2s^4 + 3s^3 + 4s^2 + 5s + 6$$

we get the following Routh table

s^5	1	3	5	0
s^4	2	4	6	0
s^3	1	2	0	
s^2	ϵ	6	0	
s^1	$\frac{2\epsilon-6}{\epsilon}$	0		
s^0	6	0		

In the limit when $\epsilon \to 0^+$, we have

$$\lim_{\epsilon \to 0^+} \frac{2\epsilon - 6}{\epsilon} \to -\infty$$

so that there are two sign changes in the first column of the Routh table and the given polynomial has two unstable poles.

$\diamond$

Singular case 2 with an all-zero row indicates that while performing a series of long divisions, as presented in (4.40)–(4.42), the polynomials $\Delta_i(s)$ and $\Delta_{i+1}(s)$ have a common divisor $d(s)$. That common divisor is also part of $\Delta(s)$. The zeros of $d(s)$ correspond to the cases (a), (b), and (c) of the singular case 2, presented in Figure 4.4, so that the polynomial $d(s)$ is even. Routh proposed the following scheme to resolve singular case 2:

(A) Form an auxiliary polynomial $d(s)$, as an even polynomial whose coefficients are obtained from the row above the all-zero row.
(B) Take the derivative of $d(s)$ and use the coefficients of the polynomial $\frac{d}{ds}d(s)$ instead of the all-zero row.
(C) Proceed with the standard way of forming the Routh table.

The next two examples demonstrate the procedure of handling singular case 2.

Example 4.19: For the polynomial

$$\Delta(s) = s^5 + s^4 + 2s^3 + 2s^2 + s + 1$$

we get all zeros in the third row, that is

s^5	1	2	1	0
s^4	1	2	1	0
s^3	0	0	0	

An auxiliary *even* polynomial is formed by using coefficients above the all-zero row, that is

$$d(s) = s^4 + 2s^2 + 1$$

whose derivative is given by

$$\frac{d(s)}{ds} = 4s^3 + 4s$$

so that the coefficients 4, 4 are used in the table instead of the all-zero row and the procedure is continued in the standard way as shown below

s^5	1	2	1	0
s^4	1	2	1	0
s^3	4	4	0	
s^2	1	1	0	
s^1	0	0		

Since again an all-zero row is encountered, a new auxiliary even polynomial is formed as

$$d_1(s) = s^2 + 1 \quad \Rightarrow \quad \frac{d}{ds}d_1(s) = 2s$$

and the Routh table is completed as follows

s^5	1	2	1	0
s^4	1	2	1	0
s^3	4	4	0	
s^2	1	1	0	
s^1	2	0		
s^0	1	0		

It can be seen that since there are no sign changes in the first column of the Routh table, then there are no poles in the right-hand half of the complex plane, but the polynomial is not asymptotically stable as well since this is a *singular case and the original Routh criterion for asymptotic stability is not applicable.* In order to find the location of the poles, we can solve two auxiliary equations

$$d(s) = s^4 + 2s^2 + 1 = 0, \quad d_1(s) = s^2 + 1 = 0$$

whose solutions are $s_{1,2} = j$, $s_{3,4} = -j$, i.e. this polynomial has a pair of double poles at $\pm j$.

$\diamond$

Example 4.20: Consider the polynomial

$$\Delta(s) = s^7 + 2s^6 + 3s^5 + 4s^4 + 4s^3 + 3s^2 + 2s + 1$$

Its Routh table is given by

s^7	1	3	4	2	0
s^6	2	4	3	1	0
s^5	1	5/2	3/2	0	
s^4	−1	0	1	0	
s^3	5/2	5/2	0		
s^2	1	1	0		
s^1	0 (2)	0			
s^0	1	0			

An auxiliary second-order polynomial has been formed as $d(s) = s^2 + 1$. Note that the polynomial under consideration has two unstable poles, due to two sign changes in the first column of the Routh table, and a pair of poles on the imaginary axis at $\pm j$.

$\diamond$

It has been shown in the literature that even for complex cases, such as the simultaneous appearance of a row with a zero element in the first column and an all-zero row, the ϵ-method, subject to minor modifications, is efficient (Fahmy and O'Reilly, 1982). Note that singular cases can be eliminated by forming an extended Routh table according to the procedure presented in Benidir and Picinbono (1990). However, the extended Routh table is more complicated than the standard Routh table.

Finally, we want to point out that the Routh–Hurwitz test can be used to obtain a measure of system relative stability, i.e. to determine how far system poles are to the left of the imaginary axis. This can be achieved by working with the shifted characteristic polynomial obtained by replacing the complex frequency s by $s - \gamma$, where $\gamma > 0$ indicates the distance from the imaginary axis. In Section 4.7 a laboratory experiment examining the system stability measure is designed. In addition, several problems dealing with the shifted characteristic polynomial are formulated in Section 4.8.

4.5 Algebraic Stability Tests for Discrete Systems

In this section we study the stability of time invariant linear discrete-time systems and present two algebraic methods: Jury's test and the bilinear transformation method.

4.5.1 Jury's Test

An algebraic method corresponding to the Routh–Hurwitz test, for examining stability of time invariant linear discrete systems, was developed in a series of papers by Jury in the early sixties (Jury, 1964). A simplified form of that method is presented in Raible (1974). An elegant proof of Jury's stability method, following the results of Raible (1974) and by using the Lyapunov stability theory in the spirit of the previous section, has been derived by Harn and Chen (1981) and Mansour (1982).

Consider a polynomial represented in the z-domain by the following general expression

$$a_n z^n + a_{n-1} z^{n-1} + \cdots + a_1 z + a_0 \tag{4.45}$$

The algebraic test to be presented is based on Table 4.2, the simplified Jury's table (Harn and Chen, 1981; Chen, 1984), which can be easily obtained by playing simple algebra with the polynomial's coefficients.

a_n	a_{n-1}	a_{n-2}	$\ldots$	a_2	a_1	a_0
b_{n-1}	b_{n-2}	$\ldots$	$\ldots$	b_1	b_0	
c_{n-2}	c_{n-3}	$\ldots$	$\ldots$	c_0		
$\ldots$	$\ldots$	$\ldots$	$\ldots$			
$\ldots$	$\ldots$	$\ldots$				
u_1	u_0					
w_0						

Table 4.2: Simplified Jury's table

The newly defined coefficients are obtained by a simple pattern given below

$$b_{n-1} = a_n - k_0 a_0, \quad b_{n-2} = a_{n-1} - k_0 a_1, \quad \ldots, \quad b_0 = a_1 - k_0 a_{n-1}$$

$$k_0 = \frac{a_0}{a_n}$$

$$c_{n-2} = b_{n-1} - k_1 b_0, \quad c_{n-3} = b_{n-2} - k_1 b_1, \quad \ldots, \quad c_0 = b_1 - k_1 b_{n-2}$$

$$k_1 = \frac{b_0}{b_{n-1}}$$

$$\ldots \qquad\qquad \ldots \qquad \ldots \tag{4.46}$$

The following theorem, corresponding to the result from the previous section, can be stated (Raible, 1974; Harn and Chen, 1981; Chen, 1984).

Theorem 4.13 *Assume that $a_n > 0$. Then, the polynomial under consideration is asymptotically stable if and only if all coefficients in the first column of Table 4.2 are positive. In addition, the number of negative coefficients in the first column indicates the number of poles outside the unit circle.*

The proof of this theorem is omitted. It follows a similar pattern to that presented in the proof of the Routh–Hurwitz criterion. Readers interested in it can find the complete proof in Harn and Chen (1981), Mansour (1982), and Chen (1984).

In the next example we demonstrate the procedure for forming Table 4.2.

Example 4.21: The polynomial under consideration is given by

$$z^3 + 0.5z^2 + 0.3z + 0.1$$

The simplified Jury's table for this example has the form

1	0.5	0.3	0.1
0.99	0.47	0.25	
0.93	0.35		
0.80			

with coefficients $k_0 = 0.1$, $k_1 = 0.25$, $k_2 = 0.24$. From this table and Theorem 4.13, we conclude that the considered polynomial is stable. Using MATLAB, we find the eigenvalues as $p_1 = -0.3893$, $p_{2,3} = -0.0554 \pm j0.53038$, i.e. all of them are inside the unit circle.

◇

For clarification of singular cases in studying the stability of time invariant linear discrete-time systems, the reader is referred to Jury (1964) and Raible (1974).

Note that a new stability table for discrete-time linear systems that is very similar to the Routh table has been recently proposed by Hu (1994).

4.5.2 Bilinear Transformation and the Routh–Hurwitz Method

It important to point out that the Routh–Hurwitz method can be used for studying the stability of discrete-time linear systems as well. That is, the very well known bilinear transformation defined by

$$z = \frac{s+1}{s-1} \qquad (4.47)$$

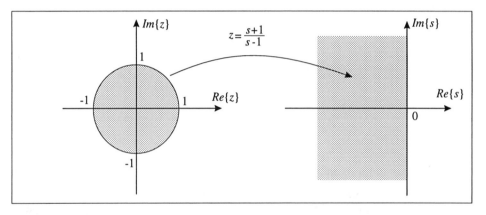

Figure 4.5: Mapping by a bilinear transformation

maps the unit circle in the z-domain into the left complex plane in the s-domain (see Figure 4.5).

For the given characteristic equation $\Delta(z) = 0$, by using the bilinear transformation (4.47), we get another characteristic equation in the s-domain, $\Delta_d(s) = 0$, so that the Routh–Hurwitz criterion can be directly applied to $\Delta_d(s)$. The stability conclusion reached for the polynomial $\Delta_d(s)$ is valid also for the polynomial $\Delta(z)$. The following example demonstrates this procedure.

Example 4.22: Consider the discrete-time characteristic equation

$$\Delta(z) = z^3 + z^2 + z + 2 = 0$$

By using the bilinear transformation, this is mapped into

$$\left(\frac{s+1}{s-1}\right)^3 + \left(\frac{s+1}{s-1}\right)^2 + \left(\frac{s+1}{s-1}\right) + 2 = 0$$

which implies

$$\Delta_d(s) = 5s^3 - 3s^2 + 7s - 1 = 0$$

Using the knowledge from the previous section, we immediately conclude that this polynomial is unstable (it has coefficients of opposite signs). The same instability conclusion is valid for the polynomial $\Delta(z)$. If we form the Routh table we get

s^3	5	7	0
s^2	−3	−1	0
s^1	16/3	0	
s^0	−1	0	

The first column of this table indicates the existence of three s-domain unstable roots (three sign changes), which means also that in the z-domain there are three roots outside of the unit circle. This can be confirmed by MATLAB, which produces the following roots: $p_1 = -1.3532$, $p_{2,3} = 0.1766 \pm j1.2028$. Also, by applying the Jury test, we get

1	1	1	2
−3	−1	−1	
−8/3	−2/3		
−15/6			

so that by Theorem 4.13 all three roots are outside the unit circle.

◇

4.6 Frequency Domain Stability Study

According to the definitions of stability given in Section 4.1, system stability is a pure time domain phenomenon. Interestingly enough, a stability test for time invariant linear systems can also be derived in the frequency domain. It is based on the complex analysis result known as *Cauchy's principle of argument*. Note that the system transfer function is a complex function. By applying Cauchy's principle of argument to the *open-loop system* transfer function, we will get information about stability of the closed-loop system transfer function and arrive at the Nyquist stability criterion (Nyquist, 1932).

The Nyquist stability criterion is extremely inefficient for testing the stability of even low-order dimensional systems since it is based on the polar plot of system transfer function over an infinite range of frequencies. However, its importance lies in the fact that it can be used to determine the relative degree of system stability by producing the so-called phase and gain stability margins. These stability margins will be needed in Chapter 9 for frequency domain controller design techniques. That is why, in this section, we present only the essence of the Nyquist stability criterion, solve several low-order examples, and define the phase and gain stability margins. It should be pointed out that the Nyquist method is efficient for studying the stability of linear systems with pure time delay (see Kuo, 1991).

For a single-input single-output feedback system (see Figure 2.2a) the closed-loop transfer function is defined in (2.15) by

$$M(s) = \frac{G(s)}{1 + H(s)G(s)} \tag{4.48}$$

where $G(s)$ represents the system and $H(s)$ is the feedback element. Since the system eigenvalues (poles) are determined as those values at which its transfer function becomes infinity, it follows from (4.48) that the closed-loop system poles are obtained by solving the following equation

$$1 + H(s)G(s) = 0 = \Delta(s) \tag{4.49}$$

which, in fact, represents the *system characteristic equation*.

In the following we consider the complex function

$$D(s) = 1 + H(s)G(s) \tag{4.50}$$

whose zeros are the closed-loop poles of the transfer function given in (4.48). In addition, it is easy to see that the poles of $D(s)$ are the zeros of $M(s)$. At the same time the poles of $D(s)$ are the open-loop control system poles since they are contributed by the poles of $H(s)G(s)$, which can be considered as the open-loop control system transfer function—obtained when the feedback loop is open at some point. The Nyquist stability test is obtained by applying the Cauchy principle of argument to the complex function $D(s)$. First, we state Cauchy's principle of argument.

Cauchy's Principle of Argument

Let $F(s)$ be an analytic function in a closed region of the complex plane s given in Figure 4.6 except at a finite number of points (namely, the poles of $F(s)$). It is also assumed that $F(s)$ is analytic at every point on the contour. Then, *as s travels around the contour in the s-plane in the clockwise direction, the function $F(s)$ encircles the origin in the $(Re\{F(s)\}, Im\{F(s)\})$-plane in the same direction N times (see Figure 4.6), with N given by*

$$N = Z - P \tag{4.51}$$

where Z and P stand for the number of zeros and poles (including their multiplicities) of the function $F(s)$ inside the contour.

The above result can be also written as

$$\arg\{F(s)\} = (Z - P)2\pi = 2\pi N \tag{4.52}$$

which justifies the terminology used, "the principle of argument".

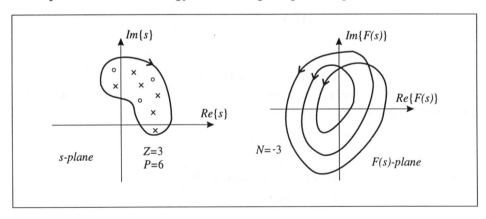

Figure 4.6: Cauchy's principle of argument

Nyquist Plot

The Nyquist plot is a polar plot of the function $D(s) = 1 + G(s)H(s)$ when the complex frequency s travels around the contour given in Figure 4.7.

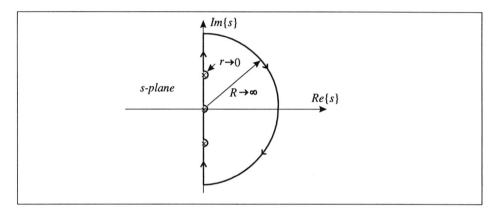

Figure 4.7: Contour in the s-plane

Note that the contour in this figure covers the whole unstable half plane of the complex plane s, $R \to \infty$. In addition, since the function $D(s)$, according to Cauchy's principle of argument, must be analytic at every point on the contour, the poles of $D(s)$ on the imaginary axis must be encircled by infinitesimally small semicircles.

Now we are ready to formulate the Nyquist stability criterion.

Nyquist Stability Criterion

The Nyquist criterion says: the number of unstable closed-loop poles is equal to the number of unstable open-loop poles plus the number of encirclements of the origin of the Nyquist plot of the complex function $D(s)$.

This can be easily justified by applying Cauchy's principle of argument to the function $D(s)$ with the s-plane contour given in Figure 4.7. Note that Z and P represent the numbers of zeros and poles, respectively, of $D(s)$ in the unstable part of the complex plane. At the same time, *the zeros of $D(s)$ are the closed-loop system poles*, and *the poles of $D(s)$ are the open-loop system poles* (closed-loop zeros).

The above criterion can be slightly simplified if instead of plotting the function $D(s) = 1 + G(s)H(s)$, we plot only the function $G(s)H(s)$ and count encirclement of the Nyquist plot of $G(s)H(s)$ around the point $(-1, j0)$, so that the modified Nyquist criterion has the following form.

The number of unstable closed-loop poles (Z) is equal to the number of unstable open-loop poles (P) plus the number of encirclements (N) of the point $(-1, j0)$ of the Nyquist plot of $G(s)H(s)$, that is

$$Z = P + N \tag{4.53}$$

Two important notions can be derived from the Nyquist diagram: *phase and gain stability margins.* The phase and gain stability margins are presented in Figure 4.8.

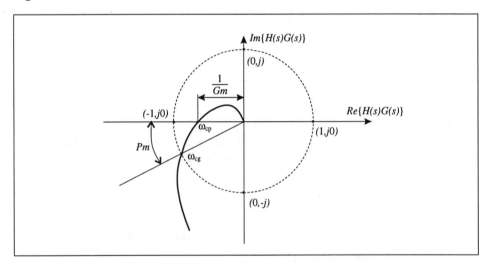

Figure 4.8: Phase and gain stability margins

They give the degree of relative stability; in other words, they tell how far the given system is from the instability region. Their formal definitions are given by

$$Pm = 180° + \arg\left\{G(j\omega_{cg})H(j\omega_{cg})\right\} \tag{4.54}$$

$$Gm\,[dB] = 20\log\frac{1}{|G(j\omega_{cp})H(j\omega_{cp})|}\,[dB] \tag{4.55}$$

where ω_{cg} and ω_{cp} stand for, respectively, the *gain and phase crossover frequencies*, which from Figure 4.8 are obtained as

$$|G(j\omega_{cg})H(j\omega_{cg})| = 1 \Rightarrow \omega_{cg} \tag{4.56}$$

and

$$\arg \{G(j\omega_{cp})H(j\omega_{cp})\} = 180° \Rightarrow \omega_{cp} \qquad (4.57)$$

These two notions will be used in Chapter 9 as the specifications for the frequency domain controller design technique.

In the following we solve two examples in order to demonstrate the Nyquist stability criterion and use of the formula (4.53).

Example 4.23: Consider a control system represented by

$$G(s)H(s) = \frac{1}{s(s+1)}$$

Since this system has a pole at the origin, the contour in the s-plane should encircle it with a semicircle of an infinitesimally small radius. This contour has three parts (a), (b), and (c). Mappings for each of them are considered below.

(a) On this semicircle the complex variable s is represented in the polar form by $s = Re^{j\Psi}$ with $R \to \infty$, $-\frac{\pi}{2} \le \Psi \le \frac{\pi}{2}$. Substituting $s = Re^{j\Psi}$ into $G(s)H(s)$, we easily see that $G(s)H(s) \to 0$. Thus, the huge semicircle from the s-plane maps into the origin in the $G(s)H(s)$-plane (see Figure 4.9).

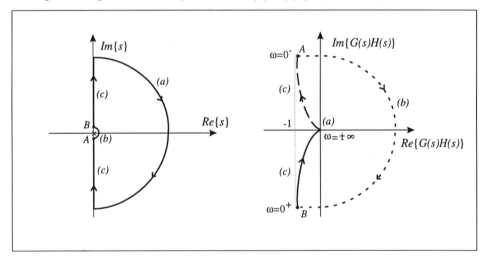

Figure 4.9: Nyquist plot for Example 4.23

(b) On this semicircle the complex variable s is represented in the polar form by $s = re^{j\Phi}$ with $r \to 0$, $-\frac{\pi}{2} \le \Phi \le \frac{\pi}{2}$, so that we have

$$G(s)H(s) \to \frac{1}{re^{j\Phi}} \to \infty \times \arg(-\Phi)$$

Since Φ changes from $-\frac{\pi}{2}$ at point A to $\frac{\pi}{2}$ at point B, $\arg\{G(s)H(s)\}$ will change from $\frac{\pi}{2}$ to $-\frac{\pi}{2}$. We conclude that the infinitesimally small semicircle at the origin in the s-plane is mapped into a semicircle of infinite radius in the $G(s)H(s)$-plane.

(c) On this part of the contour s takes pure imaginary values, i.e. $s = j\omega$ with ω changing from $-\infty$ to $+\infty$. Due to symmetry, it is sufficient to study only mapping along $0^+ \le \omega \le +\infty$. We can find the real and imaginary parts of the function $G(j\omega)H(j\omega)$, which are given by

$$Re\{G(j\omega)H(j\omega)\} = \frac{-1}{\omega^2 + 1}, \quad Im\{G(j\omega)H(j\omega)\} = \frac{-1}{\omega(\omega^2 + 1)}$$

From these expressions we see that neither the real nor the imaginary parts can be made zero, and hence the Nyquist plot has no points of intersection with the coordinate axis. For $\omega = 0^+$ we are at point B and since the plot at $\omega = +\infty$ will end up at the origin, the Nyquist diagram corresponding to part (c) has the form as shown in Figure 4.9. Note that the vertical asymptote of the Nyquist plot in Figure 4.9 is given by $Re\{G(j0^\pm)H(j0^\pm)\} = -1$ since at those points $Im\{G(j0^\pm)H(j0^\pm)\} = \mp\infty$.

From the Nyquist diagram we see that $N = 0$ and since there are no open-loop poles in the left half of the complex plane, i.e. $P = 0$, we have $Z = 0$ so that the corresponding closed-loop system has no unstable poles.

The Nyquist plot is drawn by using the MATLAB function `nyquist` as follows

```
num=1;
den=[1  1  0];
nyquist(num,den);
% axis scaling
axis([-1.5  0.5  -10  10]);
axis([-1.2  0.2  1  1]);
```

The MATLAB Nyquist plot is presented in Figure 4.10. It can be seen from Figures 4.8 and 4.9 that $1/Gm = 0$, which implies that $Gm = \infty$. Also, from the same figures it follows that $\omega_{cp} = \infty$. In order to find the phase margin and the corresponding gain crossover frequency we use the MATLAB function `margin` as follows

```
[Gm,Pm,wcp,wcg]=margin(num,den)
```

producing, respectively, gain margin, phase margin, phase crossover frequency, and gain crossover frequency. The required phase margin and gain crossover frequency are obtained as $Pm = 53.4108°$, $\omega_{cg} = 0.7862\,\mathrm{rad/s}$.

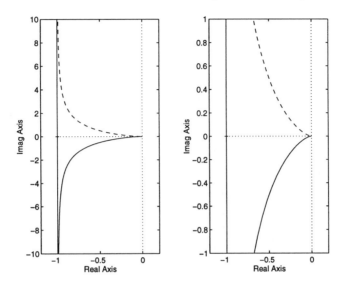

Figure 4.10: MATLAB Nyquist plot for Example 4.23

◇

Example 4.24: Consider now the following system, obtained from the one in the previous example by adding a pole, that is

$$G(s)H(s) = \frac{1}{s(s+1)(s+2)}$$

The contour in the s-plane is the same as in the previous example. For cases (a) and (b) we have the same analyses and conclusions. It remains to examine case (c). If we find the real and imaginary parts of $G(j\omega)H(j\omega)$, we get

$$Re\{G(j\omega)H(j\omega)\} = \frac{-3}{9\omega^2 + (2-\omega^2)^2}$$

$$Im\{G(j\omega)H(j\omega)\} = \frac{-(2-\omega^2)}{\omega\left[9\omega^2 + (2-\omega^2)^2\right]}$$

It can be seen that an intersection with the real axis happens at $\omega = \sqrt{2}$ at the point $R_e\{G(j\sqrt{2})H(j\sqrt{2})\} = -1/6$. The Nyquist plot is given in Figure 4.11. The corresponding Nyquist plot obtained by using MATLAB is given in Figure 4.12.

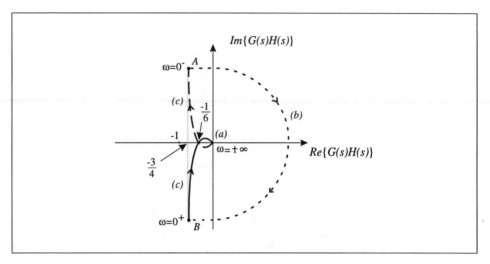

Figure 4.11: Nyquist plot for Example 4.24

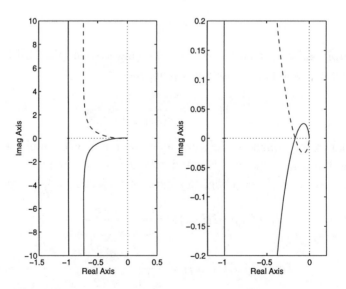

Figure 4.12: MATLAB Nyquist plot for Example 4.24

Note that the vertical asymptote is given by $Re\{G(j0)H(j0)\} = -3/4$. Thus, we have $N = 0, P = 0$, and $Z = 0$ so that the closed-loop system is stable. The MATLAB function `margin` produces the following values for relative stability parameters

$$Gm = 6\,\text{dB}, \quad Pm = 53.4108°, \quad \omega_{cg} = 0.4457\,\text{rad/s}, \quad \omega_{cp} = 1.4142\,\text{rad/s}$$

◇

Several problems involving the Nyquist plots and finding the gain and phase margins are given in the laboratory experiment and the problem section.

4.7 MATLAB Laboratory Experiment

Part 1. (a) Using the MATLAB function `poly`, find the characteristic polynomial for an F-8 aircraft whose continuous-time model is given in Section 3.5.2. Use the Routh–Hurwitz test to examine the stability of the polynomial obtained. Check the answer obtained by the MATLAB function `eig`, that is, by finding the system eigenvalues.

(b) For a fifth-order continuous-time model of an industrial reactor given in Problem 3.23, examine the stability by directly finding system eigenvalues and by using the Lyapunov method. Hint: Use $\mathbf{Q} = \mathbf{I}_5$.

Part 2. (a) A discrete-time model of a chemical plant (Gomathi *et al.*, 1980; Gajić and Shen, 1993) is given by the following system matrix

$$\mathbf{A} = \begin{bmatrix} 0.9541 & 0.0196 & 0.0036 & 0.0007 & 0.0002 \\ 0.4085 & 0.4132 & 0.1608 & 0.0447 & 0.0120 \\ 0.1222 & 0.2633 & 0.3615 & 0.1593 & 0.1238 \\ 0.0411 & 0.1286 & 0.2721 & 0.2144 & 0.4098 \\ 0.0013 & 0.0058 & 0.0188 & 0.0362 & 0.9428 \end{bmatrix}$$

Using MATLAB, find its characteristic polynomial. Examine the stability of this system by using (a) the eigenvalue approach, (b) Jury's test, and (c) Lyapunov's method.

(b) Repeat the analysis done in Part 2a for a discrete-time model of a distillation column (Kautsky *et al.*, 1985) given by

$$A = \begin{bmatrix} 0.9895 & 0.0056 & 0.0003 & 0.0000 & 0.0000 \\ 0.1172 & 0.8145 & 0.0760 & 0.0056 & 0.0004 \\ 0.0088 & 0.1239 & 0.7502 & 0.1080 & 0.0112 \\ 0.0009 & 0.0180 & 0.1838 & 0.6683 & 0.1508 \\ 0.0000 & 0.0003 & 0.0017 & 0.0133 & 0.9852 \end{bmatrix}$$

Part 3. (a) Use MATLAB and its function `nyquist(num,den,w)` to plot the Nyquist diagram for the following system

$$G(s)H(s) = \frac{100}{(s-1)(s+10)^2}$$

Determine the stability of the closed-loop system using the Nyquist criterion. Hint: Use MATLAB axis scaling function as `axis([-1.4 0.2 -0.6 0.6])`.

(b) Find the gain and phase stability margins by using MATLAB function

`[Gm,Pm,wcp,wcg]=margin(num,den)`

for the problem studied in Example 4.24, now parametrized by a constant K, that is

$$G(s)H(s) = \frac{K}{s(s+1)(s+2)}$$

Take $K = 2, 3, 4, 5, 6, 10, 100$ and comment on the results obtained.

Part 4. Use the MATLAB function `conv` to obtain a shifted characteristic polynomial of the characteristic polynomial

$$s^6 + 2s^5 + s^4 + s^3 + 2s^2 + 2s + 1$$

i.e. replace s by $s_1 - 1$. Apply the Routh–Hurwitz method to both polynomials. Find the exact location of the roots of the original characteristic polynomial and comment on the results obtained from the Routh–Hurwitz tests.

Part 5.[7] (a) For the antenna control system whose transfer function is given in Problem 3.1, find the state space form for $K = 1$. Examine system stability by finding the system minimal polynomial. Hint: Use MATLAB to find n_i from formula (4.20).

---

[7]   This part is based on the material presented in Subsections 4.2.4–4.2.6.

(b) Justify the answer obtained in (a) by finding and plotting the system response due to the initial condition $\mathbf{x}^T(0) = [1 \quad 1 \quad 1 \quad 1 \quad 1 \quad 1]$. Hint: Take `t=0:1:100`, calculate `[y,x,t]=initial(A,B,C,D,x0,t)`, and then use `plot(x)`.

4.8 References

Barnett, S. and D. Siljak, "Routh's algorithm: a centennial survey," *SIAM Review*, vol. 19, 472–489, 1977.

Benidir, M. and B. Picinbono, "Extended table for eliminating singularities in Routh's array," *IEEE Transactions on Automatic Control*, vol. AC-35, 218–222, 1990.

Calise, A. and D. Moerder, "Optimal output feedback design of systems with ill-conditioned dynamics," *Automatica*, vol. 21, 271–276, 1985.

Chen, T., *Linear Systems Theory and Design*, Holt, Rinehart and Winston, New York, 1984.

Fahmy, M. and O'Reilly, "A note on the Routh–Hurwitz test," *IEEE Transactions on Automatic Control*, vol. AC-27, 483–485, 1982.

Gajić, Z. and X. Shen, *Parallel Algorithms for Optimal Control of Large Scale Linear Systems*, Springer-Verlag, London, 1993.

Gajić, Z. and M. Qureshi, *Lyapunov Matrix Equation in System Stability and Control*, Academic Press, San Diego, California, 1995.

Gantmacher, F., *The Theory of Matrices*, Chelsea Publishing Company, New York, 1960.

Gomathi, K., S. Prabhu, and M. Pai, "A suboptimal controller for minimum sensitivity of closed-loop eigenvalues to parameter variations," *IEEE Transactions on Automatic Control*, vol. AC-25, 587–588, 1980.

Harn, Y. and C. Chen, "A proof of discrete stability test via the Lyapunov theorem," *IEEE Transactions on Automatic Control*, vol. AC-26, 733–734, 1981.

Hu, X., "A new stability table for discrete-time systems," *Systems & Control Letters*, vol. 22, 385–392, 1994.

Jury, E., *The Theory and Application of the Z-Transform Method*, Wiley, New York, 1964.

Kailath, T., *Linear Systems*, Prentice Hall, Englewood Cliffs, New Jersey, 1980.

Kando, H. and T. Iwazumi, and H. Ukai, "Singular perturbation modeling of large-scale systems with multi-time scale property," *International Journal of Control*, vol. 48, 2361–2387, 1988.

Kautsky, J., N. Nichols, and P. Van Doren, "Robust pole assignment in linear state feedback," *International Journal of Control*, vol. 41, 1129–1155, 1985.

Khalil, H., *Nonlinear Systems*, Macmillan, New York, 1992.

Kuo, B., *Automatic Control Systems*, Prentice Hall, Englewood Cliffs, New Jersey, 1991.

Lancaster, P. and M. Tismenetsky, *The Theory of Matrices*, Academic Press, Orlando, Florida, 1985.

Longhi, S. and R. Zulli, "A robust periodic pole assignment algorithm," *IEEE Transactions on Automatic Control*, vol. AC-40, 890–894, 1995.

Mahmoud, M., "Order reduction and control of discrete systems," *Proc. IEE, Part D*, vol. 129, 129–135, 1982.

Mansour, M., "A note on the stability of linear discrete systems and Lyapunov method," *IEEE Transactions on Automatic Control*, vol. AC-27, 707–708, 1982.

Meinsma, G., "Elementary proof of the Routh–Hurwitz test," *Systems & Control Letters*, vol. 25, 237–242, 1995.

Nyquist, H., "Regeneration theory," *Bell System Technical Journal*, vol. 11, 126–147, 1932.

Ogata, K., *Discrete Time Control Systems*, Prentice Hall, Englewood Cliffs, New Jersey, 1987.

Parks, P., "A new proof of the Routh–Hurwitz stability criterion using Liapunov's second method," *Proc. Cambridge Philosophical Society*, vol. 58, 694–702, 1962.

Raible, R., "A simplification of Jury's tabular form," *IEEE Transactions on Automatic Control*, vol. AC-19, 248–250, 1974.

Routh, E., *A Treatise on the Stability of a Given State of Motion,* Macmillan, London, 1887.

Strang, G., *Linear Algebra and Its Applications*, Saunders College Publishing, Orlando, Florida, 1988.

4.9 Problems

4.1 Find the response of the following systems given by

$$\dot{x} = A_i x, \qquad x(t_0) = \begin{bmatrix} 1 \\ 1 \\ 1 \end{bmatrix}$$

where

$$(a)\ A_1 = \begin{bmatrix} 0 & 1 & 0 \\ 0 & 0 & 1 \\ 0 & 0 & 0 \end{bmatrix}, \quad (b)\ A_2 = \begin{bmatrix} 0 & 1 & 0 \\ 0 & 0 & 0 \\ 0 & 0 & 0 \end{bmatrix}, \quad (c)\ A_3 = \begin{bmatrix} 0 & 0 & 0 \\ 0 & 0 & 0 \\ 0 & 0 & 0 \end{bmatrix}$$

and comment on the stability of these systems.

4.2 Assuming that the input and output system matrices are given by

$$B = [1 \quad 1 \quad 1]^T, \qquad C = [1 \quad 1 \quad 1]$$

find the system transfer functions for the three cases defined in Problem 4.1.

4.3 Find the Jordan form and the minimal polynomial for the linear time invariant system

$$\dot{x}(t) = \begin{bmatrix} 0 & -1 & 0 & 3 \\ 0 & 0 & -2 & 0 \\ 0 & 0 & 0 & 2 \\ 0 & 0 & 0 & 0 \end{bmatrix} x(t)$$

4.4 By using rules for finding the Jordan form outlined in Section 4.2.4 and MATLAB, find the minimal polynomial for the system

$$\dot{x}(t) = \begin{bmatrix} -1 & 0 & 1 & 0 & 0 & 1 \\ 0 & -1 & 2 & 1 & -1 & 0 \\ 0 & 0 & 0 & -1 & 0 & -1 \\ 0 & 0 & 0 & 0 & 0 & 0 \\ 0 & 0 & 0 & 0 & -2 & 2 \\ 0 & 0 & 0 & 0 & 0 & -2 \end{bmatrix} x(t)$$

Comment on the system stability.

4.5 Consider the discrete-time linear system

$$x(k+1) = \begin{bmatrix} 1 & 2 & 1 & 0 \\ 0 & 1 & 2 & 1 \\ 0 & 0 & -1 & 2 \\ 0 & 0 & 0 & -1 \end{bmatrix} x(k)$$

Examine the stability of this system using the statement of Theorem 4.6.

4.6 Find the response of the following discrete-time systems

$$x(k+1) = A_i x(k), \qquad x(0) = \begin{bmatrix} 2 \\ 2 \\ 2 \end{bmatrix}$$

with

$$A_1 = \begin{bmatrix} 0 & 1 & 0 \\ 0 & 0 & 1 \\ 0 & 0 & 0 \end{bmatrix}, \quad A_2 = \begin{bmatrix} 0 & 1 & 0 \\ 0 & 0 & 0 \\ 0 & 0 & 0 \end{bmatrix}, \quad A_3 = \begin{bmatrix} 0 & 0 & 0 \\ 0 & 0 & 0 \\ 0 & 0 & 0 \end{bmatrix}$$

and comment on their stability.

4.7 Use the Lyapunov test and MATLAB to examine the stability of the following continuous-time system and find a Lyapunov function if such a function exists

$$A = \begin{bmatrix} -5 & 0 & 1 & 2 \\ 0 & -4 & 2 & -1 \\ 1 & 0 & -3 & 0 \\ 0 & -1 & 1 & -5 \end{bmatrix}$$

4.8 Consider the synchronous machine from Problem 3.28. Use MATLAB to find its eigenvalues, and comment on the system stability. Verify the answer obtained by applying the Lyapunov stability test with $Q = 0.1I_7$.

4.9 Given a discrete-time model of a catalytic cracker (Kando *et al.*, 1988)

$$A = \begin{bmatrix} 0.012 & 0.047 & 0.097 & 0.072 & -0.019 \\ 0.014 & 0.056 & 0.115 & 0.085 & -0.023 \\ 0.066 & 0.252 & 0.581 & 0.431 & -0.116 \\ 0.028 & 0.105 & 0.240 & 0.178 & -0.049 \\ 0.001 & 0.003 & 0.003 & 0.002 & -0.001 \end{bmatrix}$$

Use MATLAB to examine its stability by the Lyapunov test. Hint: Take $\mathbf{Q} = \mathbf{I}_5$.

4.10 Using the Routh–Hurwitz stability criterion, show that the given characteristic equation describes an unstable system

$$s^8 + 2s^7 + 3s^6 + 4s^5 + 5s^4 + 4s^3 + 3s^2 + 2s + 1 = 0$$

Do not complete the whole Routh table. Just reach the point at which you can claim system instability.

4.11 Using the Routh–Hurwitz method, examine stability of the following systems represented by their characteristic polynomials

$$s^2 + 2s + 1$$
$$s^3 + 2s^2 + 2s + 1$$
$$s^4 + 2s^3 + 3s^2 + 2s + 1$$
$$s^5 + 2s^4 + 3s^3 + 3s^2 + 2s + 1$$
$$s^6 + 2s^5 + 3s^4 + 4s^3 + 3s^2 + 2s + 1$$

4.12 Given the following fifth-order characteristic polynomial

$$a_5 s^5 + a_4 s^4 + a_3 s^3 + a_2 s^2 + a_1 s + a_0$$

Form the Routh table for this polynomial. Show that the same coefficients (as those in the first column of the Routh table) are obtained by forming a sequence of long divisions as done in (4.32)–(4.34) and Example 4.14.

4.13 Draw the block diagram for Example 4.14 of the form given in Figure 4.3. From the block diagram obtained, get the state space form as the one given in (4.36) and (4.37).

4.14 Using instability Theorem 4.11 determine whether or not the following polynomials are unstable

$$s^4 + 2s^3 - 2s^2 + s + 2$$
$$s^5 + 3s^3 + 2s^2 + s + 1$$
$$s^6 + s^5 + s^4 + 3s^3 + 2s^2 + s$$
$$s^7 - s^5 + s^3 - s + 1$$

4.15 Find relationships among elements R_1, R_2, and C_1 of the electrical network given in Figure 2.28 such that it represents an asymptotically stable system.

4.16 Examine whether or not given systems can be made asymptotically stable by properly choosing the values of unknown parameters

$$s^5 + 3s^4 + s^3 + Ks^2 + 2s + 1$$
$$s^6 + 2s^5 + 3s^4 + 6s^3 + 2s^2 + s + K$$
$$s^4 + s^3 + Ks^2 + 2s + 1$$

4.17 Examine the stability of the following polynomials and determine the number of unstable poles

$$s^6 + 2s^5 + 3s^4 + s^3 + 3s^2 + 2s + 1$$
$$s^7 + s^6 + 2s^5 + 3s^4 + 6s^3 + 2s^2 + s + 1$$
$$s^7 + 2s^6 + 2s^5 + s^4 + s^3 + 3s^2 + 2s + 3$$
$$s^8 - 2s^7 + s^6 - 3s^5 + 3s^4 - s^3 + 2s^2 - s + 1$$

4.18 For the polynomials defined by

$$s^3 + 2s^2 + 3s + 4$$
$$s^4 + s^3 + 2s^2 + 3s + 6$$

find the number of poles located to the right of the vertical line parallel to the imaginary axis passing through the point $(-1, j0)$, in other words examine the relative stability of the above polynomials. Hint: Form the shifted polynomials by using the change of variables as $s = s_1 - 1$ and apply the Routh–Hurwitz test to the shifted polynomials obtained.

4.19 Using the Routh–Hurwitz test, examine the stability of the single-link robotic manipulator from Example 1.3 whose characteristic equation is given by

$$s^4 + 4.49s^2 + 0.3464$$

4.20 Repeat Problem 4.18 for the inverted pendulum from Section 4.2.3 whose characteristic equation is

$$s^4 - 21.56s^2$$

4.21 A discrete-time model of a steam power system (Mahmoud, 1982) is given by the following system matrix

$$A = \begin{bmatrix} 0.915 & 0.051 & 0.038 & 0.015 & 0.038 \\ -0.030 & 0.889 & -0.001 & 0.046 & 0.111 \\ -0.006 & 0.468 & 0.247 & 0.014 & 0.048 \\ -0.715 & -0.022 & -0.021 & 0.240 & -0.024 \\ -0.148 & -0.003 & -0.004 & 0.090 & 0.026 \end{bmatrix}$$

By using MATLAB, its characteristic polynomial is obtained as follows

$$z^5 - 2.317z^4 + 1.833z^3 - 0.570z^2 + 0.067z - 0.002$$

Examine the stability of this system by using Jury's test.

4.22 A linearized and discretized model of an underwater vehicle is described by the following system matrix (Longhi and Zulli, 1995)

$$A = \begin{bmatrix} 0.995 & 0.485 & 0 & -0.004 & 0 & 0 \\ -0.017 & 0.942 & 0 & -0.018 & 0 & 0 \\ 0 & -0.003 & 0.995 & 0.485 & 0 & 0 \\ 0 & -0.014 & -0.017 & 0.942 & 0 & 0 \\ 0 & 0 & 0 & 0 & 1 & 0.5 \\ 0 & 0 & 0 & 0 & 0.002 & 1 \end{bmatrix}$$

Use any method to examine its stability.

4.23 Using the simplified Jury's table and Theorem 4.13, examine stability of the following polynomial (Raible, 1974)

$$z^4 + 2.8z^3 - 4.35z^2 + 1.75z - 0.2$$

4.24 Draw the Nyquist plots for

$$G(s)H(s) = \frac{5}{s(s+2)}, \quad G(s)H(s) = \frac{10}{s(s+5)(s+10)}$$

$$G(s)H(s) = \frac{3}{(s+1)(s+2)}, \quad G(s)H(s) = \frac{s^2+1}{s(s+1)}$$

and examine the stability of the corresponding closed-loop systems.

4.25 Use MATLAB to find the gain and phase stability margins of the open-loop control systems defined in Problem 4.24.

Chapter Five

Controllability and Observability

Controllability and observability represent two major concepts of modern control system theory. These originally theoretical concepts, introduced by R. Kalman in 1960, are particularly important for practical implementations. They can be roughly defined as follows.

Controllability: *In order to be able to do whatever we want with the given dynamic system under control input, the system must be controllable.*

Observability: *In order to see what is going on inside the system under observation, the system must be observable.*

Even though the concepts of controllability and observability are almost abstractly defined, we now intuitively understand their meanings. The remaining problem is to produce some mathematical check up tests and to define controllability and observability more rigorously. Our intention is to reduce mathematical derivations and the number of definitions, but at the same time to derive and define very clearly both of them. In that respect, in Section 5.1, we start with observability derivations for linear discrete-time invariant systems and give the corresponding definition. The observability of linear discrete systems is very naturally introduced using only elementary linear algebra. This approach will be extended to continuous-time system observability, where the derivatives of measurements (observations) have to be used, Section 5.2. Next, in Sections 5.3 and 5.4 we define controllability for both discrete- and continuous-time linear systems.

In this chapter we show that the concepts of controllability and observability are related to linear systems of algebraic equations. It is well known that a solvable system of linear algebraic equations has a solution if and only if the rank of the system matrix is full (see Appendix C). Observability and controllability tests will be connected to the rank tests of ceratin matrices, known as the controllability and observability matrices.

At the end of this chapter, in Section 5.5, we will introduce the concepts of system stabilizability (detectability), which stand for controllability (observability) of unstable system modes. Also, we show that both controllability and observability are invariant under nonsingular transformations. In addition, in the same section the concepts of controllability and observability are clarified using different canonical forms, where they become more obvious.

The study of observability is closely related to observer (estimator) design, a simple, but extremely important technique used to construct another dynamic system, the observer (estimator), which produces estimates of the system state variables using information about the system inputs and outputs. The estimator design is presented in Section 5.6. Techniques for constructing both full-order and reduced-order estimators are considered. A corresponding problem to observer design is the so-called pole placement problem. It can be shown that for a controllable linear system, the system poles (eigenvalues) can be arbitrarily located in the complex plane. Since this technique can be used for system linear feedback stabilization and for controller design purposes, it will be independently presented in Section 8.2.

Several examples are included in order to demonstrate procedures for examining system controllability and observability. All of them can be checked by MATLAB. Finally, we have designed the corresponding laboratory experiment by using the MATLAB package, which can contribute to better and deeper understanding of these important modern control concepts.

Chapter Objectives

This chapter introduces definitions of system controllability and observability. Testing controllability and observability is replaced by linear algebra problems of finding ranks of certain matrices known as the controllability and observability matrices. After mastering the above concepts and tests, students will be able to determine system initial conditions from system output measurements, under the assumption that the given system is observable. As the highlight of this chapter, students will learn how to construct a system's observer (estimator), which for an observable system produces the estimates of state variables at any time instant.

5.1 Observability of Discrete Systems

Consider a linear, time invariant, discrete-time system in the state space form

$$x(k+1) = A_d x(k), \quad x(0) = x_0 = \text{unknown} \tag{5.1}$$

with output measurements

$$y(k) = C_d x(k) \tag{5.2}$$

where $x(k) \in \Re^n$, $y(k) \in \Re^p$. A_d and C_d are constant matrices of appropriate dimensions. The natural question to be asked is: can we learn everything about the dynamical behavior of the state space variables defined in (5.1) by using only information from the output measurements (5.2). If we know x_0, then the recursion (5.1) apparently gives us complete knowledge about the state variables at any discrete-time instant. Thus, the only thing that we have to determine from the state measurements is the initial state vector $x(0) = x_0$.

Since the n-dimensional vector $x(0)$ has n unknown components, it is expected that n measurements are sufficient to determine x_0. Take $k = 0, 1, ..., n-1$ in (5.1) and (5.2), i.e. generate the following sequence

$$y(0) = C_d x(0)$$
$$y(1) = C_d x(1) = C_d A_d x(0)$$
$$y(2) = C_d x(2) = C_d A_d x(1) = C_d A_d^2 x(0) \tag{5.3}$$
$$\vdots$$
$$y(n-1) = C_d x(n-1) = C_d A_d^{n-1} x(0)$$

or, in matrix form

$$\begin{bmatrix} y(0) \\ y(1) \\ y(2) \\ \vdots \\ y(n-1) \end{bmatrix}^{(np) \times 1} = \begin{bmatrix} C_d \\ C_d A_d \\ C_d A_d^2 \\ \vdots \\ C_d A_d^{n-1} \end{bmatrix}^{(np) \times n} \times x(0) \tag{5.4}$$

We know from elementary linear algebra that the system of linear algebraic equations with n unknowns, (5.4), has a unique solution if and only if the system matrix has rank n. In this case we need

$$
\text{rank}
\begin{bmatrix}
\mathbf{C}_d \\
\mathbf{C}_d \mathbf{A}_d \\
\mathbf{C}_d \mathbf{A}_d^2 \\
\vdots \\
\mathbf{C}_d \mathbf{A}_d^{n-1}
\end{bmatrix}
= n
\tag{5.5}
$$

Thus, the initial condition $\mathbf{x}_0$ is completely determined if the so-called *observability matrix*, defined by

$$
\mathcal{O}(\mathbf{A}_d, \mathbf{C}_d) =
\begin{bmatrix}
\mathbf{C}_d \\
\mathbf{C}_d \mathbf{A}_d \\
\mathbf{C}_d \mathbf{A}_d^2 \\
\vdots \\
\mathbf{C}_d \mathbf{A}_d^{n-1}
\end{bmatrix}^{(np)\times n}
\tag{5.6}
$$

has rank n, that is

$$
\text{rank}\mathcal{O} = n
\tag{5.7}
$$

The previous derivations can be summarized in the following theorem.

Theorem 5.1 *The linear discrete-time system (5.1) with measurements (5.2) is observable if and only if the observability matrix (5.6) has rank equal to n.*

A simple second-order example demonstrates the procedure for examining the observability of linear discrete-time systems. More complex examples corresponding to real physical control systems will be considered in Sections 5.7 and 5.8.

Example 5.1: Consider the following system with measurements

$$
\begin{bmatrix} x_1(k+1) \\ x_2(k+1) \end{bmatrix} =
\begin{bmatrix} 1 & 2 \\ 3 & 4 \end{bmatrix}
\begin{bmatrix} x_1(k) \\ x_2(k) \end{bmatrix}
$$

$$
y(k) = \begin{bmatrix} 1 & 2 \end{bmatrix}
\begin{bmatrix} x_1(k) \\ x_2(k) \end{bmatrix}
$$

The observability matrix for this second-order system is given by

$$\mathcal{O} = \begin{bmatrix} \mathbf{C}_d \\ \mathbf{C}_d\mathbf{A}_d \end{bmatrix} = \begin{bmatrix} 1 & 2 \\ 7 & 10 \end{bmatrix}$$

Since the rows of the matrix $\mathcal{O}$ are linearly independent, then $\mathrm{rank}\mathcal{O} = 2 = n$, i.e. the system under consideration is observable. Another way to test the completeness of the rank of square matrices is to find their determinants. In this case

$$\det\mathcal{O} = -4 \neq 0 \Leftrightarrow \text{full rank} = n = 2$$

◇

Example 5.2: Consider a case of an unobservable system, which can be obtained by slightly modifying Example 5.1. The corresponding system and measurement matrices are given by

$$\mathbf{A}_d = \begin{bmatrix} 1 & -2 \\ -3 & -4 \end{bmatrix}, \quad \mathbf{C}_d = \begin{bmatrix} 1 & 2 \end{bmatrix}$$

The observability matrix is

$$\mathcal{O} = \begin{bmatrix} 1 & 2 \\ -5 & -10 \end{bmatrix}$$

so that $\mathrm{rank}\mathcal{O} = 1 < 2$, and the system is unobservable.

◇

5.2 Observability of Continuous Systems

A linear, time invariant, continuous system in the state space form was studied in Chapter 3. For the purpose of studying its observability, we consider an input-free system

$$\dot{\mathbf{x}}(t) = \mathbf{A}\mathbf{x}(t), \quad \mathbf{x}(t_0) = \mathbf{x}_o = \text{unknown} \tag{5.8}$$

with the corresponding measurements

$$\mathbf{y}(t) = \mathbf{C}\mathbf{x}(t) \tag{5.9}$$

of dimensions $\mathbf{x}(k) \in \Re^n$, $\mathbf{y}(k) \in \Re^p$, $\mathbf{A} \in \Re^{n \times n}$, and $\mathbf{C} \in \Re^{p \times n}$. Following the same arguments as in the previous section, we can conclude that the knowledge of $\mathbf{x}_0$ is sufficient to determine $\mathbf{x}(t)$ at any time instant, since from (5.8) we have

$$\mathbf{x}(t) = e^{\mathbf{A}(t - t_0)} \mathbf{x}(t_0) \tag{5.10}$$

The problem that we are faced with is to find $\mathbf{x}(t_0)$ from the available measurements (5.9). In Section 5.1 we have solved this problem for discrete-time systems by generating the sequence of measurements at discrete-time instants $k = 0, 1, 2, ..., n - 1$, i.e. by producing relations given in (5.3). Note that a time shift in the discrete-time corresponds to a derivative in the continuous-time. Thus, an analogous technique in the continuous-time domain is obtained by taking derivatives of the continuous-time measurements (5.9)

$$\begin{aligned}
\mathbf{y}(t_0) &= \mathbf{C}\mathbf{x}(t_0) \\
\dot{\mathbf{y}}(t_0) &= \mathbf{C}\dot{\mathbf{x}}(t_0) = \mathbf{C}\mathbf{A}\mathbf{x}(t_0) \\
\ddot{\mathbf{y}}(t_0) &= \mathbf{C}\ddot{\mathbf{x}}(t_0) = \mathbf{C}\mathbf{A}^2\mathbf{x}(t_0) \\
&\vdots \\
\mathbf{y}^{(n-1)}(t_0) &= \mathbf{C}\mathbf{x}^{(n-1)}(t_0) = \mathbf{C}\mathbf{A}^{n-1}\mathbf{x}(t_0)
\end{aligned} \tag{5.11}$$

Our goal is to generate n linearly independent algebraic equations in n unknowns of the state vector $\mathbf{x}(t_0)$. Equations (5.11) comprise a system of np linear algebraic equations. They can be put in matrix form as

$$\begin{bmatrix} \mathbf{y}(t_0) \\ \dot{\mathbf{y}}(t_0) \\ \ddot{\mathbf{y}}(t_0) \\ \vdots \\ \mathbf{y}^{(n-1)}(t_0) \end{bmatrix}^{(np) \times 1} = \begin{bmatrix} \mathbf{C} \\ \mathbf{C}\mathbf{A} \\ \mathbf{C}\mathbf{A}^2 \\ \vdots \\ \mathbf{C}\mathbf{A}^{n-1} \end{bmatrix}^{(np) \times n} \times \mathbf{x}(t_0) = \mathcal{O}\mathbf{x}(t_0) = Y(t_0) \tag{5.12}$$

where $\mathcal{O}$ is the observability matrix already defined in (5.6) and where the definition of $Y(t_0)$ is obvious. Thus, the initial condition $\mathbf{x}(t_0)$ can be determined uniquely from (5.12) if and only if the observability matrix has full rank, i.e. $\text{rank}\mathcal{O} = n$.

As expected, we have obtained the same observability result for both continuous- and discrete-time systems. The continuous-time observability theorem, dual to Theorem 5.1, can be formulated as follows.

Theorem 5.2 *The linear continuous-time system* (5.8) *with measurements* (5.9) *is observable if and only if the observability matrix has full rank.*

It is important to notice that adding higher-order derivatives in (5.12) cannot increase the rank of the observability matrix since by the Cayley–Hamilton theorem (see Appendix C) for $k \geq n$ we have

$$\mathbf{A}^k = \sum_{i=0}^{n-1} \alpha_i \mathbf{A}^i \tag{5.13}$$

so that the additional equations would be linearly dependent on the previously defined n equations (5.12). The same applies to the discrete-time domain and the corresponding equations given in (5.4).

There is no need to produce a test example for the observability study of continuous-time systems since the procedure is basically the same as in the case of discrete-time systems studied in the previous section. Thus, Examples 5.1 and 5.2 demonstrate the presented procedure in this case also; however, we have to keep in mind that the corresponding matrices $\mathbf{A}$ and $\mathbf{C}$ describe systems which operate in different time domains. Fortunately, the algebraic procedures are exactly the same in both cases.

5.3 Controllability of Discrete Systems

Consider a linear discrete-time invariant control system defined by

$$\mathbf{x}(k+1) = \mathbf{A}_d \mathbf{x}(k) + \mathbf{B}_d \mathbf{u}(k), \quad \mathbf{x}(0) = \mathbf{x}_0 \tag{5.14}$$

The system controllability is roughly defined as an ability to do whatever we want with our system, or in more technical terms, the ability to transfer our system from any initial state $\mathbf{x}(0) = \mathbf{x}_0$ to any desired final state $\mathbf{x}(k_1) = \mathbf{x}_f$ in a finite time, i.e. for $k_1 < \infty$ (it makes no sense to achieve that goal at $k_1 = \infty$). Thus, the question to be answered is: can we find a control sequence $\mathbf{u}(0), \mathbf{u}(1), \ldots, \mathbf{u}(n-1)$, such that $\mathbf{x}(k) = \mathbf{x}_f$?

Let us start with a simplified problem, namely let us assume that the input $\mathbf{u}(k)$ is a scalar, i.e. the input matrix $\mathbf{B}_d$ is a vector denoted by $\mathbf{b}_d$. Thus, we have

$$\mathbf{x}(k+1) = \mathbf{A}_d \mathbf{x}(k) + \mathbf{b}_d u(k), \quad \mathbf{x}(0) = \mathbf{x}_0 \tag{5.15}$$

Taking $k = 0, 1, 2, ..., n$ in (5.15), we obtain the following set of equations

$$\mathbf{x}(1) = \mathbf{A}_d\mathbf{x}(0) + \mathbf{b}_d u(0)$$
$$\mathbf{x}(2) = \mathbf{A}_d\mathbf{x}(1) + \mathbf{b}_d u(1) = \mathbf{A}_d^2\mathbf{x}(0) + \mathbf{A}_d\mathbf{b}_d u(0) + \mathbf{b}_d u(1)$$
$$\vdots \qquad\qquad\qquad\qquad\qquad\qquad\qquad\qquad (5.16)$$
$$\mathbf{x}(n) = \mathbf{A}_d^n\mathbf{x}(0) + \mathbf{A}_d^{n-1}\mathbf{b}_d u(0) + \cdots + \mathbf{b}_d u(n-1)$$

The last equation in (5.16) can be written in matrix form as

$$\mathbf{x}(n) - \mathbf{A}_d^n\mathbf{x}(0) = \left[\mathbf{b}_d \vdots \mathbf{A}_d\mathbf{b}_d \vdots \cdots \vdots \mathbf{A}_d^{n-1}\mathbf{b}_d\right]\begin{bmatrix} u(n-1) \\ u(n-2) \\ \vdots \\ u(1) \\ u(0) \end{bmatrix} \qquad (5.17)$$

Note that $\left[\mathbf{b}_d \vdots \mathbf{A}_d\mathbf{b}_d \vdots \cdots \vdots \mathbf{A}_d^{n-1}\mathbf{b}_d\right]$ is a square matrix. We call it the *controllability matrix* and denote it by C. If the controllability matrix C is nonsingular, equation (5.17) produces the unique solution for the input sequence given by

$$\begin{bmatrix} u(n-1) \\ u(n-2) \\ \vdots \\ u(1) \\ u(0) \end{bmatrix} = C^{-1}(\mathbf{x}(n) - \mathbf{A}_d^n\mathbf{x}(0)) \qquad (5.18)$$

Thus, for any $\mathbf{x}(n) = \mathbf{x}_f$, the expression (5.18) determines the input sequence that transfers the initial state $\mathbf{x}_0$ to the desired state $\mathbf{x}_f$ in n steps. It follows that the controllability condition, in this case, is equivalent to nonsingularity of the controllability matrix C.

In a general case, when the input $\mathbf{u}(k)$ is a vector of dimension r, the repetition of the same procedure as in (5.15)–(5.17) leads to

$$\mathbf{x}(n) - \mathbf{A}_d^n\mathbf{x}(0) = \left[\mathbf{B}_d \vdots \mathbf{A}_d\mathbf{B}_d \vdots \cdots \vdots \mathbf{A}_d^{n-1}\mathbf{B}_d\right]\begin{bmatrix} \mathbf{u}(n-1) \\ \mathbf{u}(n-2) \\ \vdots \\ \mathbf{u}(1) \\ \mathbf{u}(0) \end{bmatrix} \qquad (5.19)$$

The controllability matrix, in the general vector input case, defined by

$$C(\mathbf{A}_d, \mathbf{B}_d) = \left[\mathbf{B}_d \vdots \mathbf{A}_d\mathbf{B}_d \vdots \cdots \vdots \mathbf{A}_d^{n-1}\mathbf{B}_d \right] \qquad (5.20)$$

is of dimension $n \times r \cdot n$.   The corresponding system of n linear alge-braic equations in $r \cdot n$ unknowns for n r-dimensional vector components of $\mathbf{u}(0), \mathbf{u}(1), \ldots, \mathbf{u}(n-1)$, given by

$$C^{n\times(nm)} \begin{bmatrix} \mathbf{u}(n-1) \\ \mathbf{u}(n-2) \\ \vdots \\ \mathbf{u}(1) \\ \mathbf{u}(0) \end{bmatrix}^{(nr)\times1} = \mathbf{x}(n) - \mathbf{A}_d^n\mathbf{x}(0) = \mathbf{x}_f - \mathbf{A}_d^n\mathbf{x}(0) \qquad (5.21)$$

will have a solution for any $\mathbf{x}_f$ if and only if the matrix C has full rank, i.e. rank$C = n$ (see Appendix C).

The controllability theorem is as follows.

Theorem 5.3 *The linear discrete-time system (5.14) is controllable if and only if*

$$\text{rank}C = n \qquad (5.22)$$

where the controllability matrix C is defined by (5.20).

5.4 Controllability of Continuous Systems

Studying the concept of controllability in the continuous-time domain is more challenging than in the discrete-time domain. At the beginning of this section we will first apply the same strategy as in Section 5.3 in order to indicate difficulties that we are faced with in the continuous-time domain. Then, we will show how to find a control input that will transfer our system from any initial state to any final state.

A linear continuous-time system with a scalar input is represented by

$$\dot{\mathbf{x}} = \mathbf{A}\mathbf{x} + \mathbf{b}u, \quad \mathbf{x}(t_0) = \mathbf{x}_o \qquad (5.23)$$

Following the discussion and derivations from Section 5.3, we have, for a scalar input, the following set of equations

$$\dot{\mathbf{x}} = \frac{d}{dt}\mathbf{x} = \mathbf{A}\mathbf{x} + \mathbf{b}u$$

$$\ddot{\mathbf{x}} = \frac{d^2}{dt^2}\mathbf{x} = \mathbf{A}^2\mathbf{x} + \mathbf{A}\mathbf{b}u + \mathbf{b}\dot{u}$$

$$\vdots$$ (5.24)

$$\mathbf{x}^{(n)} = \frac{d^n}{dt^n}\mathbf{x} = \mathbf{A}^n\mathbf{x} + \mathbf{A}^{n-1}\mathbf{b}u + \mathbf{A}^{n-2}\mathbf{b}\dot{u} + \cdots + \mathbf{b}u^{(n-1)}$$

The last equation in (5.24) can be written as

$$\mathbf{x}^{(n)}(t) - \mathbf{A}^n\mathbf{x}(t) = C\begin{bmatrix} u^{(n-1)}(t) \\ u^{(n-2)}(t) \\ \vdots \\ \dot{u}(t) \\ u(t) \end{bmatrix} \qquad (5.25)$$

Note that (5.25) is valid for any $t \in [t_0, t_f]$ with t_f free but finite. Thus, the nonsingularity of the controllability matrix C implies the existence of the scalar input function $u(t)$ and its $n - 1$ derivatives, for any $t < t_f < \infty$.

For a vector input system dual to (5.23), the above discussion produces the same relation as (5.25) with the controllability matrix C given by (5.20) and with the input vector $\mathbf{u}(t) \in \mathfrak{R}^r$, that is

$$C^{n \times m \cdot n}\begin{bmatrix} \mathbf{u}^{(n-1)}(t) \\ \mathbf{u}^{(n-2)}(t) \\ \vdots \\ \dot{\mathbf{u}}(t) \\ \mathbf{u}(t) \end{bmatrix}^{r \cdot n \times 1} = \mathbf{x}^{(n)}(t) - \mathbf{A}^n\mathbf{x}(t) = \gamma(t) \qquad (5.26)$$

It is well known from linear algebra that in order to have a solution of (5.26), it is sufficient that

$$\mathrm{rank}C = \mathrm{rank}\left[C \vdots \gamma(t) \right] \qquad (5.27)$$

Also, a solution of (5.26) exists for any $\gamma(t)$—any desired state at t—if and only if

$$\text{rank}\mathcal{C} = n \tag{5.28}$$

Equations (5.25) and (5.26) establish relationships between the state and control variables. However, from (5.25) and (5.26) we do not have an explicit answer about a control function that is transferring the system from any initial state $\mathbf{x}(t_0)$ to any final state $\mathbf{x}(t_1) = \mathbf{x}_f$. Thus, elegant and simple derivations for the discrete-time controllability problem cannot be completely extended to the continuous-time domain. Another approach, which is mathematically more complex, is required in this case. It will be presented in the remaining part of this section.

From Section 3.2 we know that the state space equation with the control input has the following solution

$$\mathbf{x}(t) = e^{\mathbf{A}(t-t_0)}\mathbf{x}(t_0) + \int_{t_0}^{t} e^{\mathbf{A}(t-\tau)}\mathbf{B}\mathbf{u}(\tau)d\tau$$

At the final time t_1 we have

$$\mathbf{x}(t_1) = \mathbf{x}_f = e^{\mathbf{A}(t_1-t_0)}\mathbf{x}(t_0) + \int_{t_0}^{t_1} e^{\mathbf{A}(t_1-\tau)}\mathbf{B}\mathbf{u}(\tau)d\tau$$

or

$$e^{-\mathbf{A}t_1}\mathbf{x}_f - e^{-\mathbf{A}t_0}\mathbf{x}(t_0) = \int_{t_0}^{t_1} e^{-\mathbf{A}\tau}\mathbf{B}\mathbf{u}(\tau)d\tau$$

Using the Cayley–Hamilton theorem (see Appendix C), that is

$$e^{-\mathbf{A}\tau} = \sum_{i=0}^{n-1} \alpha_i(\tau)\mathbf{A}^i \tag{5.29}$$

where $\alpha_i(\tau)$, $i = 0, 1, ..., n-1$, are scalar time functions, the previous equation can be rewritten as

$$e^{-\mathbf{A}t_1}\mathbf{x}_f - e^{-\mathbf{A}t_0}\mathbf{x}(t_0) = \sum_{i=0}^{n-1} \mathbf{A}^i\mathbf{B} \int_{t_0}^{t_1} \alpha_i(\tau)\mathbf{u}(\tau)d\tau$$

or

$$e^{-At_1}\mathbf{x}_f - e^{-At_0}\mathbf{x}(t_0) = \left[\mathbf{B} \vdots \mathbf{AB} \vdots \cdots \vdots \mathbf{A}^{n-1}\mathbf{B}\right] \begin{bmatrix} \int_{t_0}^{t_1} \alpha_0(\tau)\mathbf{u}(\tau)d\tau \\ \int_{t_0}^{t_1} \alpha_1(\tau)\mathbf{u}(\tau)d\tau \\ \vdots \\ \int_{t_0}^{t_1} \alpha_{n-1}(\tau)\mathbf{u}(\tau)d\tau \end{bmatrix}$$

On the left-hand side of this equation all quantities are known, i.e. we have a constant vector. On the right-hand side the controllability matrix is multiplied by a vector whose components are functions of the required control input. Thus, we have a functional equation in the form

$$\mathbf{const}^{n \times 1} = \mathcal{C}(\mathbf{A}, \mathbf{B})^{n \times rn} \begin{bmatrix} \mathbf{f}_1(\mathbf{u}(\tau)) \\ \mathbf{f}_2(\mathbf{u}(\tau)) \\ \vdots \\ \mathbf{f}_{n-1}(\mathbf{u}(\tau)) \end{bmatrix}^{rn \times 1} , \quad \tau \in (t_0, t_1) \qquad (5.30)$$

A solution of this equation exists if and only if $\operatorname{rank} \mathcal{C}(\mathbf{A}, \mathbf{B}) = n$, which is the condition already established in (5.28). In general, it is very hard to solve this equation. One of the many possible solutions of (5.30) will be given in Section 5.8 in terms of the controllability Grammian. The controllability Grammian is defined by the following integral

$$\mathbf{W}(t_0, t_1) = \int_{t_0}^{t_1} e^{\mathbf{A}(t_0 - \tau)} \mathbf{B}\mathbf{B}^T e^{\mathbf{A}^T(t_0 - \tau)} d\tau \qquad (5.31)$$

The results presented in this section can be summarized in the following theorem.

Theorem 5.4 *The linear continuous-time system is controllable if and only if the controllability matrix $\mathcal{C}$ has full rank, i.e. $\operatorname{rank}\mathcal{C} = n$.*

We have seen that controllability of linear continuous- and discrete-time systems is given in terms of the controllability matrix (5.20). Examining the rank of the controllability matrix comprises an algebraic criterion for testing system controllability. The example below demonstrates this procedure.

Example 5.3: Given the linear continuous-time system

$$\dot{x} = \begin{bmatrix} 0 & 1 & -2 \\ 3 & -4 & 5 \\ -6 & 7 & 8 \end{bmatrix} x + \begin{bmatrix} 0 & -1 \\ 2 & -3 \\ 4 & -5 \end{bmatrix} u$$

The controllability matrix for this third-order system is given by

$$C = \begin{bmatrix} B \vdots AB \vdots A^2B \end{bmatrix}$$

$$= \begin{bmatrix} 0 & -1 & \vdots & -6 & 7 & \vdots & \\ 2 & -3 & \vdots & 12 & -10 & \vdots & A^2B \\ 4 & -5 & \vdots & 46 & -55 & \vdots & \end{bmatrix}$$

Since the first three columns are linearly independent we can conclude that rank$C = 3$. Hence there is no need to compute A^2B since it is well known from linear algebra that the row rank of the given matrix is equal to its column rank. Thus, rank$C = 3 = n$ implies that the system under consideration is controllable.

◇

5.5 Additional Controllability/Observability Topics

In this section we will present several interesting and important results related to system controllability and observability.

Invariance Under Nonsingular Transformations

In Section 3.4 we introduced the similarity transformation that transforms a given system from one set of coordinates to another. Now we will show that both system controllability and observability are invariant under similarity transformation.

Consider the vector input form of (5.23) and the similarity transformation

$$\hat{x} = Px \tag{5.32}$$

such that

$$\dot{\hat{x}} = \hat{A}\hat{x} + \hat{B}u$$

where $\hat{\mathbf{A}} = \mathbf{PAP}^{-1}$ and $\hat{\mathbf{B}} = \mathbf{PB}$. Then the following theorem holds.

Theorem 5.5 *The pair* $(\mathbf{A}, \mathbf{B})$ *is controllable if and only if the pair* $\left(\hat{\mathbf{A}}, \hat{\mathbf{B}}\right)$ *is controllable.*

This theorem can be proved as follows

$$\mathcal{C}\left(\hat{\mathbf{A}}, \hat{\mathbf{B}}\right) = \left[\hat{\mathbf{B}} \vdots \hat{\mathbf{A}}\hat{\mathbf{B}} \vdots \cdots \vdots \hat{\mathbf{A}}^{n-1}\hat{\mathbf{B}}\right]$$

$$= \left[\mathbf{PB} \vdots \mathbf{PAP}^{-1}\mathbf{PB} \vdots \cdots \vdots \mathbf{PA}^{n-1}\mathbf{P}^{-1}\mathbf{PB}\right]$$

$$= \mathbf{P}\left[\mathbf{B} \vdots \mathbf{AB} \vdots \cdots \vdots \mathbf{A}^{n-1}\mathbf{B}\right] = \mathbf{P}\mathcal{C}(\mathbf{A}, \mathbf{B})$$

Since $\mathbf{P}$ is a nonsingular matrix (it cannot change the rank of the product $\mathbf{P}\mathcal{C}$), we get

$$\text{rank}\mathcal{C}\left(\hat{\mathbf{A}}, \hat{\mathbf{B}}\right) = \text{rank}\mathcal{C}(\mathbf{A}, \mathbf{B})$$

which proves the theorem and establishes controllability invariance under a similarity transformation.

A similar theorem is valid for observability. The similarity transformation (5.32) applied to (5.8) and (5.9) produces

$$\dot{\hat{\mathbf{x}}} = \hat{\mathbf{A}}\hat{\mathbf{x}}$$
$$\mathbf{y} = \hat{\mathbf{C}}\hat{\mathbf{x}}$$

where

$$\hat{\mathbf{C}} = \mathbf{CP}^{-1}$$

Then, we have the following theorem

Theorem 5.6 *The pair* $(\mathbf{A}, \mathbf{C})$ *is observable if and only if the pair* $\left(\hat{\mathbf{A}}, \hat{\mathbf{C}}\right)$ *is observable.*

The proof of this theorem is as follows

$$\mathcal{O}\left(\hat{\mathbf{A}}, \hat{\mathbf{C}}\right) = \begin{bmatrix} \hat{\mathbf{C}} \\ \hat{\mathbf{C}}\hat{\mathbf{A}} \\ \hat{\mathbf{C}}\hat{\mathbf{A}}^2 \\ \vdots \\ \hat{\mathbf{C}}\hat{\mathbf{A}}^{n-1} \end{bmatrix} = \begin{bmatrix} \mathbf{CP}^{-1} \\ \mathbf{CP}^{-1}\mathbf{PAP}^{-1} \\ \mathbf{CP}^{-1}\mathbf{PA}^2\mathbf{P}^{-1} \\ \vdots \\ \mathbf{CP}^{-1}\mathbf{PA}^{n-1}\mathbf{P}^{-1} \end{bmatrix} = \begin{bmatrix} \mathbf{C} \\ \mathbf{CA} \\ \mathbf{CA}^2 \\ \vdots \\ \mathbf{CA}^{n-1} \end{bmatrix}\mathbf{P}^{-1}$$

that is,

$$\mathcal{O}\left(\hat{\mathbf{A}}, \hat{\mathbf{C}}\right) = \mathcal{O}(\mathbf{A}, \mathbf{C})\mathbf{P}^{-1}$$

The nonsingularity of $\mathbf{P}$ implies

$$\text{rank}\mathcal{O}\left(\hat{\mathbf{A}}, \hat{\mathbf{C}}\right) = \text{rank}\mathcal{O}(\mathbf{A}, \mathbf{C})$$

which proves the stated observability invariance.

Note that Theorems 5.5 and 5.6 are applicable to both continuous- and discrete-time linear systems.

Frequency Domain Controllability and Observability Test

Controllability and observability have been introduced in the state space domain as pure time domain concepts. It is interesting to point out that in the frequency domain there exists a very powerful and simple theorem that gives a single condition for both the controllability and the observability of a system. It is given below.

Let $H(s)$ be the transfer function of a single-input single-output system

$$H(s) = \mathbf{c}(s\mathbf{I} - \mathbf{A})^{-1}\mathbf{b}$$

Note that $H(s)$ is defined by a ratio of two polynomials containing the corresponding system poles and zeros. The following controllability–observability theorem is given without a proof.

Theorem 5.7 *If there are no zero-pole cancellations in the transfer function of a single-input single-output system, then the system is both controllable and observable. If the zero-pole cancellation occurs in $H(s)$, then the system is either uncontrollable or unobservable or both uncontrollable and unobservable.*

A similar theorem can be formulated for discrete linear time invariant systems.

Example 5.4: Consider a linear continuous-time dynamic system represented by its transfer function

$$H(s) = \frac{(s+3)}{(s+1)(s+2)(s+3)} = \frac{s+3}{s^3 + 6s^2 + 11s + 6}$$

Theorem 5.7 indicates that any state space model for this system is either uncontrollable or/and unobservable. To get the complete answer we have to go to a state space form and examine the controllability and observability matrices. One of the possible many state space forms of $H(s)$ is as follows

$$\begin{bmatrix} \dot{x}_1 \\ \dot{x}_2 \\ \dot{x}_3 \end{bmatrix} = \begin{bmatrix} -6 & -11 & -6 \\ 1 & 0 & 0 \\ 0 & 1 & 0 \end{bmatrix} \begin{bmatrix} x_1 \\ x_2 \\ x_3 \end{bmatrix} + \begin{bmatrix} 1 \\ 0 \\ 0 \end{bmatrix} u$$

$$y = \begin{bmatrix} 0 & 1 & 3 \end{bmatrix} \begin{bmatrix} x_1 \\ x_2 \\ x_3 \end{bmatrix}$$

It is easy to show that the controllability and observability matrices are given by

$$\mathcal{C} = \begin{bmatrix} 1 & -6 & 25 \\ 0 & 1 & -6 \\ 0 & 0 & 1 \end{bmatrix}, \quad \mathcal{O} = \begin{bmatrix} 0 & 1 & 3 \\ 1 & 3 & 0 \\ -3 & -11 & -6 \end{bmatrix}$$

Since

$$\det \mathcal{C} = 1 \neq 0 \Rightarrow \text{rank}\mathcal{C} = 3 = n$$

and

$$\det \mathcal{O} = 0 \Rightarrow \text{rank}\mathcal{O} < 3 = n$$

this system is controllable, but unobservable.

Note that, due to a zero-pole cancellation at $s = -3$, the system transfer function $H(s)$ is reducible to

$$H(s) = H_r(s) = \frac{1}{(s+1)(s+2)} = \frac{1}{s^2 + 3s + 2}$$

so that the equivalent system of order $n = 2$ has the corresponding state space form

$$\begin{bmatrix} \dot{x}_{1r} \\ \dot{x}_{2r} \end{bmatrix} = \begin{bmatrix} -2 & -3 \\ 1 & 0 \end{bmatrix} \begin{bmatrix} x_{1r} \\ x_{2r} \end{bmatrix} + \begin{bmatrix} 1 \\ 0 \end{bmatrix} u$$

$$y = \begin{bmatrix} 0 & 1 \end{bmatrix} \begin{bmatrix} x_{1r} \\ x_{2r} \end{bmatrix}$$

For this reduced-order system we have

$$\mathcal{C} = \begin{bmatrix} 1 & -2 \\ 0 & 1 \end{bmatrix}, \quad \mathcal{O} = \begin{bmatrix} 0 & 1 \\ 1 & 0 \end{bmatrix}$$

and therefore the system is both controllable and observable.

Interestingly enough, the last two mathematical models of dynamic systems of order $n = 3$ and $n = 2$ represent exactly the same physical system. Apparently, the second one ($n = 2$) is preferred since it can be realized with only two integrators.

◇

It can be concluded from Example 5.4 that Theorem 5.7 gives an answer to the problem of dynamic system reducibility. It follows that a single-input single-output dynamic system is irreducible if and only if it is both controllable and observable. Such a system realization is called the *minimal realization*. If the system is either uncontrollable and/or unobservable it can be represented by a system whose order has been reduced by removing uncontrollable and/or unobservable modes. It can be seen from Example 5.4 that the reduced system with $n = 2$ is both controllable and observable, and hence it cannot be further reduced. This is also obvious from the transfer function $H_r(s)$.

Theorem 5.7 can be generalized to multi-input multi-output systems, where it plays very important role in the procedure of testing whether or not a given system is in the minimal realization form. The procedure requires the notion of the characteristic polynomial for proper rational matrices which is beyond the scope of this book. Interested readers may find all details and definitions in Chen (1984).

It is important to point out that the similarity transformation does not change the transfer function as was shown in Section 3.4.

Controllability and Observability of Special Forms

In some cases, it is easy to draw conclusions about system controllability and/or observability by examining directly the state space equations. In those cases there is no need to find the corresponding controllability and observability matrices and check their ranks.

Consider the phase variable canonical form with

$$\dot{x} = Ax + Bu$$
$$y = Cx$$

where

$$A = \begin{bmatrix} 0 & 1 & 0 & \cdots & 0 \\ 0 & 0 & 1 & \cdots & 0 \\ \vdots & \vdots & \vdots & \ddots & \vdots \\ 0 & 0 & 0 & \cdots & 1 \\ -a_0 & -a_1 & -a_2 & \cdots & -a_{n-1} \end{bmatrix}, \quad B = \begin{bmatrix} 0 \\ 0 \\ \vdots \\ 0 \\ 1 \end{bmatrix}$$

$$C = \begin{bmatrix} 1 & 0 & 0 & \cdots & 0 \end{bmatrix}$$

This form is both controllable and observable due to an elegant chain connection of the state variables. The variable $x_1(t)$ is directly measured, so that $x_2(t)$ is known from $x_2(t) = \dot{x}_1(t)$. Also, $x_3(t) = \dot{x}_2(t) = \ddot{x}_1(t)$, and so on, $x_n(t) = x_1^{(n-1)}(t)$. Thus, this form is observable. The controllability follows from the fact that all state variables are affected by the control input, i.e. x_n is affected directly by $u(t)$ and then $\dot{x}_{n-1}(t)$ by $x_n(u(t))$ and so on. The control input is able to indirectly move all state variables into the desired positions so that the system is controllable. This can be formally verified by forming the corresponding controllability matrix and checking its rank. This is left as an exercise for students (see Problem 5.13).

Another example is the modal canonical form. Assuming that all eigenvalues of the system matrix are distinct, we have

$$\dot{x} = \Lambda x + \Gamma u$$
$$y = \mathfrak{D} x$$

where

$$\Lambda = \begin{bmatrix} \lambda_1 & 0 & \cdots & 0 \\ 0 & \lambda_2 & \cdots & 0 \\ \vdots & \vdots & \ddots & \vdots \\ 0 & 0 & \cdots & \lambda_n \end{bmatrix}, \quad \Gamma = \begin{bmatrix} \gamma_1 \\ \gamma_2 \\ \vdots \\ \gamma_n \end{bmatrix}$$

$$\mathfrak{D} = \begin{bmatrix} \delta_1 & \delta_2 & \cdots & \delta_n \end{bmatrix}$$

We are apparently faced with n completely decoupled first-order systems. Obviously, for controllability all γ_i, $i = 1, \ldots, n$, must be different from zero, so that each state variable can be controlled by the input $u(t)$. Similarly, $\delta_i \neq 0$, $i = 1, \ldots, n$, ensures observability since, due to the state decomposition, each system must be observed independently.

The Role of Observability in Analog Computer Simulation

In addition to applications in control system theory and practice, the concept of observability is useful for analog computer simulation. Consider the problem of solving an nth-order differential equation given by

$$y^{(n)} + \sum_{i=1}^{n} a_{n-i} y^{(n-i)} = \sum_{i=0}^{m} b_{m-i} u^{(m-i)}$$

with known initial conditions for $y(0), \dot{y}(0), ..., y^{(n-1)}(0)$. This system can be solved by an analog computer by using n integrators. The outputs of these n integrators represent the state variables $x_1, x_2, ..., x_n$, so that this system has the state space form

$$\dot{x} = \mathbf{A}x + \mathbf{b}u, \qquad x(0) = \text{unknown}$$

$$y = \mathbf{c}x$$

However, the initial condition for $x(0)$ is not given. In other words, the initial conditions for the considered system of n integrators are unknown. They can be determined from $y(0), \dot{y}(0), ..., y^{(n-1)}(0)$ by following the observability derivations performed in Section 5.2, namely

$$y(0) = \mathbf{c}x(0)$$
$$\dot{y}(0) = \mathbf{c}\dot{x}(0) = \mathbf{c}\mathbf{A}x(0) + \mathbf{c}\mathbf{b}u(0)$$
$$\ddot{y}(0) = \mathbf{c}\ddot{x}(0) = \mathbf{c}\mathbf{A}^2 x(0) + \mathbf{c}\mathbf{A}\mathbf{b}u(0) + \mathbf{c}\mathbf{b}\dot{u}(0)$$
$$\vdots$$
$$y^{(n-1)}(0) = \mathbf{c}x^{(n-1)}(0) = \mathbf{c}\mathbf{A}^{n-1}x(0) + \mathbf{c}\mathbf{A}^{n-2}\mathbf{b}u(0)$$
$$+ \mathbf{c}\mathbf{A}^{n-3}\mathbf{b}\dot{u}(0) + \cdots + \mathbf{c}\mathbf{A}\mathbf{b}u^{(n-3)}(0) + \mathbf{c}\mathbf{b}u^{(n-2)}(0)$$

This system can be written in matrix form as follows

$$\begin{bmatrix} y(0) \\ \dot{y}(0) \\ \vdots \\ y^{(n-1)}(0) \end{bmatrix} = \mathcal{O} \cdot x(0) + \mathcal{T} \begin{bmatrix} 0 \\ u(0) \\ \vdots \\ u^{(n-2)}(0) \end{bmatrix} \tag{5.33}$$

where $\mathcal{O}$ is the observability matrix and $\mathcal{T}$ is a known matrix. Since $u(0), \dot{u}(0), ..., u^{(n-1)}(0)$ are known, it follows that a unique solution for $x(0)$

exists if and only if the observability matrix, which is square in this case, is invertible, i.e. the pair $(\mathbf{A}, \mathbf{c})$ is observable.

Example 5.5: Consider a system represented by the differential equation

$$\frac{d^2 y}{dt^2} + 4\frac{dy}{dt} + 4y = \frac{du}{dt} + u, \quad y(0) = 2, \quad \dot{y}(0) = 1, \quad u(t) = e^{-4t}, \quad t \geq 0$$

Its state space form is given by

$$\dot{\mathbf{x}} = \mathbf{A}\mathbf{x} + \mathbf{b}u = \begin{bmatrix} 0 & 1 \\ -4 & -4 \end{bmatrix} \mathbf{x} + \begin{bmatrix} 0 \\ 1 \end{bmatrix} u$$

$$y = \mathbf{c}\mathbf{x} = \begin{bmatrix} 1 & 1 \end{bmatrix} \mathbf{x}$$

The initial condition for the state space variables is obtained from (5.33) as

$$\begin{bmatrix} \mathbf{c} \\ \mathbf{c}\mathbf{A} \end{bmatrix} \mathbf{x}(0) = \begin{bmatrix} y(0) \\ \dot{y}(0) \end{bmatrix} - \begin{bmatrix} 0 \\ \mathbf{c}\mathbf{b}u(0) \end{bmatrix} = \begin{bmatrix} 2 \\ 1 \end{bmatrix} - \begin{bmatrix} 0 \\ 1 \end{bmatrix}$$

leading to

$$\begin{bmatrix} 1 & 1 \\ -4 & -3 \end{bmatrix} \mathbf{x}(0) = \begin{bmatrix} 2 \\ 0 \end{bmatrix} \Rightarrow \mathbf{x}(0) = \begin{bmatrix} x_1(0) \\ x_2(0) \end{bmatrix} = \begin{bmatrix} -6 \\ 8 \end{bmatrix}$$

This means that if analog computer simulation is used to solve the above second-order differential equation, the initial conditions for integrators should be set to -6 and 8.

◇

Stabilizability and Detectability

So far we have defined and studied observability and controllability of the complete state vector. We have seen that the system is controllable (observable) if all components of the state vector are controllable (observable). The natural question to be asked is: do we really need to control and observe all state variables? In some applications, it is sufficient to take care only of the unstable components of the state vector. This leads to the definition of stabilizability and detectability.

Definition 5.1 *A linear system (continuous or discrete) is stabilizable if all unstable modes are controllable.*

Definition 5.2 *A linear system (continuous or discrete) is detectable if all unstable modes are observable.*

The concepts of stabilizability and detectability play very important roles in optimal control theory, and hence are studied in detail in advanced control theory courses. For the purpose of this course, it is enough to know their meanings.

5.6 Observer (Estimator) Design[1]

Sometimes all state space variables are not available for measurements, or it is not practical to measure all of them, or it is too expensive to measure all state space variables. In order to be able to apply the state feedback control to a system, *all of its state space variables must be available at all times.* Also, in some control system applications, one is interested in having information about system state space variables at any time instant. Thus, one is faced with the problem of estimating system state space variables. This can be done by constructing another dynamical system called the observer or estimator, connected to the system under consideration, whose role is to produce good estimates of the state space variables of the original system.

The theory of observers started with the work of Luenberger (1964, 1966, 1971) so that observers are very often called Luenberger observers. According to Luenberger, any system driven by the output of the given system can serve as an observer for that system. Two main techniques are available for observer design. The first one is used for the full-order observer design and produces an observer that has the same dimension as the original system. The second technique exploits the knowledge of some state space variables available through the output algebraic equation (system measurements) so that a reduced-order dynamic system (observer) is constructed only for estimating state space variables that are not directly obtainable from the system measurements.

5.6.1 Full-Order Observer Design

Consider a linear time invariant continuous system

$$\dot{\mathbf{x}}(t) = \mathbf{A}\mathbf{x}(t) + \mathbf{B}\mathbf{u}(t), \quad \mathbf{x}(t_0) = \mathbf{x}_0 = \text{unknown}$$
$$\mathbf{y}(t) = \mathbf{C}\mathbf{x}(t)$$
(5.34)

---

[1]   This section may be skipped without loss of continuity.

where $\mathbf{x} \in \mathfrak{R}^n$, $\mathbf{u} \in \mathfrak{R}^r$, $\mathbf{y} \in \mathfrak{R}^p$ with constant matrices $\mathbf{A}, \mathbf{B}, \mathbf{C}$ having appropriate dimensions. Since from the system (5.34) only the output variables, $\mathbf{y}(t)$, are available at all times, we may construct another artificial dynamic system of order n (built, for example, of capacitors and resistors) having the same matrices $\mathbf{A}, \mathbf{B}, \mathbf{C}$

$$\dot{\hat{\mathbf{x}}}(t) = \mathbf{A}\hat{\mathbf{x}}(t) + \mathbf{B}\mathbf{u}(t), \quad \hat{\mathbf{x}}(t_0) = \hat{\mathbf{x}}_0$$

$$\hat{\mathbf{y}}(t) = \mathbf{C}\hat{\mathbf{x}}(t)$$
(5.35)

and compare the outputs $\mathbf{y}(t)$ and $\hat{\mathbf{y}}(t)$. Of course these two outputs will be different since in the first case the system's initial condition is unknown, and in the second case it has been chosen arbitrarily. The difference between these two outputs will generate an error signal

$$\mathbf{y}(t) - \hat{\mathbf{y}}(t) = \mathbf{C}\mathbf{x}(t) - \mathbf{C}\hat{\mathbf{x}}(t) = \mathbf{C}\mathbf{e}(t) \qquad (5.36)$$

which can be used as the feedback signal to the artificial system such that the estimation (observation) error $\mathbf{e}(t) = \mathbf{x}(t) - \hat{\mathbf{x}}(t)$ is reduced as much as possible. This can be physically realized by proposing the system-observer structure as given in Figure 5.1.

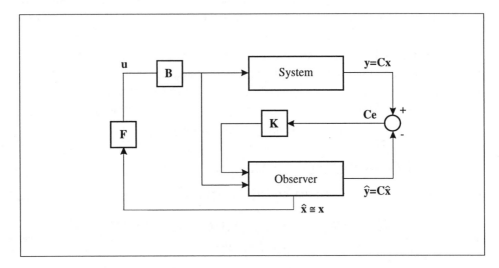

Figure 5.1: System-observer structure

In this structure $\mathbf{K}$ represents the observer gain and has to be chosen such that the observation error is minimized. The observer alone from Figure 5.1 is given by

$$\dot{\hat{\mathbf{x}}}(t) = \mathbf{A}\hat{\mathbf{x}}(t) + \mathbf{B}\mathbf{u}(t) + \mathbf{K}(\mathbf{y}(t) - \hat{\mathbf{y}}(y)) = \mathbf{A}\hat{\mathbf{x}}(t) + \mathbf{B}\mathbf{u}(t) + \mathbf{K}\mathbf{e}(t) \quad (5.37)$$

From (5.34) and (5.37) it is easy to derive an expression for dynamics of the estimation (observation) error as

$$\dot{\mathbf{e}}(t) = (\mathbf{A} - \mathbf{KC})\mathbf{e}(t) \quad (5.38)$$

If the observer gain $\mathbf{K}$ is chosen such that the feedback matrix $\mathbf{A} - \mathbf{KC}$ *is asymptotically stable*, then the estimation error $\mathbf{e}(t)$ will decay to zero for any initial condition $\mathbf{e}(t_0)$. *This can be achieved if the pair $(\mathbf{A}, \mathbf{C})$ is observable.* More precisely, by taking the transpose of the estimation error feedback matrix, i.e. $\mathbf{A}^T - \mathbf{C}^T\mathbf{K}^T$, we see that if the pair $(\mathbf{A}^T, \mathbf{C}^T)$ is controllable, then we can do whatever we want with the system, and thus we can locate its poles in arbitrarily asymptotically stable positions. Note that controllability of the pair $(\mathbf{A}^T, \mathbf{C}^T)$ is equal to observability of the pair $(\mathbf{A}, \mathbf{C})$, see expressions for the observability and controllability matrices.

In practice the observer poles should be chosen to be about ten times faster than the system poles. This can be achieved by setting the minimal real part of observer eigenvalues to be ten times bigger than the maximal real part of system eigenvalues, that is

$$|Re\{\lambda_{min}\}|_{observer} > 10|Re\{\lambda_{max}\}|_{system}$$

Theoretically, the observer can be made arbitrarily fast by pushing its eigenvalues far to the left in the complex plane, but very fast observers generate noise in the system. A procedure suggesting an efficient choice of the observer initial condition is discussed in Johnson (1988).

It is important to point out that the system-observer structure preserves the closed-loop system poles that would have been obtained if the linear perfect state feedback control had been used. The system (5.34) under the perfect state feedback control, i.e. $\mathbf{u}(t) = -\mathbf{Fx}(t)$ has the closed-loop form as

$$\dot{\mathbf{x}}(t) = (\mathbf{A} - \mathbf{BF})\mathbf{x}(t) \quad (5.39)$$

so that the eigenvalues of the matrix $\mathbf{A} - \mathbf{BF}$ are the closed-loop system poles under perfect state feedback. In the case of the system-observer structure, as given in Figure 5.1, we see that the actual control applied to both the system and the observer is given by

$$\mathbf{u}(t) = -\mathbf{F}\hat{\mathbf{x}}(t) = -\mathbf{F}\mathbf{x}(t) + \mathbf{F}\mathbf{e}(t) \tag{5.40}$$

so that from (5.34) and (5.38) we have

$$\begin{bmatrix} \dot{\mathbf{x}} \\ \dot{\mathbf{e}} \end{bmatrix} = \begin{bmatrix} \mathbf{A} - \mathbf{BF} & \mathbf{BF} \\ 0 & \mathbf{A} - \mathbf{KC} \end{bmatrix} \begin{bmatrix} \mathbf{x} \\ \mathbf{e} \end{bmatrix} \tag{5.41}$$

Since the state matrix of this system is upper block triangular, its eigenvalues are equal to the eigenvalues of matrices $\mathbf{A} - \mathbf{BF}$ and $\mathbf{A} - \mathbf{KC}$. A very simple relation among $\mathbf{x}, \mathbf{e}$, and $\hat{\mathbf{x}}$ can be written from the definition of the estimation error as

$$\begin{bmatrix} \mathbf{x} \\ \mathbf{e} \end{bmatrix} = \begin{bmatrix} \mathbf{I} & 0 \\ \mathbf{I} & -\mathbf{I} \end{bmatrix} \begin{bmatrix} \mathbf{x} \\ \hat{\mathbf{x}} \end{bmatrix} = \mathbf{T} \begin{bmatrix} \mathbf{x} \\ \hat{\mathbf{x}} \end{bmatrix} \tag{5.42}$$

Note that the matrix $\mathbf{T}$ is nonsingular. In order to go from xe-coordinates to $\mathbf{x}\hat{\mathbf{x}}$-coordinates we have to use the similarity transformation defined in (5.42), which by the main property of the similarity transformation indicates that the same eigenvalues, i.e. $\lambda(\mathbf{A} - \mathbf{BF})$ and $\lambda(\mathbf{A} - \mathbf{KC})$, are obtained in the $\mathbf{x}\hat{\mathbf{x}}$-coordinates.

This important observation that the system-observer configuration has closed-loop poles separated into the original system closed-loop poles under perfect state feedback and the actual observer closed-loop poles is known as the *separation principle*.

5.6.2 Reduced-Order Observer (Estimator)

In this section we show how to construct an observer of reduced dimension by exploiting knowledge of the output measurement equation. Assume that the output matrix $\mathbf{C}$ has rank l, which means that the output equation represents l linearly independent algebraic equations. Thus, equation

$$\mathbf{y}(t) = \mathbf{C}\mathbf{x}(t) \tag{5.43}$$

produces l equations for n unknowns of the state space vector $\mathbf{x}(t)$. Our goal is to construct an observer of order $n - l$ for estimation of the remaining $n - l$ state space variables.

The reduced-order observer design will be presented according to the results of Cumming (1969) and Gopinath (1968, 1971). The procedure for obtaining this observer is not unique, which is obvious from the next step. Assume that a matrix C_1 exists such that

$$\text{rank}\begin{bmatrix} C \\ C_1 \end{bmatrix} = n \qquad (5.44)$$

and introduce a vector $p \in \Re^l$ as

$$p(t) = C_1 x(t) \qquad (5.45)$$

From equations (5.43) and (5.45) we have

$$x(t) = \begin{bmatrix} C \\ C_1 \end{bmatrix}^{-1} \begin{bmatrix} y(t) \\ p(t) \end{bmatrix} \qquad (5.46)$$

Since the vector $p(t)$ is unknown, we will construct an observer to estimate it. Introduce the notation

$$\begin{bmatrix} C \\ C_1 \end{bmatrix}^{-1} = [L_1 \quad L_2] \qquad (5.47)$$

so that from (5.46) we get

$$x(t) = L_1 y(t) + L_2 p(t) \qquad (5.48)$$

An observer for $p(t)$ can be constructed by finding first a differential equation for $p(t)$ from (5.45), that is

$$\dot{p} = C_1 \dot{x} = C_1 Ax + C_1 Bu = C_1 AL_2 p + C_1 AL_1 y + C_1 Bu \qquad (5.49)$$

Note that from (5.49) we are not able to construct an observer for $p(t)$ since $y(t)$ does not contain explicit information about the vector $p(t)$, but if we differentiate the output variable we get from (5.34) and (5.48)

$$\dot{y} = C\dot{x} = CAx + CBu = CAL_2 p + CAL_1 y + CBu \qquad (5.50)$$

i.e. $\dot{y}(t)$ carries information about $p(t)$.

An observer for $p(t)$, according to the observer structure given in (5.37), is obtained from the last two equations as

$$\dot{\hat{p}} = C_1 AL_2 \hat{p} + C_1 AL_1 y + C_1 Bu + K_1 \left(\dot{y} - \dot{\hat{y}} \right) \tag{5.51}$$

where K_1 is the observer gain. If in equation (5.50) we replace $p(t)$ by its estimate, we will have

$$\dot{\hat{y}} = CAL_2 \hat{p} + CAL_1 y + CBu \tag{5.52}$$

so that

$$\dot{\hat{p}} = C_1 AL_2 \hat{p} + C_1 AL_1 y + C_1 Bu + K_1 (\dot{y} - CAL_2 \hat{p} - CAL_1 y - CBu) \tag{5.53}$$

Since it is impractical and undesirable to differentiate $y(t)$ in order to get $\dot{y}(t)$ (this operation introduces noise in practice), we take the change of variables

$$\hat{q} = \hat{p} - K_1 y \tag{5.54}$$

This leads to an observer for $\hat{q}(t)$ of the form

$$\dot{\hat{q}}(t) = A_q \hat{q}(t) + B_q u(t) + K_q y(t) \tag{5.55}$$

where

$$A_q = C_1 AL_2 - K_1 CL_2, \quad B_q = C_1 B - K_1 CB$$
$$K_q = C_1 AL_2 K_1 + C_1 AL_1 - K_1 CAL_1 - K_1 CAL_2 K_1 \tag{5.56}$$

It is left as an exercise to students (see Problem 5.18) to derive (5.55) and (5.56). The estimates of the original system state space variables are now obtained from (5.48) and (5.53) as

$$\hat{x}(t) = L_1 y(t) + L_2 \hat{p}(t) = L_2 \hat{q}(t) + (L_1 + L_2 K_1) y \tag{5.57}$$

The obtained system-reduced-observer structure is presented in Figure 5.2.

There are other ways of constructing the system observers (Luenberger, 1971; Chen, 1984). The reader particularly interested in observers is referred to a specialized book on observers for linear systems (O'Reilly, 1983).

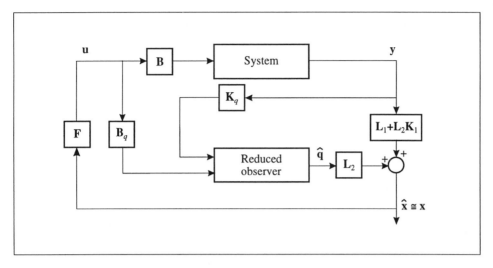

Figure 5.2: System-reduced-observer structure

5.7 MATLAB Case Study: F-8 Aircraft

In the case of high-order systems $(n > 3)$, obtaining the controllability and observability matrices is computationally very involved. The MATLAB package for computer-aided control system design and its CONTROL toolbox help to overcome this problem. Moreover, use of MATLAB enables a deeper understanding of controllability and observability concepts. Consider the following fourth-order model of an F-8 aircraft studied in Teneketzis and Sandell (1977), Khalil and Gajić (1984), Gajić and Shen (1993). The aircraft dynamics in continuous-time is described by the following matrices

$$A = \begin{bmatrix} -0.0135700 & -32.2 & -46.300 & 0.0000 \\ 0.0001200 & 0.0 & 1.214 & 0.0000 \\ -0.0001212 & 0.0 & -1.214 & 1.0000 \\ 0.0005700 & 0.0 & -9.010 & -0.6696 \end{bmatrix}$$

$$B = \begin{bmatrix} -0.433 & 0.1394 & -0.1394 & -0.1577 \end{bmatrix}^T$$

$$C = \begin{bmatrix} 0 & 0 & 0 & 1 \\ 1 & 0 & 0 & 0 \end{bmatrix}$$

By using the MATLAB function ctrb (for calculation of the controllability matrix C) and obsv (for calculation of the observability matrix $\mathcal{O}$), it can be verified that this system is both controllable and observable, namely

$$\text{rank}\, C^{4\times 4} = 4, \quad \text{rank}\, \mathcal{O}^{8\times 4} = 4$$

By using the MATLAB function det (to calculate a matrix determinant), we get

$$\det C = -6.8690$$

Since $det C$ is far from zero it seems that this system is well controllable (the controllability margin is big).

If we discretize the continuous-time matrices $\mathbf{A}$ and $\mathbf{B}$ by using the sampling period $\Delta T = 0.1$, we get a somewhat surprising result. Namely

$$\det C(\mathbf{A}_d, \mathbf{B}_d)|_{\Delta t=0.1} = -4.5381 \times 10^{-10}$$

Thus, this discrete system is almost uncontrollable. Theoretically, it is still controllable but we need an enormous amount of energy in order to control it. For example, let the initial condition be $\mathbf{x}(0) = \begin{bmatrix} 1 & 1 & 1 & 1 \end{bmatrix}$ and let the final state $\mathbf{x}(4)$ be the coordinate origin. Then, by (5.18), the control sequence that solves the problem of transferring the system from $\mathbf{x}(0)$ to $\mathbf{x}(4) = \mathbf{0}$, obtained by using MATLAB is

$$\begin{bmatrix} u(1) \\ u(2) \\ u(3) \\ u(4) \end{bmatrix} = 10^7 \times \begin{bmatrix} 3.1682 \\ -9.7464 \\ 9.9944 \\ 3.4163 \end{bmatrix}$$

Apparently, this result is unacceptable and this discrete system is practically uncontrollable.

Note that the eigenvalues of the continuous-time controllability Grammian (5.31), obtained by using the MATLAB function gram, have values 7.48×10^4, 0.42, 0.037, 0.0068. *The eigenvalues of the controllability Grammian are the best indicators of the controllability measure.* Since two of them are very close to zero, the original system is very badly conditioned from the controllability point of view even though $\det C(\mathbf{A}, \mathbf{B})$ is far from zero. The interested reader can find more about controllability and observability measures in a very comprehensive paper by Muller and Weber (1972).

5.8 MATLAB Laboratory Experiment

Part 1. The controllability staircase form of the system

$$\dot{x} = Ax + Bu$$
$$y = Cx$$

clearly distinguishes controllable and uncontrollable parts of a control system. It can be obtained by the similarity transformation, and is defined by

$$\begin{bmatrix} \dot{x}_c \\ \dot{x}_{nc} \end{bmatrix} = \begin{bmatrix} A_c & A_{12} \\ 0 & A_{nc} \end{bmatrix} \begin{bmatrix} x_c \\ x_{nc} \end{bmatrix} + \begin{bmatrix} B_c \\ 0 \end{bmatrix} u$$

$$y = [C_c \; C_{nc}] \begin{bmatrix} x_c \\ x_{nc} \end{bmatrix}$$

(5.58)

where x_c are controllable modes, and x_{nc} are uncontrollable modes. Apparently, in this structure the input u cannot influence the state variables x_{nc}; hence these are uncontrollable. Similarly, one can define the observability staircase form as

$$\begin{bmatrix} \dot{x}_o \\ \dot{x}_{no} \end{bmatrix} = \begin{bmatrix} A_o & 0 \\ A_{21} & A_{no} \end{bmatrix} \begin{bmatrix} x_o \\ x_{no} \end{bmatrix} + \begin{bmatrix} B_o \\ B_{no} \end{bmatrix} u$$

$$y = [C_o \; 0] \begin{bmatrix} x_o \\ x_{no} \end{bmatrix}$$

(5.59)

with x_o observable and x_{no} unobservable. Due to the fact that only x_o appears in the output and that x_o and x_{no} are not coupled through the state equations, the state variables x_{no} cannot be observed.

Use the MATLAB functions `ctrbf` (controllable staircase form) and `obsvf` (observable staircase form) to get the corresponding forms for the system

$$A = \begin{bmatrix} 1 & 2 & 3 & 4 & 5 & 6 \\ -1 & -2 & -3 & -4 & -5 & -6 \\ 0 & 0 & 0 & 0 & 0 & 0 \\ 1 & 1 & 1 & 1 & 1 & 1 \\ 2 & 2 & 2 & 2 & 2 & 2 \\ -1 & 0 & 2 & -1 & 0 & 2 \end{bmatrix}, \quad B = \begin{bmatrix} 1 & 1 \\ -1 & 1 \\ 1 & -1 \\ 0 & 0 \\ 1 & 2 \\ -2 & 1 \end{bmatrix}$$

$$C = [-1 \quad 1 \quad -1 \quad 0 \quad -2 \quad 2]$$

Identify the corresponding similarity transformation.

Part 2. Derive analytically that the transfer function of (5.58) is given in terms of the controllable parts, i.e. it is equal to

$$H_c(s) = C_c(sI - A_c)^{-1}B_c = H(s) = C(sI - A)^{-1}B \qquad (5.60)$$

Clarify your answer by using the MATLAB function for the transfer function ss2zp, i.e. show that both transfer functions have the same gains, poles, and zeros (subject to zero-pole cancellation).

Do the same for the observable staircase form, i.e. show that

$$H_o(s) = C_o(sI - A_o)^{-1}B_o = H(s) \qquad (5.61)$$

and justify this identity by using the MATLAB function ss2zp.

Part 3. Examine the controllability and observability of the power system composed of two interconnected areas considered in Geromel and Peres (1985) and Shen and Gajić (1990)

$$A = \begin{bmatrix}
0 & 0.55 & 0 & 0 & 0 & -5.5 & 0 & 0 & 0 \\
0 & 0 & 1 & 0 & 0 & 0 & 0 & 0 & 0 \\
0 & -3.3 & -0.05 & 6 & 0 & 3.3 & 0 & 0 & 0 \\
0 & 0 & 0 & -3.3 & 3.3 & 0 & 0 & 0 & 0 \\
0 & 0 & -5.2 & 0 & -13 & 0 & 0 & 0 & 0 \\
0 & 0 & 0 & 0 & 0 & 0 & 1 & 0 & 0 \\
0 & 3.3 & 0 & 0 & 0 & -3.3 & -0.05 & 6 & 0 \\
0 & 0 & 0 & 0 & 0 & 0 & 0 & -3.3 & 3.3 \\
0 & 0 & 0 & 0 & 0 & 0 & -5.2 & 0 & -13
\end{bmatrix}$$

$$B = \begin{bmatrix}
0 & 0 & 0 & 0 & 13 & 0 & 0 & 0 & 0 \\
0 & 0 & 0 & 0 & 0 & 0 & 0 & 0 & 13
\end{bmatrix}^T$$

$$C = \begin{bmatrix}
1 & 0.43 & 0 & 0 & 0 & 0 & 0 & 0 & 0 \\
0 & 0 & 0 & 1 & 0 & 0 & 0 & 0 & 0 \\
-1 & 0 & 0 & 0 & 0 & 0.43 & 0 & 0 & 0 \\
0 & 0 & 0 & 0 & 0 & 0 & 0 & 1 & 0
\end{bmatrix}$$

Part 4. Follow the steps used in Section 5.7, but this time for the F-15 aircraft, whose state space model was presented in Example 1.4. Consider both the subsonic and supersonic flight conditions. Comment on the results obtained.

Part 5. The controllability Grammian is defined in (5.31) as

$$\mathbf{W}(t_0, t_1) = \int_{t_0}^{t_1} e^{\mathbf{A}(t_0-\tau)} \mathbf{B}\mathbf{B}^T e^{\mathbf{A}^T(t_0-\tau)} d\tau$$

(a) Show analytically that the control input given by

$$\mathbf{u}(t) = -\mathbf{B}^T e^{\mathbf{A}^T(t_0-t)} \mathbf{W}^{-1}(t_0, t_1) \left[\mathbf{x}(t_0) - e^{\mathbf{A}(t_0-t_1)} \mathbf{x}(t_1) \right] \qquad (5.62)$$

will drive any initial state $\mathbf{x}(t_0)$ into any desired final state $\mathbf{x}(t_1) = \mathbf{x}_f$. Note that under the controllability assumption many control inputs can be found to transfer the system from the initial to the final state. The expression given in (5.62) is also known as the minimum energy control (Klamka, 1991) since in addition to driving the system from $\mathbf{x}(t_0)$ to $\mathbf{x}(t_1) = \mathbf{x}_f$, it also minimizes an integral of the square of the input (energy), $\mathbf{u}^T(t)\mathbf{u}(t)$, in the time interval (t_0, t_1).

(b) Using the MATLAB function gram, find the controllability Grammian for the system defined in Part 4 for $t_0 = 0$ and $t_1 = 1$. One of several known controllability tests states that the *system is controllable if and only if its controllability Grammian is positive definite* (Chen, 1984; Klamka, 1991). Verify whether or not the controllability Grammian for this problem is positive definite.

(c) Find the control input $\mathbf{u}(t)$ that drives the system defined in Part 4 from the initial condition $\mathbf{x}(0) = \begin{bmatrix} 0 & 0 & 0 & 0 & 0 \end{bmatrix}^T$ to the final state $\mathbf{x}(1) = \begin{bmatrix} 1 & 1 & 1 & 1 & 1 \end{bmatrix}^T$.

Part 6. By duality to the controllability Grammian, the observability Grammian is defined as

$$\mathbf{V}(t_0, t_1) = \int_{t_0}^{t_1} e^{-\mathbf{A}^T(t_0-\tau)} \mathbf{C}^T \mathbf{C} e^{-\mathbf{A}(t_0-\tau)} d\tau \qquad (5.63)$$

Note that the observability Grammian is in general a positive semidefinite matrix. It is known in the literature on observability that *if and only if the observability Grammian is positive definite, the system is observable* (Chen, 1984). Check the observability of the system given in Part 3 by using the observability Grammian test.

5.9 References

Chen, C., *Introduction to Linear System Theory*, Holt, Rinehart and Winson, New York, 1984.

Chow, J. and P. Kokotović, "A decomposition of near-optimum regulators for systems with slow and fast modes," *IEEE Transactions on Automatic Control*, vol. AC-21, 701–705, 1976.

Cumming, S., "Design of observers of reduced dynamics," *Electronic Letters*, vol. 5, 213–214, 1969.

Gajić, Z. and X. Shen, *Parallel Algorithms for Optimal Control of Large Scale Linear Systems*, Springer-Verlag, London, 1993.

Geromel, J. and P. Peres, "Decentralized load-frequency control," *IEE Proc., Part D*, vol. 132, 225–230, 1985.

Gopinath, B., *On the Identification and Control of Linear Systems*, Ph.D. Dissertation, Stanford University, 1968.

Gopinath, B., "On the control of linear multiple input–output systems," *Bell Technical Journal*, vol. 50, 1063–1081, 1971.

Johnson, C., "Optimal initial conditions for full-order observers," *International Journal of Control*, vol. 48, 857–864, 1988.

Kalman, R., "Contributions to the theory of optimal control," *Boletin Sociedad Matematica Mexicana*, vol. 5, 102–119, 1960.

Khalil, H. and Z. Gajić, "Near optimum regulators for stochastic linear singularly perturbed systems," *IEEE Transactions on Automatic Control*, vol. AC-29, 531–541, 1984.

Klamka, J., *Controllability of Dynamical Systems*, Kluwer, Warszawa, 1991.

Longhi, S. and R. Zulli, "A robust pole assignment algorithm," *IEEE Transactions on Automatic Control*, vol. AC-40, 890–894, 1995.

Luenberger, D., "Observing the state of a linear system," *IEEE Transactions on Military Electronics*, vol. 8, 74–80, 1964.

Luenberger, D., "Observers for multivariable systems," *IEEE Transactions on Automatic Control*, vol. AC-11, 190–197, 1966.

Luenberger, D., "An introduction to observers," *IEEE Transactions on Automatic Control*, vol. AC-16, 596–602, 1971.

Mahmoud, M., "Order reduction and control of discrete systems," *IEE Proc., Part D*, vol. 129, 129–135, 1982.

Muller, P. and H. Weber, "Analysis and optimization of certain qualities of controllability and observability of linear dynamical systems," *Automatica*, vol. 8, 237–246, 1972.

O'Reilly, J., *Observers for Linear Systems*, Academic Press, New York, 1983.

Petkov, P., N. Christov, and M. Konstantinov, "A computational algorithm for pole assignment of linear multiinput systems," *IEEE Transactions on Automatic Control*, vol. AC-31, 1004–1047, 1986.

Teneketzis, D. and N. Sandell, "Linear regulator design for stochastic systems by multiple time-scale method," *IEEE Transactions on Automatic Control*, vol. AC-22, 615–621, 1977.

Shen, X. and Z. Gajić, "Near optimum steady state regulators for stochastic linear weakly coupled systems," *Automatica*, vol. 26, 919-923, 1990.

5.10 Problems

5.1   Test the controllability and observability of the following systems

$$A = \begin{bmatrix} 1 & -2 \\ -1 & 0 \end{bmatrix}, \quad B = \begin{bmatrix} 1 \\ 0 \end{bmatrix}, \quad C = \begin{bmatrix} 0 & -1 \end{bmatrix}$$

$$A = \begin{bmatrix} 2 & -1 & 0 \\ 0 & 1 & 0 \\ 1 & 0 & -1 \end{bmatrix}, \quad B = \begin{bmatrix} -1 \\ 0 \\ 1 \end{bmatrix}, \quad C = \begin{bmatrix} 1 & 0 & 0 \end{bmatrix}$$

5.2   Find the values for parameters b_1, b_2, and b_3 such that the given system is controllable

$$A = \begin{bmatrix} 1 & 0 & -1 \\ 0 & 2 & 0 \\ -2 & 0 & 3 \end{bmatrix}, \quad B = \begin{bmatrix} b_1 & 0 \\ 0 & b_2 \\ b_3 & 0 \end{bmatrix}$$

5.3   Find the values for parameters c_1 and c_2 such that the following system is observable

$$A = \begin{bmatrix} -1 & 1 \\ 1 & -1 \end{bmatrix}, \quad C = \begin{bmatrix} c_1 & c_2 \end{bmatrix}$$

If the output vector of the corresponding discrete system is given by $[y(0) \ y(1)] = [1 \ 2]$, find the system's initial condition.

5.4 Verify that all columns of the matrix

$$
\mathbf{A}^3 = \begin{bmatrix} 0 & 1 & 0 \\ -2 & 3 & 1 \\ -1 & 0 & 1 \end{bmatrix}^3
$$

can be expressed as a linear combination of the columns forming matrices **I**, **A**, and $\mathbf{A}^2$ (see 5.13).

5.5 Assuming that the desired final state of a discrete system represented by

$$
\mathbf{A} = \begin{bmatrix} 0 & 1 & 0 \\ -2 & 3 & 1 \\ -1 & 0 & 1 \end{bmatrix}, \quad \mathbf{B} = \begin{bmatrix} 0 \\ 1 \\ 2 \end{bmatrix}, \quad \mathbf{x}(0) = \begin{bmatrix} 1 \\ 1 \\ 1 \end{bmatrix}
$$

is $\mathbf{x}(3) = [0 \ -1 \ 1]^T$ find the control sequence that transfers the system from $\mathbf{x}(0)$ to $\mathbf{x}(3)$.

5.6 Find a solution to Problem 5.5 in the case of a two-input system that has the input matrix

$$
\mathbf{B} = \begin{bmatrix} 0 & 1 \\ -1 & 0 \\ 0 & 2 \end{bmatrix}
$$

The remaining elements are the same as in Problem 5.5.

5.7 Determine conditions on b_1, b_2, c_1, and c_2 such that the following system is both controllable and observable

$$
\mathbf{A} = \begin{bmatrix} 1 & 1 \\ 0 & 1 \end{bmatrix}, \quad \mathbf{B} = \begin{bmatrix} b_1 \\ b_2 \end{bmatrix}, \quad \mathbf{C} = [c_1 \ c_2]
$$

Assume that the input to this system is known. Find the initial conditions of this system in terms of the given input in the case when the measured output is $y(t) = e^{-t}\cos t$.

5.8 Using the frequency domain criterion, check the joint controllability and observability of the system

$$
\mathbf{A} = \begin{bmatrix} 1 & 0 & 2 \\ -1 & 0 & 0 \\ 1 & 2 & 0 \end{bmatrix}, \quad \mathbf{B} = \begin{bmatrix} 1 \\ 0 \\ -1 \end{bmatrix}, \quad \mathbf{C} = [1 \ 1 \ 0]
$$

5.9 Find the initial conditions of all integrators in an analog computer simulation of the following differential equation

$$\frac{d^3y}{dt^3} + 3\frac{d^2y}{dt^2} + 3\frac{dy}{dt} + y = \frac{d^2u}{dt^2} + \frac{du}{dt} + u$$

$$y(0) = \frac{dy(0)}{dt} = \frac{d^2y(0)}{dt^2} = 1$$

5.10 The transfer function of a system given by

$$\frac{C(s)}{R(s)} = \frac{10(s+1)(s+3)(s+5)}{s(s+2)(s+4)(s+5)}$$

indicates that this system is either uncontrollable or unobservable. Check by the rank test, after a zero-pole cancellation takes place, that the remaining system is both controllable and observable.

5.11 A discrete model of a steam power system was considered in Mahmoud (1982) and Gajić and Shen (1993).

(a) Using MATLAB, examine the controllability and observability of this system, represented by

$$\mathbf{A}_d = \begin{bmatrix} 0.915 & 0.051 & 0.038 & 0.015 & 0.038 \\ -0.030 & 0.889 & -0.001 & 0.046 & 0.111 \\ -0.006 & 0.648 & 0.247 & 0.014 & 0.048 \\ -0.715 & -0.022 & -0.021 & 0.240 & -0.024 \\ -0.148 & -0.003 & -0.004 & 0.090 & 0.026 \end{bmatrix}$$

$$\mathbf{B}_d = \begin{bmatrix} 0.010 & 0.122 & 0.036 & 0.562 & 0.115 \end{bmatrix}^T$$

$$\mathbf{C}_d = \begin{bmatrix} 1 & 1 & 0 & 0 & 0 \\ 0 & 0 & 1 & 1 & 1 \end{bmatrix}$$

(b) Find the system transfer function and justify the answer obtained in (a).

5.12 Using MATLAB, examine the controllability of the magnetic type control system considered in Chow and Kokotović (1976)

$$\mathbf{A} = \begin{bmatrix} 0 & 0.40 & 0.000 & 0.00 \\ 0 & 0.00 & 0.349 & 0.00 \\ 0 & -5.24 & -4.65 & 2.62 \\ 0 & 0.00 & 0.000 & -10.00 \end{bmatrix}, \quad \mathbf{B} = \begin{bmatrix} 0 \\ 0 \\ 0 \\ 10 \end{bmatrix}$$

5.13 Find the controllability matrix of the system in the phase variable canonical form and show that its rank is always equal to n.

5.14 Linearize the given system at the nominal point $(x_1, x_2, u) = (1, 0, k)$ and examine system controllability and observability in terms of k

$$\dot{x}_1 = x_1^2 u, \qquad x_1(0) = 1$$
$$\dot{x}_2 = x_1 x_2 + u, \qquad x_2(0) = 0$$
$$y = x_1$$

5.15 Find the state space form of a system given by

$$\frac{d^3 y}{dt^3} + \frac{d^2 y}{dt^2} + k\frac{dy}{dt} + 2y = \frac{du}{dt} + u$$

and examine system controllability and observability in terms of k. Do they depend on the choice of the state space form?

5.16 Given a linear system described by

$$\frac{d^2 y}{dt^2} + 2\frac{dy}{dt} + y = \frac{du}{dt} + u, \quad y(0) = 0, \quad \frac{dy(0)}{dt} = 2$$

Transfer this differential equation into a state space form and determine the initial conditions for the state space variables. Can you solve this problem by using an unobservable state space form? Justify your answer.

5.17 Check that the matrix $\mathbf{A}$ given in (4.36) and the matrix $\mathbf{C} = \sqrt{\mathbf{Q}}$ defined in (4.39) form an observable pair.

5.18 Derive formulas (5.55) and (5.56) for the reduced-order observer design.

5.19 Using MATLAB, examine the controllability of a fifth-order distillation column considered in Petkov *et al.* (1986)

$$\mathbf{A} = \begin{bmatrix} -0.194 & 0.0628 & 0 & 0 & 0 \\ 1.306 & -2.132 & 0.9807 & 0 & 0 \\ 0 & 1.595 & -3.149 & 1.547 & 0 \\ 0 & 0.0355 & 2.632 & -4.257 & 1.855 \\ 0 & 0.00227 & 0 & 0.1636 & -0.1625 \end{bmatrix}$$

$$\mathbf{B} = \begin{bmatrix} 0 & 0.0632 & 0.0838 & 0.1004 & 0.0063 \\ 0 & 0 & -0.1396 & -0.2060 & -0.0128 \end{bmatrix}^T$$

5.20 Examine both the controllability and observability of the robotic manipula-tor acrobot whose state space matrices are given in Problem 3.2.

5.21 Repeat problem 5.20 for the industrial reactor defined in Problem 3.26.

5.22 Consider the state space model of the flexible beam given in Example 3.2. Find the system transfer function and determine its poles and zeros. Use Theorem 5.7 to check the controllability and observability of this linear control system.

5.23 The system matrix for a linearized model of the inverted pendulum studied in Section 1.6 is given in Section 4.2.3. Using the same data as in Section 4.2.3, the input matrix is obtained as

$$\mathbf{B} = \begin{bmatrix} 0 & 1 & 0 & -2 \end{bmatrix}^T$$

Examine the controllability of this inverted pendulum.

5.24 A system matrix of a discrete-time model of an underwater vehicle is given in Problem 4.22. Its input matrix is given by Longhi and Zulli (1995)

$$\mathbf{B} = \begin{bmatrix} 0.0258 & -0.0002 & 0 \\ 0.1023 & -0.0010 & 0 \\ -0.0001 & 0.0258 & 0 \\ -0.0008 & 0.1023 & 0 \\ 0 & 0 & 0.0055 \\ 0 & 0 & 0.0221 \end{bmatrix}$$

Check the controllability of this system.

Part II

ANALYSIS AND DESIGN

Chapter Six

Transient and Steady State Responses

In control system analysis and design it is important to consider the complete system response and to design controllers such that a satisfactory response is obtained for all time instants $t \geq t_0$, where t_0 stands for the initial time. It is known that the system response has two components: transient response and steady state response, that is

$$y(t) = y_{tr}(t) + y_{ss}(t) \tag{6.1}$$

The transient response is present in the short period of time immediately after the system is turned on. If the system is asymptotically stable, the transient response disappears, which theoretically can be recorded as

$$\lim_{t \to \infty} y_{tr}(t) = 0 \tag{6.2}$$

However, if the system is unstable, the transient response will increase very quickly (exponentially) in time, and in the most cases the system will be practically unusable or even destroyed during the unstable transient response (as can occur, for example, in some electrical networks). Even if the system is asymptotically stable, the transient response should be carefully monitored since some undesired phenomena like high-frequency oscillations (e.g. in aircraft during landing and takeoff), rapid changes, and high magnitudes of the output may occur.

Assuming that the system is asymptotically stable, then the system response in the long run is determined by its steady state component only. For control systems it is important that steady state response values are as close as possible to desired ones (specified ones) so that we have to study the corresponding errors, which represent the difference between the actual and desired system outputs at steady state, and examine conditions under which these errors can be reduced or even eliminated.

In Section 6.1 we find analytically the response of a second-order system due to a unit step input. The obtained result is used in Section 6.2 to define important parameters that characterize the system transient response. Of course, these parameters can be exactly defined and determined only for second-order systems. For higher-order systems, only approximations for the transient response parameters can be obtained by using computer simulation. Several cases of real control systems and the corresponding MATLAB simulation results for the system transient response are presented in Sections 6.3 and 6.5. The steady state errors of linear control systems are defined in Section 6.4, and the feedback elements which help to reduce the steady state errors to zero are identified. In this section we also give a simplified version of the basic linear control problem originally defined in Section 1.1. Section 6.6 presents a summary of the main control system specifications and introduces the concept of control system sensitivity function. In Section 6.7 a laboratory experiment is formulated.

Chapter Objectives

The chapter has the main objective of introducing and explaining the concepts that characterize system transient and steady state responses. In addition, system dominant poles and the system sensitivity function are introduced in this chapter.

6.1 Response of Second-Order Systems

Consider the second-order feedback system represented, in general, by the block diagram given in Figure 6.1, where K represents the system static gain and T is the system time constant. It is quite easy to find the closed-loop transfer function of this system, that is

$$M(s) = \frac{Y(s)}{U(s)} = \frac{\frac{K}{T}}{s^2 + \frac{1}{T}s + \frac{K}{T}} \tag{6.3}$$

The closed-loop transfer function can be written in the following form

$$\frac{Y(s)}{U(s)} = \frac{\omega_n^2}{s^2 + 2\zeta\omega_n s + \omega_n^2} \tag{6.4}$$

where from (6.3) and (6.4) we have

$$\zeta = \frac{1}{2\omega_n T}, \qquad \omega_n^2 = \frac{K}{T} \tag{6.5}$$

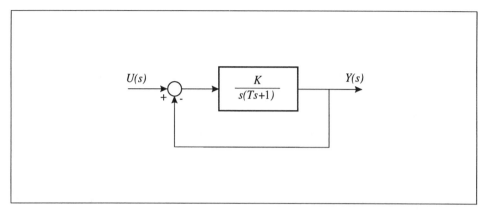

Figure 6.1: Block diagram of a general second-order system

Quantities ζ and ω_n are called, respectively, the *system damping ratio* and the *system natural frequency*. The system eigenvalues obtained from (6.4) are given by

$$\lambda_{1,2} = -\zeta\omega_n \pm j\omega_n\sqrt{1 - \zeta^2} = -\zeta\omega_n \pm j\omega_d \tag{6.6}$$

where ω_d is the *system damped frequency*. The location of the system poles and the relation between damping ratio, natural and damped frequencies are given in Figure 6.2.

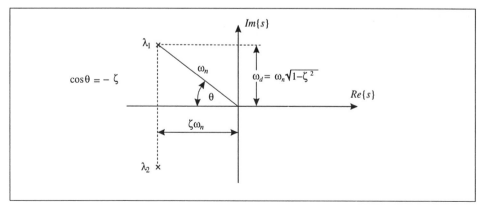

Figure 6.2: Second-order system eigenvalues in terms of parameters $\zeta, \omega_n, \omega_d$

In the following we find the closed-loop response of this second-order system due to a unit step input. Since the Laplace transform of a unit step is $1/s$ we have

$$Y(s) = \frac{w_n^2}{s(s^2 + 2\zeta w_n s + w_n^2)} \tag{6.7}$$

Depending on the value of the damping ratio ζ three interesting cases appear: (a) the critically damped case, $\zeta = 1$; (b) the over-damped case, $\zeta > 1$; and (c) the under-damped case, $\zeta < 1$. All of them are considered below. These cases are distinguished by the nature of the system eigenvalues. In case (a) the eigenvalues are multiple and real, in (b) they are real and distinct, and in case (c) the eigenvalues are complex conjugate.

(a) Critically Damped Case

For $\zeta = 1$, we get from (6.6) a double pole at $-w_n$. The corresponding output is obtained from

$$Y(s) = \frac{w_n^2}{s(s + w_n)^2} = \frac{1}{s} - \frac{1}{s + w_n} - \frac{w_n}{(s + w_n)^2}$$

which after taking the Laplace inverse produces

$$y(t) = 1 - e^{-w_n t} - w_n t e^{-w_n t} \tag{6.8}$$

The shape of this response is given in Figure 6.3a, where the location of the system poles ($\lambda_1 = p_1, \lambda_2 = p_2$) is also presented.

(b) Over-Damped Case

For the over-damped case, we have two real and asymptotically stable poles at $-\zeta w_n \pm w_d$. The corresponding closed-loop response is easily obtained from

$$Y(s) = \frac{1}{s} + \frac{k_1}{s + \zeta w_n + w_d} + \frac{k_2}{s + \zeta w_n - w_d}$$

as

$$y(t) = 1 + k_1 e^{-(\zeta w_n + w_d)t} + k_2 e^{-(\zeta w_n - w_d)t} \tag{6.9}$$

It is represented in Figure 6.3b.

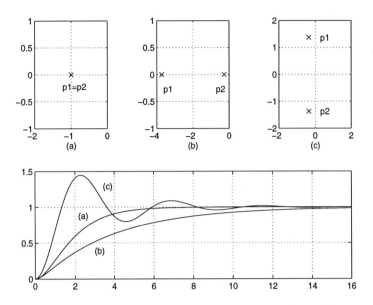

Figure 6.3: Responses of second-order systems and locations of system poles

(c) Under-Damped Case

This case is the most interesting and important one. The system has a pair of complex conjugate poles so that in the s-domain we have

$$Y(s) = \frac{k_1}{s} + \frac{k_2}{s + \zeta\omega_n + j\omega_d} + \frac{k_2^*}{s + \zeta\omega_n - j\omega_d} \qquad (6.10)$$

Applying the Laplace transform it is easy to show (see Problem 6.1) that the system output in the time domain is given by

$$y(t) = 1 + \frac{e^{-\zeta\omega_n t}}{\sqrt{1 - \zeta^2}} \sin\left[\left(\omega_n \sqrt{1 - \zeta^2}\right)t - \theta\right] \qquad (6.11)$$

where from Figure 6.2 we have

$$\cos\theta = -\zeta, \quad \sin\theta = \sqrt{1 - \zeta^2}, \quad \tan\theta = \frac{\sqrt{1 - \zeta^2}}{-\zeta} \qquad (6.12)$$

The response of this system is presented in Figure 6.3c.

The under-damped case is the most common in control system applications. A magnified figure of the system step response for the under-damped case is presented in Figure 6.4. It will be used in the next section in order to define the transient response parameters. These parameters are important for control system analysis and design.

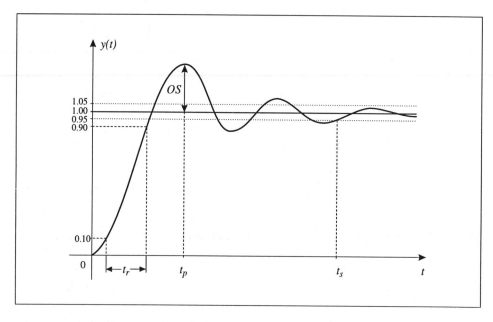

Figure 6.4: Response of an under-damped second-order system

6.2 Transient Response Parameters

The most important transient response parameters are denoted in Figure 6.4. These parameters are: response overshoot, settling time, peak time, and rise time.

The response overshoot can be obtained by finding the maximum of the function $y(t)$, as given by (6.11), with respect to time. This leads to

$$\frac{dy(t)}{dt} = -\frac{\zeta\omega_n}{\sqrt{1-\zeta^2}}e^{-\zeta\omega_n t}\sin{(\omega_d t - \theta)} + \frac{\omega_d}{\sqrt{1-\zeta^2}}e^{-\zeta\omega_n t}\cos{(\omega_d t - \theta)} = 0$$

or

$$\zeta\omega_n \sin{(\omega_d t - \theta)} - \omega_d \cos{(\omega_d t - \theta)} = 0$$

which by using relations (6.12) and Figure 6.2 implies

$$\sin \omega_d t = 0 \tag{6.13}$$

It is left as an exercise to students to derive (6.13) (see Problem 6.2). From this equation we have

$$\omega_d t = i\pi, \quad i = 0, 1, 2, ... \tag{6.14}$$

The *peak time* is obtained for $i = 1$, i.e. as

$$t_p = \frac{\pi}{\omega_d} = \frac{\pi}{\omega_n \sqrt{1 - \zeta^2}} \tag{6.15}$$

and times for other minima and maxima are given by

$$t_{ip} = \frac{i\pi}{\omega_d} = \frac{i\pi}{\omega_n \sqrt{1 - \zeta^2}}, \quad i = 2, 3, 4, ... \tag{6.16}$$

Since the steady state value of $y(t)$ is $y_{ss}(t) = 1$, it follows that the *response overshoot* is given by

$$OS = y(t_p) - y_{ss}(t) = 1 + e^{-\zeta \omega_n t_p} - 1 = e^{-\zeta \omega_n t_p} = e^{-\frac{\zeta \pi}{\sqrt{1 - \zeta^2}}} \tag{6.17}$$

Overshoot is very often expressed in percent, so that we can define the *maximum percent overshoot* as

$$MPOS = OS(\%) = e^{-\frac{\zeta \pi}{\sqrt{1 - \zeta^2}}} \frac{1}{100}(\%) \tag{6.18}$$

From Figure 6.4, the expression for the response 5 percent *settling time* can be obtained as

$$y(t_s) = 1 + \frac{e^{-\zeta \omega_n t_s}}{\sqrt{1 - \zeta^2}} = 1.05 \tag{6.19}$$

which for the standard values of ζ leads to

$$t_s = -\frac{1}{\zeta \omega_n} \ln \left(0.05 \sqrt{1 - \zeta^2} \right) \approx \frac{3}{\zeta \omega_n} \tag{6.20}$$

Note that in practice $0.5 < \zeta < 0.8$.

The response *rise time* is defined as the time required for the unit step response to change from 0.1 to 0.9 of its steady state value. The rise time is inversely proportional to the system bandwidth, i.e. the wider bandwidth, the smaller the rise time. However, designing systems with wide bandwidth is costly, which indicates that systems with very fast response are expensive to design.

Example 6.1: Consider the following second-order system

$$\frac{Y(s)}{U(s)} = \frac{4}{s^2 + 2s + 4}$$

Using (6.4) and (6.5) we get

$$\omega_n^2 = 4 \Rightarrow \omega_n = 2\,\text{rad/s}, \quad 2\zeta\omega_n = 2 \Rightarrow \zeta = 0.5$$

$$\omega_d = \omega_n\sqrt{1 - \zeta^2} = \sqrt{3}\ \text{rad/s}$$

The peak time is obtained from (6.15) as

$$t_p = \frac{\pi}{\omega_d} = \frac{\pi}{\sqrt{3}} = 1.82\,\text{s}$$

and the settling time, from (6.20), is found to be

$$t_s \approx \frac{3}{\zeta\omega_n} = 3\,\text{s}$$

The maximum percent overshoot is equal to

$$MPOS = e^{-\frac{\zeta\pi}{\sqrt{1-\zeta^2}}} 100(\%) = 16.3\%$$

The step response of this system obtained by the MATLAB function `[y,x]=step(num,den,t)` with `t=0:0.1:5` is presented in Figure 6.5. It can be seen that the analytically obtained results agree with the results presented in Figure 6.5. From Figure 6.5 we are able to estimate the rise time, which in this case is approximately equal to $t_r \approx 0.8\,\text{s}$.

Note that the response rise time can be very precisely determined by using MATLAB (see Problem 6.15). Also, MATLAB can be used to find accurately the transient response settling time (see Problem 6.14).

◇

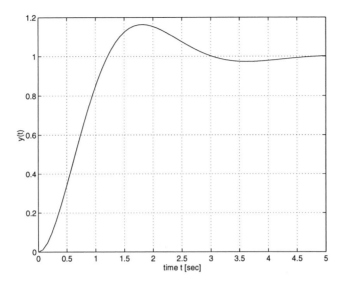

Figure 6.5: System step response for Example 6.1

6.3 Transient Response of High-Order Systems

In the previous section we have been able to precisely define and determine parameters that characterize the system transient response. This has been possible due to the fact that the system under consideration has been of order two only. For higher-order systems, analytical expressions for the system response are not generally available. However, in some cases of high-order systems one is able to determine approximately the transient response parameters.

A particularly important is the case in which an asymptotically stable system has a pair of complex conjugate poles (eigenvalues) much closer to the imaginary axis than the remaining poles. This situation is represented in Figure 6.6. The system poles far to the left of the imaginary axis have large negative real parts so that they decay very quickly to zero (as a matter of fact, they decay exponentially with $e^{\sigma_i t}$, where σ_i are negative real parts of the corresponding poles). Thus, the system response is dominated by the pair of complex conjugate poles closest to the imaginary axis since they decay slowest, as they have relatively small real parts. Hence, these poles are called the *dominant system poles*.

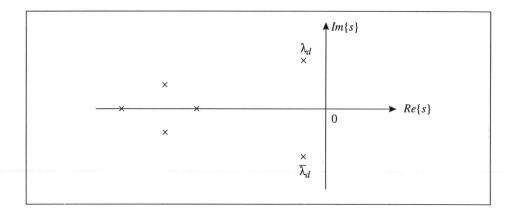

Figure 6.6: Complex conjugate dominant system poles

This analysis can be also justified by using the closed-loop system transfer function. Consider, for example, a system described by its transfer function as

$$M(s) = \frac{Y(s)}{U(s)} = \frac{12600(s+1)}{(s+3)(s+10)(s+60)(s+70)}$$

Since the poles at –60 and –70 are far to the left, their contribution to the system response is negligible (they decay very quickly to zero as e^{-60t} and e^{-70t}). The transfer function can be formally simplified as follows

$$M(s) = \frac{12600(s+1)}{(s+3)(s+10)60\left(\frac{s}{60}+1\right)70\left(\frac{s}{70}+1\right)}$$

$$\approx \frac{3(s+1)}{(s+3)(s+10)} = M_r(s)$$

(6.21)

Example 6.2: In this example we use MATLAB to compare the step responses of the original and reduced-order systems whose transfer functions are given in (6.21). The results obtained for $y(t)$ and $y_r(t)$ are given in Figure 6.7. It can be seen from this figure that step responses for the original and reduced-order (approximate) systems almost overlap.

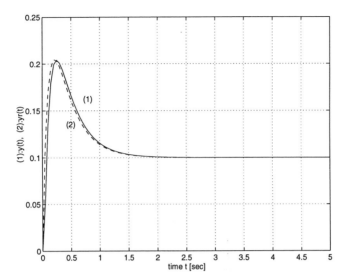

Figure 6.7: System step responses for the original
(1) and reduced-order approximate (2) systems

The corresponding responses are obtained by the following sequence of
MATLAB functions

```
z=-1;
p=[-3 -10 -60 -70];
k=12600;
[num,den]=zp2tf(z,p,k);
t=0:0.05:5;
[y,x]=step(num,den,t);
zr=-1;
pr=[-3 -10];
kr=3;
[numr,denr]=zp2tf(zr,pr,kr);
[yr,xr]=step(numr,denr,t);
plot(t,y,t,yr,'- -');
xlabel('time t [sec]');
ylabel('(1):y(t), (2):yr(t)');
grid;
text(0.71,0.16,'(1)');
```

```
text(0.41,0.13,'(2)');
```

◇

Similarly one can neglect the complex conjugate non-dominant poles, as is demonstrated in the next example.

Example 6.3: Consider the following transfer function containing two pairs of complex conjugate poles

$$M(s) = \frac{20(s+2)}{(s+1-j)(s+1+j)(s+10-j10)(s+10+j10)}$$

and the corresponding approximate reduced-order transfer function obtained by

$$M(s) = \frac{20(s+2)}{(s^2+2s+2)(s^2+20s+200)}$$

$$= \frac{20(s+2)}{(s^2+2s+2)200\left(\frac{s^2}{200}+\frac{20s}{200}+1\right)} \approx \frac{(s+2)}{10(s^2+2s+2)} = M_r(s)$$

The step responses of the original and approximate reduced-order systems are presented in Figure 6.8.

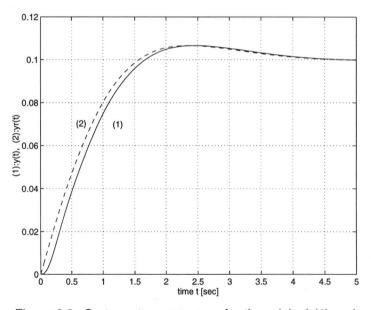

Figure 6.8: System step responses for the original (1) and approximate (2) systems with complex conjugate poles

It can be seen from this figure that a very good approximation for the step response is obtained by using the approximate reduced-order model.

◇

However, the above technique is rather superficial. In addition, for multi-input multi-output systems this procedure becomes computationally cumbersome. In that case we need a more systematic method. In the control literature one is able to find several techniques used for the system order reduction. One of them, the *method of singular perturbations* (Kokotović and Khalil, 1986; Kokotović *et al.*, 1986), is presented below. The method systematically generalizes the previously explained idea of dominant poles.

The eigenvalues of certain systems (having large and small time constants, or slow and fast system modes) are clustered in two or several groups (see Figure 6.9). According to the theory of singular perturbations, if it is possible to find an isolated group of poles (eigenvalues) closest to the imaginary axis, then the system response will be predominantly determined by that group of eigenvalues.

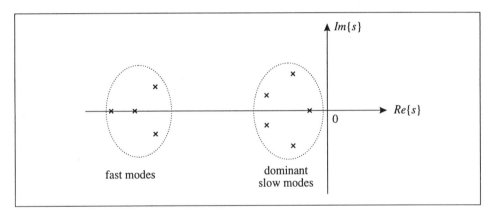

Figure 6.9: System eigenvalues clustered in two disjoint groups

The state space form of such systems is given by

$$\begin{bmatrix} \dot{x}_1 \\ \dot{x}_2 \end{bmatrix} = \begin{bmatrix} A_1 & A_2 \\ \frac{1}{\epsilon}A_3 & \frac{1}{\epsilon}A_4 \end{bmatrix} \begin{bmatrix} x_1 \\ x_2 \end{bmatrix} + \begin{bmatrix} B_1 \\ \frac{1}{\epsilon}B_2 \end{bmatrix} u \qquad (6.22)$$

$$y = C_1 x_1 + C_2 x_2$$

where ϵ is a small positive parameter. It indicates that the time derivatives for state variables x_2 are large, so that variables x_2 change quickly, in contrast to variables x_1, which are slow. If the state variables x_2 are asymptotically stable, then they decay very quickly, so that after the fast dynamics disappear ($\dot{x}_2 = 0$), we get an approximation for the fast subsystem as

$$0 = \mathbf{A}_3 \mathbf{x}_{1app} + \mathbf{A}_4 \mathbf{x}_{2app} + \mathbf{B}_2 \mathbf{u} \tag{6.23}$$

From this equation we are able to find x_{2app} (assuming that the matrix $\mathbf{A}_4$ in nonsingular, which is the standard assumption in the theory of singular perturbations; Kokotović et al., 1986) as

$$\mathbf{x}_{2app} = -\mathbf{A}_4^{-1}(\mathbf{A}_3 \mathbf{x}_{1app} + \mathbf{B}_2 \mathbf{u}) \tag{6.24}$$

Substituting this approximation in (6.22), we get an approximate reduced-order slow subsystem as

$$\dot{\mathbf{x}}_{1app} = \mathbf{A}_s \mathbf{x}_{1app} + \mathbf{B}_s \mathbf{u}$$
$$\mathbf{y}_{app} = \mathbf{C}_s \mathbf{x}_{1app} + \mathbf{D}_s \mathbf{u}$$

$$\tag{6.25}$$

$$\mathbf{A}_s = \mathbf{A}_1 - \mathbf{A}_2 \mathbf{A}_4^{-1} \mathbf{A}_3, \quad \mathbf{B}_s = \mathbf{B}_1 - \mathbf{A}_2 \mathbf{A}_4^{-1} \mathbf{B}_2$$
$$\mathbf{C}_s = \mathbf{C}_1 - \mathbf{C}_2 \mathbf{A}_4^{-1} \mathbf{A}_3, \quad \mathbf{D}_s = -\mathbf{C}_2 \mathbf{A}_4^{-1} \mathbf{B}_2$$

From the theory of singular perturbations it is known that $x_1(t)$ is close to $x_{1app}(t)$ for every $t \geq t_0$, and $y_{app}(t)$ is a good approximation for $y(t)$ for $t \geq t_1 > t_0$, where $t \geq t_1$ indicates the fact that this approximation becomes valid shortly after the fast transient disappears (Kokotović et al., 1986).

Example 6.4: Consider a mathematical model of a singularly perturbed fluid catalytic cracker considered in Arkun and Ramakrishnan (1983). The problem matrices are given by

$$\mathbf{A} = \begin{bmatrix} -16.11 & -0.39 & 27.2 & 0 & 0 \\ 0.01 & -16.99 & 0 & 0 & 12.47 \\ 15.11 & 0 & -53.6 & -16.57 & 71.78 \\ -53.36 & 0 & 0 & -107.2 & 232.11 \\ 2.27 & 69.1 & 0 & 2.273 & -102.99 \end{bmatrix}$$

$$\mathbf{B}^T = \begin{bmatrix} 11.12 & -3.61 & -21.91 & -53.6 & 69.1 \\ -12.6 & 3.36 & 0 & 0 & 0 \end{bmatrix}$$

$$\mathbf{C} = \begin{bmatrix} 0 & 0 & 0 & 0 & 1 \\ 0 & 1 & 0 & 0 & 0 \end{bmatrix}$$

The eigenvalues of this system are

$$\lambda(\mathbf{A}) = \{-2.85, -7.78, -74.32, -82.86, -129.08\}$$

which indicates that the system has two slow (–2.85 and –7.78) and three fast modes. The small parameter ϵ represents the separation of system eigenvalues into two disjoint groups. It can be roughly estimated as $\epsilon \approx 7.78/74.32 \approx 0.1$ (the ratio of the smallest and largest eigenvalues in the given slow and fast subsets). We use MATLAB to partition matrices $\mathbf{A}, \mathbf{B}, \mathbf{C}$ as follows

```
eps=0.1;
A1=A(1:2,1:2);
A2=A(1:2,3:5);
A3=A(3:5,1:2)*eps;
A4=A(3:5,3:5)*eps;
B1=B(1:2,1:2);
B2=B(3:5,1:2)*eps;
C1=C(1:2,1:2);
C2=C(1:2,3:5);
```

The slow subsystem matrices, obtained from (6.25), are given by

$$\mathbf{A}_s = \begin{bmatrix} -4.0452 & 12.4474 \\ 0.1548 & -8.2035 \end{bmatrix}, \quad \mathbf{B}_s = \begin{bmatrix} 16.8321 & -12.6 \\ 5.0320 & 3.36 \end{bmatrix}$$

$$\mathbf{C}_s = \begin{bmatrix} 0.0116 & 0.7046 \\ 0 & 1 \end{bmatrix}, \quad \mathbf{D}_s = \begin{bmatrix} 0.693 & 0 \\ 0 & 0 \end{bmatrix}$$

The eigenvalues of the slow subsystem matrix are $\lambda(\mathbf{A}_s) = \{-3.6245, -8.6243\}$. This reflects the impact of the fast modes on the slow modes so that the original slow eigenvalues located at –2.85 and –7.78 are now changed to –3.6245 and –8.6243. In Figure 6.10 the outputs of the original (solid lines) and reduced (dashed lines) systems are presented in the time interval specified by MATLAB as $\mathtt{t=0:0.025:5}$. It can be seen that the output responses of these systems are remarkably close to each other.

◇

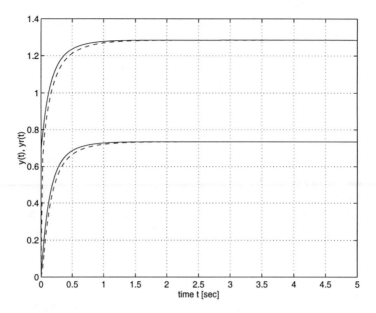

Figure 6.10: Outputs of the original fifth-order system and reduced
second-order system obtained by the method of singular perturbations

Models of many real physical linear control systems that have the singularly
perturbed structure, displaying slow and fast state variables, can be found in
Gajić and Shen (1993).

A MATLAB laboratory experiment involving system order reduction and
comparison of corresponding system trajectories and outputs of a real physical
control system by using the method of singular perturbations is formulated in
Section 6.7.

6.4 Steady State Errors

The response of an asymptotically stable linear system is in the long run determined by its steady state component. During the initial time interval the transient
response decays to zero, according to the asymptotic stability requirement (6.2),
so that in the remaining part of the time interval the system response is represented by its steady state component only. Control engineers are interested in
having steady state responses as close as possible to the desired ones so that we

define the so-called steady state errors, which represent the differences at steady state of the actual and desired system responses (outputs).

Before we proceed to steady state error analysis, we introduce a simplified version of the basic linear control system problem defined in Section 1.1.

Simplified Basic Linear Control Problem

As defined in Section 1.1 the basic linear control problem is still very difficult to solve. A simplified version of this problem can be formulated as follows. Apply to the system input a time function equal to the desired system output. This time function is known as the system's *reference input* and is denoted by $r(t)$. Note that $r(t) = u(t)$. Compare the actual and desired outputs by feeding back the actual output variable. The difference $y(t) - r(t) = e(t)$ represents the error signal. Use the error signal together with simple controllers (if necessary) to drive the system under consideration such that $e(t)$ is reduced as much as possible, at least at steady state. If a simple controller is used in the feedback loop (Figure 6.11) the error signal has to be slightly redefined, see formula (6.26).

In the following we use this simplified basic linear control problem in order to identify the structure of controllers (feedback elements) that for certain types of reference inputs (desired outputs) produce zero steady state errors.

Consider the simplest feedback configuration of a single-input single-output system given in Figure 6.11.

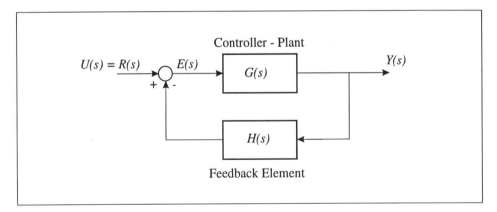

Figure 6.11: Feedback system and steady state errors

Let the input signal $U(s) = R(s)$ represent the Laplace transform of the desired output (in this feedback configuration the desired output signal is used as an input signal); then for $H(s) = 1$, we see that in Figure 6.11 the quantity $E(s)$ represents the difference between the desired output $R(s) = U(s)$ and the actual output $Y(s)$. In order to be able to reduce this error as much as possible, we allow dynamic elements in the feedback loop. Thus, $H(s)$ as a function of s has to be chosen such that for the given type of reference input, the error, now defined by

$$E(s) = R(s) - H(s)Y(s) \tag{6.26}$$

is eliminated or reduced to its minimal value at steady state.

From the block diagram given in Figure 6.11 we have

$$E(s) = R(s) - H(s)G(s)E(s)$$

so that the expression for the error is given by

$$E(s) = \frac{R(s)}{1 + H(s)G(s)} \tag{6.27}$$

The steady state error component can be obtained by using the final value theorem of the Laplace transform as

$$e_{ss} = \lim_{t \to \infty} e(t) = \lim_{s \to 0} \{sE(s)\} = \lim_{s \to 0} \left\{ \frac{sR(s)}{1 + H(s)G(s)} \right\} \tag{6.28}$$

This expression will be used in order to determine the nature of the feedback element $H(s)$ such that the steady state error is reduced to zero for different types of desired outputs. We will particularly consider step, ramp, and parabolic functions as desired system outputs.

Before we proceed to the actual steady state error analysis, we introduce one additional definition.

Definition 6.1 *The type of feedback control system* is determined by the number of poles of the open-loop feedback system transfer function located at the origin, i.e. it is equal to j, where j is obtained from

$$G(s)H(s) = \frac{K(s + z_1) \cdots (s + z_m)}{s^j (s + p_1)(s + p_2) \cdots (s + p_{n-j})} \tag{6.29}$$

Now we consider the steady state errors for different desired outputs, namely unit step, unit ramp, and unit parabolic outputs.

Unit Step Function as Desired Output

Assuming that our goal is that the system output follows as close as possible the unit step function, i.e. $U(s) = R(s) = 1/s$, we get from (6.28)

$$e_{ss} = \lim_{s \to 0} \left\{ \frac{s}{1 + H(s)G(s)} \frac{1}{s} \right\} = \frac{1}{1 + \lim_{s \to 0} \{H(s)G(s)\}} = \frac{1}{1 + K_p} \qquad (6.30)$$

where K_p is known as the *position constant* and from (6.30) is given by

$$K_p = \lim_{s \to 0} \{H(s)G(s)\} \qquad (6.31)$$

It can be seen from (6.30) that the steady state error for the unit step reference is reduced to zero for $K_p = \infty$. Examining closely (6.31), taking into account (6.29), we see that this condition is satisfied for $j \geq 1$.

Thus, we can conclude that the feedback type system of order at least one allows the system output at steady state to track the unit step function perfectly.

Unit Ramp Function as Desired Output

In this case the steady state error is obtained as

$$e_{ss} = \lim_{s \to 0} \{sE(s)\} = \lim_{s \to 0} \left\{ \frac{s}{1 + H(s)G(s)} \frac{1}{s^2} \right\} = \frac{1}{\lim_{s \to 0} \{sH(s)G(s)\}} = \frac{1}{K_v} \qquad (6.32)$$

where

$$K_v = \lim_{s \to 0} \{sH(s)G(s)\} \qquad (6.33)$$

is known as the *velocity constant*. It can be easily concluded from (6.29) and (6.33) that $K_v = \infty$, i.e. $e_{ss} = 0$ for $j \geq 2$. Thus, systems having two and more pure integrators ($1/s$ terms) in the feedback loop will be able to perfectly track the unit ramp function as a desired system output.

Unit Parabolic Function as Desired Output

For a unit parabolic function we have $R(s) = 2/s^3$ so that from (6.28)

$$e_{ss} = \lim_{s \to 0} \left\{ \frac{s}{1 + H(s)G(s)} \frac{2}{s^3} \right\} = \frac{2}{\lim_{s \to 0} \{s^2 H(s)G(s)\}} = \frac{2}{K_a} \qquad (6.34)$$

where the so-called *acceleration constant*, K_a, is defined by

$$K_a = \lim_{s \to 0} \{s^2 H(s)G(s)\} \qquad (6.35)$$

From (6.29) and (6.35), we conclude that $K_a = \infty$ for $j \geq 3$, i.e. the feedback loop must have three pure integrators in order to reduce the corresponding steady state error to zero.

Example 6.5: The steady state errors for a system that has the open-loop transfer function as

$$H(s)G(s) = \frac{20(s+1)}{s(s+2)(s+5)}$$

are

$$K_p = \infty \Rightarrow e_{ss} = 0 \quad \text{(step)}$$
$$K_v = 2 \quad \Rightarrow e_{ss} = 0.5 \quad \text{(ramp)}$$
$$K_a = 0 \quad \Rightarrow e_{ss} = \infty \quad \text{(parabolic)}$$

Since the open-loop transfer function of this system has one integrator the output of the closed-loop system can perfectly track only the unit step.

◇

Example 6.6: Consider the second-order system whose open-loop transfer function is given by

$$H(s)G(s) = \frac{(s+3)}{(s+1)(s+2)}$$

The position constant for this system is $K_p = 1.5$ so that the corresponding steady state error is

$$e_{ss} = \frac{1}{1+K_p} = \frac{1}{1+1.5} = 0.4$$

The unit step response of this system is presented in Figure 6.12, from which it can be clearly seen that the steady state output is equal to 0.6; hence the steady state error is equal to $1 - 0.6 = 0.4$.

◇

Note that the transient analysis and the study of steady state errors can be performed for discrete-time linear systems in exactly the same way as was used for continuous-time systems. The steady state errors for discrete-time systems are obtained by using the final value theorem of the $\mathcal{Z}$-transform and following the same procedure as in Section 6.4.

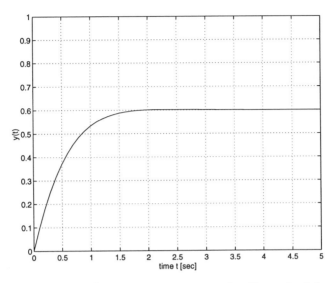

Figure 6.12: System step response for Example 6.6

6.5 Response of High-Order Systems by MATLAB

For high-order systems, analytical expressions for system step responses are quite complex. However, we are still able to determine approximately the response parameters in many cases. In this section, we plot the unit step response of a high-order control system by using MATLAB and determine approximately from the graph obtained some of transient response parameters and the corresponding steady state error.

Consider the mathematical model of a synchronous machine connected to an infinite bus. The matrix $\mathbf{A}$ of this seventh-order system is given in Problem 3.28. The remaining matrices are chosen as

$$\mathbf{B} = [0 \ \ 0 \ \ 0 \ \ 0 \ \ 0 \ \ 0 \ \ 20]^T$$
$$\mathbf{C} = [1 \ \ 1 \ \ 1 \ \ 1 \ \ 1 \ \ 1 \ \ 1], \quad \mathbf{D} = 0$$

For a system represented in the state space form, the step response is obtained by using the MATLAB function `[y,x]=step(A,B,C,D,1,t)`, where 1 indicates that the step signal is applied to the first system input and t represents time. The step response of this system is given in Figures 6.13 and 6.14.

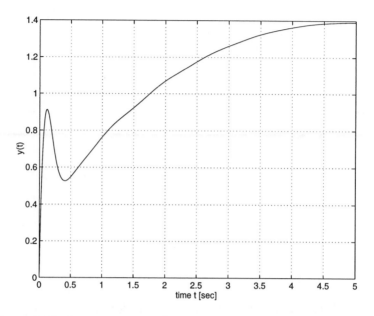

Figure 6.13: Step response of a synchronous machine for $t \in [0, 5]$

Figure 6.13 is obtained for the initial time interval of $t \in [0, 5]$. It shows the actual response shape, but it is hard to draw conclusions about the transient response parameters from this figure. However, if we plot the system step response for time interval $t \in [0, 40]$, then a response shape very similar to that in Figure 6.4 is obtained. It is pretty straightforward to read from Figure 6.14 that the peak time is $t_p \approx 5$ s, the overshot is approximately equal to 0.4, the rise time is $t_r \approx 2$ s, and the settling time is roughly equal to 12 s. By using MATLAB, it is obtained that $y_{ss} \to 1.0226$ so that the steady state error is $e_{ss} \to 0.0226$. This can be obtained either by finding $y(t)$ for some t long enough or by using the final value theorem of the Laplace transform as

$$y_{ss} = \lim_{s \to 0} \{sY(s)\} = \lim_{s \to 0} \{sM(s)U(s)\} = \lim_{s \to 0} \left\{ sM(s)\frac{1}{s} \right\} = M(0) \quad (6.36)$$

where $M(s)$ is the system closed-loop transfer function, which can be obtained by using MATLAB as [num, den]=ss2tf(A,B,C,D,1). Then, for this particular example of order seven, we have yss=num(1,8)/den(1,8). Note that num(1,8)=5048.8 and den(1,8)=4937.2.

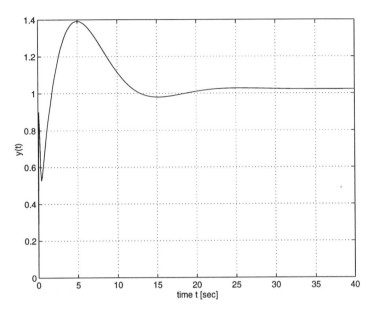

Figure 6.14: Step response of a synchronous machine for $t \in [0, 40]$

6.6 Control System Performance Specifications

Control systems should satisfy certain specifications such that systems under consideration have the desired behavior for both transient and steady state responses. If the desired specifications are not met, controllers should be designed and placed either in the forward path or in the feedback loop such that the desired specifications are obtained. The desired specifications include the required values (or upper and/or lower limits) of already defined quantities such as phase and gain margins, settling time, rise time, peak time, maximum percent overshoot, and steady state errors. Additional specifications can be defined in the frequency domain like control system frequency bandwidth, resonant frequency, and resonance peak, which will be presented in Chapter 9.

Of course, it is impossible to meet all the specifications mentioned above. Sometimes some requirements are contradictory and sometimes some of them are not affordable. Thus, control engineers have to compromise while trying to satisfy all of imposed control system requirements. Fortunately, we are able to identify the most important ones. First of all, *systems must be stable*; hence the main goal of controller design is to stabilize the system under consideration, in other words, the system phase and gain stability margins should be handled with increased care. Secondly, *systems should have limited overshoot and settling time* and *the steady state errors should be kept within admissible bounds*. In the most of cases only these specifications will be taken into account while designing controllers in Chapters 8 and 9.

In addition to the above specifications control systems should be insensitive to variation of system parameters and components. Linear models are very often obtained by performing linearization of nonlinear models, i.e. the linear models are in many cases just approximations of nonlinear systems at given operating points. That is why it is required that controllers used for control of such systems be robust, i.e. they should produce satisfactory results for broad families of linear systems that are close to linearized systems at given operating points. The importance of control system sensitivity to parameter changes has been recognized since the beginning of modern control theory (Tomović, 1963; Kokotović and Rutman, 1965; Tomović and Vukobratović, 1972). Control system robustness has been the trend of the eighties and nineties (Morari and Zafiriou, 1989; Chiang and Safonov, 1992; Grimble, 1994; Green and Limebeer, 1995). Studying these control system specifications (reduced sensitivity and increased robustness) in detail is beyond the scope of this book. Here, we just introduce the basic system sensitivity result and define the control system sensitivity function.

Consider the feedback control system given in Figure 6.11. The plant transfer function $G(s)$ is obtained through mathematical modeling either analytically or experimentally and is assumed to be known. However, due to plant parameter changes, e.g. due to components aging or parameter uncertainties, the actual plant transfer function is in fact $G_a(s) = G(s) + \Delta G(s)$, where $\Delta G(s)$ represents the absolute error of the plant transfer function, so that the corresponding relative error is $\Delta G(s)/G(s)$.

The closed-loop transfer function for the system in Figure 6.11 is given by

$$M(s) = \frac{G(s)}{1 + G(s)H(s)} \qquad (6.37)$$

and the actual closed-loop system transfer function is

$$M_a(s) = \frac{G_a(s)}{1 + G_a(s)H(s)} \tag{6.38}$$

The corresponding absolute transfer function error is obtained as

$$\Delta M(s) = M_a(s) - M(s) = M(s)\left(\frac{M_a(s)}{M(s)} - 1\right)$$

$$= M(s)\frac{1}{1 + G_a(s)H(s)}\frac{G_a(s) - G(s)}{G(s)} \tag{6.39}$$

This leads to

$$\frac{\Delta M(s)}{M(s)} = \frac{1}{1 + G_a(s)H(s)}\frac{\Delta G(s)}{G(s)} = S_a(s)\frac{\Delta G(s)}{G(s)} \tag{6.40}$$

where

$$S_a(s) = \frac{1}{1 + G_a(s)H(s)} \tag{6.41}$$

represents the so-called control *system sensitivity function*. Note that the sensitivity function depends on the complex frequency s. It follows from formula (6.40) that the magnitude of the sensitivity function should be chosen to be as small as possible over the frequency range of interest. Since from (6.41) $|S_a(s)| < 1$, it follows that the closed-loop relative transfer function error is reduced compared to the open-loop relative plant transfer function error. In conclusion, *feedback alone reduces system sensitivity to system parameter variations*.

Finally, let us point out that feedback also decreases system sensitivity to external disturbances. This problem has been already tacitly studied in Section 2.2—see the block diagram presented in Figure 2.3 and formula (2.18).

6.7 MATLAB Laboratory Experiment

Part 1. Consider a general second-order system given in (6.3). Choose values for parameters K and T such that all three cases appear (over-damped, under-damped, and critically damped). Using MATLAB, plot the unit step responses for all cases. Find the transient response parameter for the under-damped case.

Part 2. Consider the second-order system as given by (6.3) with $T = 1$. Take several values for the static gain K such that $1 = K_1 < K_2 < K_3 < K_4 = 50$ and plot the corresponding unit step responses. Draw conclusions about the impact of K on the maximum percent overshoot and the steady state errors.

Part 3. Use the method of dominant complex conjugate poles in order to approximate the step response for the second output of the F-8 aircraft, given in Section 3.5.2, by an equivalent second-order system. Hint: Find the fourth-order transfer function for the second output and reduce it to the second-order transfer function by following the procedure of Example 6.3. Note that the same reduction technique has to be applied to the transfer function zeros. In that respect eliminate the pair of complex conjugate zeros.

Part 4. Use MATLAB in order to find approximately the transient response parameters and the steady state error for the synchronous machine considered in Section 6.5, this time with the matrix C equal to

$$C = [0 \quad 1 \quad 0 \quad 1 \quad 1 \quad 1 \quad 1]$$

Hint: Use `t=0:0.5;30` while plotting the step response. Find the exact value for y_{ss} by using formula (6.36).

Part 5.[1] Use the method of singular perturbations in order to reduce the fifth-order model of a voltage regulator considered in Kokotović (1972) to an equivalent second-order slow model. The voltage regulator matrices are given by

$$A = \begin{bmatrix} -0.2 & 0.5 & 0 & 0 & 0 \\ 0 & -0.5 & 1.6 & 0 & 0 \\ 0 & 0 & -14.28 & 85.71 & 0 \\ 0 & 0 & 0 & -25 & 75 \\ 0 & 0 & 0 & 0 & -10 \end{bmatrix}, \quad B = \begin{bmatrix} 0 \\ 0 \\ 0 \\ 0 \\ 30 \end{bmatrix}$$

$$C = [1 \quad 0 \quad 0 \quad 0 \quad 0], \quad D = 0$$

Use MATLAB to partition this system as a singularly perturbed system having two slow and three fast modes with `A1=A(1:2,1:2)` and so on. Take $\epsilon = 0.05$. Show that the step responses of the original and reduced-order systems are very close to each other by plotting them on the same graph.

---

[1]   This part is optional.

6.8 References

Arkun, Y. and S. Ramakrishnan, "Bounds on the optimum quadratic cost of structure-constrained controllers," *IEEE Transactions on Automatic Control*, vol. AC-28, 924–927, 1983.

Chiang, R. and M. Safonov, *Robust Control Tool Box User's Guide*, The Math Works, Inc., Natick, Massachusetts, 1992.

Gajić, Z. and X. Shen, *Parallel Algorithms for Optimal Control of Large Scale Linear Systems*, Springer-Verlag, London, 1993.

Green, M. and D. Limebeer, *Linear Robust Control*, Prentice Hall, Englewood Cliffs, New Jersey, 1995.

Grimble, M., *Robust Industrial Control*, Prentice Hall International, Hemel Hempstead, 1994.

Kokotović, P., "Feedback design of large scale linear systems," in *Feedback Systems*, J. Cruz (ed.), McGraw-Hill, New York, 1972.

Kokotović, P. and R. Rutman, "Sensitivity of automatic control systems," *Automation and Remote Control*, vol. 26, 247–249, 1965.

Kokotović, P. and H. Khalil, *Singular Perturbations in Systems and Control*, IEEE Press, New York, 1986.

Kokotović, P., H. Khalil, and J. O'Reilly, *Singular Perturbation Methods in Control: Analysis and Design*, Academic Press, Orlando, Florida, 1986.

Morari, M. and E. Zafiriou, *Robust Process Control*, Prentice Hall, Englewood Cliffs, New Jersey, 1989.

Tomović, R. *Sensitivity Analysis of Dynamic Systems*, McGraw-Hill, New York, 1963.

Tomović, R. and M. Vukobratović, *General Sensitivity Theory*, Elsevier, New York, 1972.

6.9 Problems

6.1 Find expressions for constants k_1 and k_2 in (6.10) and derive formula (6.11).

6.2 Derive formula (6.13).

6.3 Find the transient response parameters for the following second-order systems

(a) $\dfrac{Y(s)}{U(s)} = \dfrac{5}{s^2 + 2s + 2}$

(b) $\dfrac{Y(s)}{U(s)} = \dfrac{1}{s^2 + 2s + 1}$

(c) $\dfrac{Y(s)}{U(s)} = \dfrac{10}{s^2 + 6s + 8}$

6.4 Consider the second-order system that has a zero in its transfer function, that is

$$\frac{Y(s)}{U(s)} = \frac{5(s+1)}{s^2 + 2s + 2}$$

Use the Laplace transform to obtain its step response. Find the transient response parameters and the steady state error for a unit step. Compare the step responses of this system and the system considered in Problem 6.3a. Plot the corresponding responses by using MATLAB.

6.5 Determine the steady state errors for unit step, unit ramp, and unit parabolic inputs of a unity feedback control system having the plant transfer function

$$G(s) = \frac{50(s+1)}{s^2(s+3)(s+5)(s+10)}$$

6.6 Compare the steady state errors for unit feedback control systems represented by

$$G_1(s) = \frac{10}{s(s+2)}, \quad G_2(s) = \frac{5}{s^2(s+1)(2s+1)}$$

assuming that the input signal (desired output) is given by $2t^2 - 3t - 2$, $t > 0$.

6.7 For a linear system with a unit feedback represented by

$$G(s) = \frac{10}{(s+1)(s+5)}$$

calculate steady state errors, pick time, 5 percent settling time, and maximum percent overshoot.

6.8 Find the values for the static gain K and the time constant T such that the second-order system represented by

$$H(s)G(s) = \frac{K}{s(Ts+5)}, \qquad H(s) = 1$$

has prespecified values for the maximum percent overshoot and the peak time.

6.9 Solve Problem 6.8 by requiring that the peak time and settling time be prespecified.

6.10 Find the closed-loop system transfer function(s) for the F-8 aircraft from Section 3.5.2 by using MATLAB, and calculate the steady state error(s) due to a unit step input by using formula (6.36). Note that there are two outputs in this problem.

6.11 Repeat Problem 6.10 for the ninth-order model of a power system having two inputs and four outputs. This model is given in Section 5.8, Part 3.

6.12 Repeat Problem 6.10 for the fifth-order distillation column considered in Problem 5.19.

6.13 Generalize the order-reduction procedure by the method of singular perturbations, presented in Section 6.3, to the case when the output equation has the matrix **D** different from zero.

6.14 Write a MATLAB program for finding the transient response settling time. Hint: See the MATLAB program presented in Example 8.8.

6.15 Write a MATLAB program for finding the transient response rise time. Hint: First solve Problem 6.14.

Chapter Seven

Root Locus Technique

7.1 Introduction

In this chapter we study a method for finding locations of system poles. The method is presented for a very general set-up, namely for the case when the closed-loop system poles are functions of an unknown parameter. In most cases the parameter of interest is the system static gain K satisfying $-\infty < K < +\infty$. However, any other unknown and variable system parameter affecting pole locations can be used instead of K. Even more, this method may be formulated for several unknown and variable parameters. The method is known as the root locus technique for solving polynomial equations with constant or variable parameters. It was originally presented in Ewans (1948, 1950).

The importance of the root locus method for control system theory lies in the fact that the location of the system poles determines the system stability and the system transient response. In some cases, the desired control system performance can be obtained by changing only the system static gain K. It is known from Chapter 6 that the choice of the system static gain determines the errors of the system steady state response in the sense that a bigger value for K implies smaller values for steady state errors (assuming that the system remains asymptotically stable). However, changing K causes the system transient response parameters also to change. If one is not able to achieve all the control system requirements by changing only the static gain K (the essence of the root locus method), one has to design a dynamic compensator (controller). The question of designing dynamic compensators by using the root locus method will be addressed in detail in Chapter 8.

This chapter is organized as follows. In Section 7.2, we present the main rules for constructing the root locus. Eleven rules are given with complete justification and proofs. Only elementary mathematics is used in the proofs. The time domain controller design problem based on the root locus technique is formulated in Section 7.3. Much more about controller design by using the root locus method will be said in Sections 8.5 and 8.6. In Section 7.4, the root locus technique for discrete-time control systems is discussed. At the end of the chapter, we present several case studies and formulate two MATLAB laboratory experiments on the root locus technique.

Chapter Objective

The main and only objective of this chapter is that students master rules for drawing control system root loci and develop a feeling for the shape of root loci as functions of system static gain K. This knowledge will be used in Chapter 8 to develop powerful techniques for system controller design.

Background

In the following part of this introduction we consider mathematical fundamentals and the motivation for the development of the root locus technique. The root locus technique allows adjustment of the system poles by changing the feedback system static gain. The closed-loop feedback system, in general, can be represented by a block diagram as given in Figure 7.1.

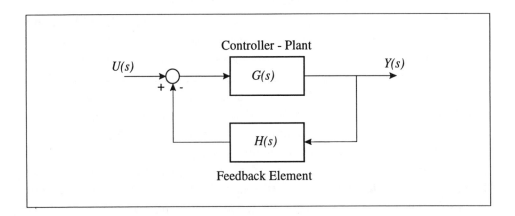

Figure 7.1: Block diagram of a feedback control system

The characteristic equation of the above closed-loop system, according to (4.48) and (4.49), is given by

$$1 + G(s)H(s) = 0 \qquad (7.1)$$

which can be written as

$$1 + KG_1(s)H_1(s) = 0 \qquad (7.2)$$

where K represents all static gains present in the loop. Usually the static gains come solely from the controller-plant element. That is, the plant $G_1(s)$ is controlled by changing the static gain K, which can be represented by a block diagram given in Figure 7.2.

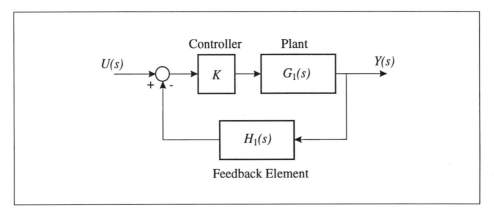

Figure 7.2: Feedback control system with a static controller in the direct path

The closed-loop transfer function of this system is given by (2.15), that is

$$M(s) = \frac{Y(s)}{U(s)} = \frac{G(s)}{1 + G(s)H(s)} = \frac{G(s)}{1 + KG_1(s)H_1(s)}$$

$$= \frac{G(s)}{1 + \dfrac{K(s^m + b_{m-1}s^{m-1} + \cdots + b_1 s + b_0)}{s^n + a_{n-1}s^{n-1} + \cdots + a_1 s + a_0}}$$

$$= \frac{G(s)(s^n + a_{n-1}s^{n-1} + \cdots + a_1 s + a_0)}{(s^n + a_{n-1}s^{n-1} + \cdots + a_0) + K(s^m + b_{m-1}s^{m-1} + \cdots + b_0)}, \quad n \geq m \qquad (7.3)$$

The corresponding characteristic equation from (7.3) is given by

$$\left(s^n + a_{n-1}s^{n-1} + \cdots + a_0\right) + K\left(s^m + b_{m-1}s^{m-1} + \cdots + b_0\right) = 0 \qquad (7.4)$$

The question to be answered by the root locus technique is: what can be achieved by changing the static gain K, theoretically, from $-\infty$ to $+\infty$? Can we find the location of system poles for all values of K? The positive answer to this question led to the development of the root locus technique. It was discovered by W. Ewans in 1948 and was mathematically formulated in 1950 in his famous paper (Ewans, 1950).

The main idea behind the root locus technique is hidden in equation (7.2). From (7.2) we have

$$G_1(s)H_1(s) = -\frac{1}{K} \qquad (7.5)$$

which is an algebraic equation involving complex numbers. It actually represents two equations (for real and imaginary parts, or for magnitudes and phase angles). In this book, *we will consider the root locus technique for $0 \le K < \infty$*. This will simplify derivations and make the root locus plots clearer. The complementary root locus for negative values of K can be similarly derived. Equation (7.5) produces the following equations for the magnitudes

$$\mid G_1(s)H_1(s) \mid = \frac{1}{K} \qquad (7.6)$$

and for the phase angles

$$\measuredangle G_1(s)H_1(s) = (2l+1)\pi, \qquad l = 0, \pm1, \pm2, \ldots \qquad (7.7)$$

If we factor $G_1(s)H_1(s)$ as

$$G_1(s)H_1(s) = \frac{(s - z_1)(s - z_2)\cdots(s - z_m)}{(s - p_1)(s - p_2)\cdots(s - p_n)} \qquad (7.8)$$

then from (7.6), by using elementary algebra with complex numbers, we get

$$\mid G_1(s)H_1(s) \mid = \frac{\prod\limits_{i=1}^{m} \mid s - z_i \mid}{\prod\limits_{i=1}^{n} \mid s - p_i \mid} = \frac{1}{K} \qquad (7.9)$$

and

$$\measuredangle G_1(s)H_1(s) = \sum_{i=1}^{m} \measuredangle(s - z_i) - \sum_{i=1}^{n} \measuredangle(s - p_i) = (2l + 1)\pi \qquad (7.10)$$

The last two equations are crucial for the development of the root locus technique.
Since the procedure is quite lengthy, we will break it down into eleven rules
(theorems), all of which are presented and proved in the next section. Before
proceeding to that, let us perform a simple exercise in order to get more familiar
with equations (7.9) and (7.10).

Example 7.1: Given the open-loop transfer function

$$G(s)H(s) = \frac{K(s + 1)}{s(s + 2)(s + 4)}$$

The locations of the open-loop poles and zeros are given in Figure 7.3.

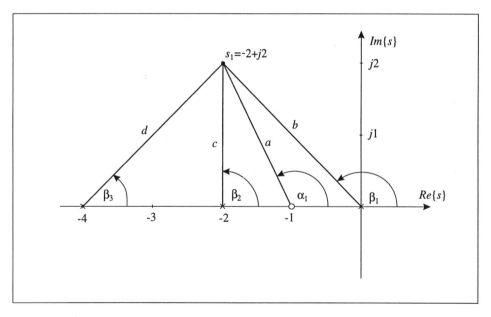

Figure 7.3: A point in the complex plain that does not lie on the root locus

Take any point s_1 in the complex plane. If that point belongs to the root
locus, it must satisfy both equations (7.9) and (7.10). For example, for the point

$s_1 = -2 + j2$ we have from (7.9)

$$\frac{|s_1 + 1|}{|s_1 + 0||s_1 + 2||s_1 + 4|} = \frac{1}{K} = \frac{a}{b \cdot c \cdot d} = \frac{\sqrt{5}}{\sqrt{8} \cdot 2 \cdot \sqrt{8}} \Rightarrow K = \frac{16}{\sqrt{5}}$$

Thus, if the point s_1 belongs to the root locus, the static gain K at that point must be equal to $16/\sqrt{5}$.

Consider now equation (7.10). It follows that for the point s_1 the following must be satisfied

$$\measuredangle G(s_1)H(s_1) = \measuredangle(s_1 + 1) - \measuredangle(s_1 + 0) - \measuredangle(s_1 + 2) - \measuredangle(s_1 + 4) = (2l + 1)\pi$$

which leads to

$$\alpha_1 - \beta_1 - \beta_2 - \beta_3 = (2l + 1)\pi, \qquad l = 0, \pm 1, \pm 2, \ldots$$
$$116.57° - 135° - 90° - 45° = -143.33° \neq (2l + 1)\pi, \quad \text{for any } l$$

We can conclude that the point $s_1 = -2 + j2$ cannot belong to the root locus since equation (7.10) is apparently not valid.

This example shows that only very selected points from the complex plane can belong to the root locus.

$\diamond$

7.2 Construction of the Root Locus

In this section, we will give eleven theorems that serve as guidelines for the root locus construction. *The root locus represents a location of the closed-loop system poles for all possible choices of the static gain $K \in [0, +\infty)$.*

The system characteristic equation is given by (7.2), that is

$$1 + KG_1(s)H_1(s) = 0 = 1 + K\frac{(s - z_1)(s - z_2) \cdots (s - z_m)}{(s - p_1)(s - p_2) \cdots (s - p_n)} = 0$$

or

$$(s - p_1)(s - p_2) \cdots (s - p_n) + K(s - z_1)(s - z_2) \cdots (s - z_m) = 0 \quad (7.11)$$

where z_i's and p_i's are the finite open-loop zeros and poles, respectively. Note that the open-loop transfer function $G_1(s)H_1(s)$, in addition to m finite zeros, has $n - m$ zeros at infinity. Setting $K = 0$ in (7.11) yields

$$(s - p_1)(s - p_2) \cdots (s - p_n) = 0 \quad (7.12)$$

This indicates that the root locus starts, for $K = 0$, at the open-loop system poles. For $K \to \infty$, equation (7.11) gives

$$(s - z_1)(s - z_2) \cdots (s - z_m) = 0 \qquad (7.13)$$

leading to the conclusion that the root locus ends at the open-loop system zeros. These two observations comprise our first two theorems on the root locus.

Theorem 7.1 *The root locus, representing the closed-loop system poles, starts for $K = 0$ at the open-loop system poles.*

Theorem 7.2 *The root locus ends for $K = +\infty$ at the open-loop system zeros, namely m poles end up at m finite zeros and the remaining $n - m$ poles escape to infinity since $n - m$ open-loop zeros are at infinity.*

The characteristic equation is of order n, so that we have to find the location of n poles for $K \in [0, +\infty)$ and thus follow n branches as K increases from zero to $+\infty$. Due to the fact that the solutions of the polynomial equations are either real or complex conjugate, it follows that the root locus must be symmetric with respect to the real axis. From these facts we obtain the next two theorems.

Theorem 7.3 *The root locus is composed of n branches.*

Theorem 7.4 *The root locus is symmetrical with respect to the real axis.*

If a point that belongs to the root locus lies on the real axis, then the angular contribution is only due to the real poles and zeros since the angular contributions of the complex conjugate poles and zeros cancel out. This is demonstrated in the following example.

Example 7.2: Let the open-loop transfer function be represented by

$$G_1(s)H_1(s) = \frac{(s + 3)}{(s + 1)(s + 4)(s^2 + 2s + 2)}$$

$$= \frac{(s - z_1)}{(s - p_1)(s - p_2)(s - p_3)(s - \bar{p}_3)}$$

Consider the point $s_1 = 2$. Then

$$\angle(s_1 + 3) = 0^\circ, \quad \angle(s_1 + 4) = 0^\circ$$

$$\angle(s_1 + 1) = 180^\circ$$

$$\angle(s_1^2 + 2s_1 + 2) = \angle(s_1 + 1 + j1) + \angle(s_1 + 1 - j1) = 45^\circ - 45^\circ = 0^\circ$$

so that

$$\angle G_1(s)H_1(s) = \angle(s_1 + 3) - \angle(s_1 + 1) - \angle(s_1 + 4)$$
$$-\angle(s_1 + 1 + j1) - \angle(s_1 + 1 - j1)$$
$$= 0° - 180° - 0° - 0° - 0° = -180°$$

◇

From this example, it can also be seen that the angular contribution of the poles and zeros to the left of the point on the real axis is 0°. Thus, it can be concluded that only real poles and zeros to the right of the point on the real axis contribute to the phase angle 180° each. Even more, since by (7.7)

$$\angle G_1(s)H_1(s) = (\text{odd number}) \times \pi$$

it follows that a point on the real axis can belong to the root locus only if the total number of poles and zeros to the right of that point is odd. This result can be formulated in the form of the following extremely useful and powerful theorem.

Theorem 7.5 *A point of the real axis belongs to the root locus if the total number of the finite open-loop poles and zeros to the right of the point on the real axis is odd.*

The power of this theorem lies in the fact that some poles are real for all values of $K \in [0, +\infty)$ so that the application of Theorem 7.5 completely determines some branches of the root locus.

Example 7.3: Consider the open-loop transfer function given in Example 7.2. The location of the open-loop poles and zeros is given in Figure 7.4.

In addition to the finite zero $z_1 = -3$, the transfer function under consideration has three zeros at infinity (the number of poles is equal to the number of zeros). According to Theorem 7.2, three branches ($n = 4$) of the root locus will end up at infinity and one of them, by Theorems 7.2 and 7.3, at $z_1 = -3$. Using Theorem 7.5, it can be concluded that some points of the real axis, i.e. $(-\infty, -4]$ and $[-3, -1]$ belong to the root locus. This completes two branches of the root locus since the pole $p_1 = -1$ (for $K = 0$) goes to the zero $z_1 = -3$ for $K = +\infty$ and the pole $p_2 = -4$ goes to zero at infinity. The partial root locus for this example is given in Figure 7.5.

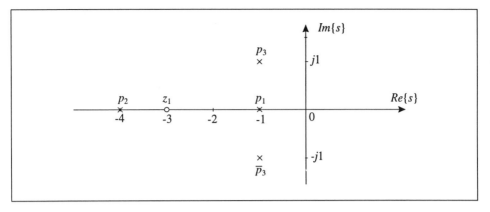

Figure 7.4: Location of the open-loop poles and zeros

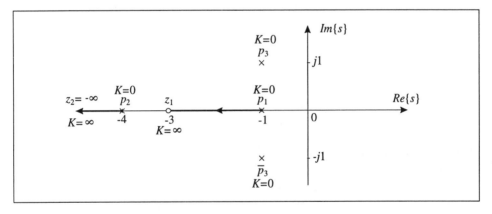

7.5: Partial root locus for Example 7.3

The remaining two branches correspond to the complex conjugate poles p_3 and $\bar{p}_3$. They should escape to infinity (according to Theorem 7.2) as K increases to $+\infty$. In order to be able to complete these two branches, we need a little bit more theory.

◇

For large values of s, i.e. for $s \to \infty$, it can be seen from (7.3) that

$$\lim_{s \to \infty} G_1(s)H_1(s) = \frac{1}{s^{n-m}}$$

which by (7.7) gives

$$\lim_{s \to \infty} \angle G_1(s)H_1(s) = (2l+1)\pi = -\angle s^{n-m} = -(n-m)\angle s \qquad (7.14)$$

Since the $n-m$ points s are theoretically at infinity, we can conclude that all finite poles and zeros produce the same angular contribution to any point at infinity lying on the root locus, say θ_l, which by (7.14) is

$$\theta_l = -\frac{(2l+1)\pi}{n-m}, \quad l = 0, \pm1, \pm2, \ldots \qquad (7.15)$$

This can be interpreted as follows: the distances among all finite poles and zeros are negligible with respect to their distance to any point at infinity lying on the root locus; hence all of them, viewed from $s = \infty$, appear to be concentrated at a single point. Thus, the angles of the asymptotes for $s \to \infty$ are obtained in (7.15), which since we have only $n - m$ branches escaping to infinity is slightly modified, that is

$$\theta_l = \frac{(2l+1)\pi}{n-m}, \quad l = 0, \pm1, \pm2, \ldots, \pm(n-m-1) \qquad (7.16)$$

Note that the total number of angles in (7.16) is $2(n-m)$. The reason for this is that the expression (7.15) is also valid for $K \to -\infty$, which corresponds to the complementary root locus ($-\infty < K \le 0$). As we have pointed before in this chapter, we study the root locus for $K \ge 0$ only. In that respect, it is sufficient to take only positive values for l in (7.16).

The previous facts are summarized in the following theorem.

Theorem 7.6 *The angles in between the asymptotes, corresponding to the $n - m$ poles at infinity, and the real axis are given by (7.16).*

Due to the fact that the root locus is symmetrical with respect to the real axis (Theorem 7.4), we can conclude that the asymptotes are also symmetrical with respect to the same axis, so that they must intersect on the real axis. The point of intersection is unique for all of them ($n - m$ asymptotes) since all finite poles and zeros viewed from $s = \infty$ behave like a single point. That point is often called the *asymptote centroid* or the *center of gravity* and it can be determined by using the following reasoning.

Consider $2(n-m)$ asymptotes represented by the corresponding vectors having a common intersection on the real axis (see Figure 7.6).

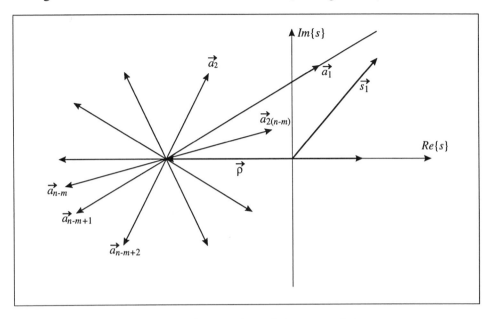

Figure 7.6: Intersection of the asymptotes

It is obvious from this figure that the vectors $\vec{a}_i$ satisfy

$$\sum_{i=1}^{2(n-m)} \vec{a}_i = 0 \tag{7.17}$$

The relations among these asymptotes and $2(n-m)$ poles at infinity (note that $n-m$ poles at infinity belong to the complementary root locus) are easily obtained by using their vector representations as given in Figure 7.6, which implies

$$\vec{a}_i = \vec{s}_i - \vec{p}, \qquad i = 1, 2, \ldots, 2(n-m) \tag{7.18}$$

In Figure 7.6 only the pole $\vec{s}_1$ is presented. Substituting (7.18) in (7.17), we get

$$\left(\sum_{i=1}^{2(n-m)} \vec{s}_i \right) - 2(n-m)\vec{p} = 0 \tag{7.19}$$

On the other hand, from the characteristic equation (7.4) we have

$$\left(s^m + b_{m-1}s^{m-1} + \cdots + b_1 s + b_0\right)\left[K + \frac{s^n + a_{n-1}s^{n-1} + \cdots + a_1 s + a_0}{s^m + b_{m-1}s^{m-1} + \cdots + b_1 s + b_0}\right] = 0$$

which, after performing long division in the square bracket, yields

$$\left(s^m + b_{m-1}s^{m-1} + \cdots + b_1 s + b_0\right)\left(K + s^{n-m} + (a_{n-1} - b_{m-1})s^{n-m-1} + \cdots\right) = 0$$

The first polynomial equation in this product gives the finite open-loop zeros (or m closed-loop poles for $K = \infty$). The remaining $n - m$ closed-loop poles are obtained from the second polynomial equation, that is

$$K + s^{n-m} + (a_{n-1} - b_{m-1})s^{n-m-1} + \cdots = 0 \qquad (7.20)$$

Using the well-known result about the roots of polynomial equations (Viete's formulas—see Appendix C), we get from (7.20)

$$\sum_{i=1}^{n-m} \vec{s}_i = -(a_{n-1} - b_{m-1}) \qquad (7.21)$$

Note that this is valid for both $K \geq 0$ and $K \leq 0$. Let us denote the complementary roots by $\vec{s}_{n-m+1}, \vec{s}_{n-m+2}, \ldots, \vec{s}_{2(n-m)}$. Then, in addition to (7.21), Viete's formulas produce the following relationship for the complementary roots

$$\sum_{i=n-m+1}^{2(n-m)} \vec{s}_i = -(a_{n-1} - b_{m-1}) \qquad (7.22)$$

Using (7.21) and (7.22) in (7.19) produces

$$-2(a_{n-1} - b_{m-1}) - 2(n - m)\rho = 0$$

which leads to

$$\rho = \frac{b_{m-1} - a_{n-1}}{n - m} \qquad (7.23)$$

Furthermore, the coefficients a_{n-1} and b_{m-1} can be replaced by using Viete's formulas and expressed in terms of the finite open-loop zeros and poles as

$$\sum_{i=1}^{m} z_i = -b_{m-1}$$

and

$$\sum_{i=1}^{n} p_i = -a_{n-1}$$

This leads to the final expression for the asymptote centroid in the form

$$\rho = \frac{\sum_{i=1}^{n} p_i - \sum_{i=1}^{m} z_i}{n - m} \tag{7.24}$$

Thus, we have proved the following theorem.

Theorem 7.7 *The intersection of the asymptotes lies on the real axis at the point given by (7.24).*

The root locus branches escaping to infinity can either stay entirely in the left-hand part of the complex plane (stability region), or move to the right (towards the instability region) and for certain values of K intersect the imaginary axis. This can be easily checked by using the Routh–Hurwitz stability criterion, which is demonstrated in the next example.

Example 7.4: Consider the characteristic equation corresponding to the open-loop transfer function studied in Example 7.2, that is

$$1 + KG_1(s)H_1(s) = 1 + \frac{K(s+3)}{(s+1)(s+4)(s^2+2s+2)} = 0$$

which implies

$$s^4 + 7s^3 + 16s^2 + (18 + K)s + (8 + 3K) = 0$$

Applying the Routh–Hurwitz test, we get

s^4	1	16	$8 + 3K$
s^3	7	$18 + K$	0
s^2	$(94 - K)/7$	$8 + 3K$	0
s^1	$(1300 - 71K - K^2)/(94 - K)$	0	0
s^0	$8 + 3K$	0	0

By examining the first column of the Routh table, the values for the static gain K that preserve system stability are obtained as

$$0 \leq K \leq 15.1$$

Note that we are studying the root locus for positive values of K. Using $K = 15.1$ produces all zeros in the row corresponding to s^1, so that the auxiliary even-power polynomial equation, obtained by using the coefficients from the previous row, is

$$A(s) = \frac{94 - 15.1}{7}s^2 + (8 + 3 \times 15.1) = 11.27s^2 + 53.3 = 0$$

From the Routh–Hurwitz stability theory, we know that solutions of this polynomial equation, given by $s_{1,2} = \pm j2.17$, represent intersections of the root locus with the imaginary axis. Using these results and Theorems 7.6 and 7.7, we can complete all branches for the root locus considered in Examples 7.2 and 7.3. We get the angles of asymptotes from (7.16) as

$$\theta_0 = \frac{\pi}{3}, \quad \theta_1 = \pi, \quad \theta_2 = \frac{5\pi}{3}$$

Either formula (7.23) or (7.24) produces the value for the intersection point of these asymptotes with the real axis, leading to

$$\rho = \frac{3 - 7}{4 - 1} = -\frac{4}{3} = \frac{-1 - 4 - 1 + j1 - 1 - j1 + 3}{4 - 1} = -\frac{4}{3}$$

The corresponding root locus is given in Figure 7.7. It completes the remaining two branches of the partial root locus presented in Figure 7.5.

$\diamond$

The next theorem gives very useful facts about the intersection of the root locus and the imaginary axis.

Theorem 7.8 *The points of intersection, if any, of the root locus and the imaginary axis are obtained from the Routh–Hurwitz stability test applied to the characteristic equation (7.4). The same test gives the values of the corresponding static gains at the points of the intersection with the imaginary axis.*

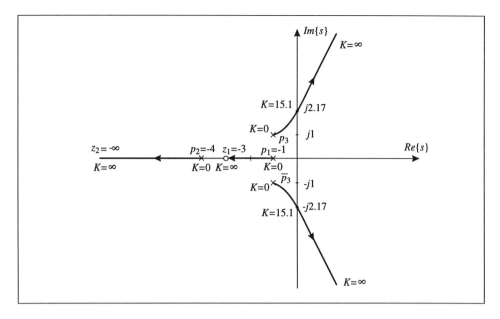

Figure 7.7: Root locus for Example 7.3

By changing the static gain K we are able to move along the branches of the root locus. If we choose a particular point on the root locus as an operating point of our control system that satisfies the required design specifications (settling time, overshoot, etc.), the following question naturally arises: how to calculate a value of the static gain at the given point on the root locus? That question has been already answered in (7.9). It can be now put in the form of the following theorem.

Theorem 7.9 *If a point s_1 belongs to the root locus, then the static gain K at that point is given by*

$$K = \frac{\prod_{i=1}^{n} |s_1 - p_i|}{\prod_{i=1}^{m} |s_1 - z_i|} \tag{7.25}$$

The distances among the point s_1 and all poles and zeros can be obtained either analytically or graphically. If we intend to obtain them graphically, the root locus must be drawn very precisely.

The next theorem will help us to get a better picture about the root locus in the neighborhood of the complex conjugate poles. In fact, that theorem can be applied to any point of the root locus since it basically rephrases the result obtained in (7.10). However, the complex conjugate poles, especially the pair of the dominant complex conjugate poles, predominantly determine the system oscillatory transient behavior so that special care has to be taken about their locations. Sometimes the desired location can be obtained by changing the static gain K only.

Theorem 7.10 *The angles of departure and angles of arrival of the complex conjugate poles are obtained from the formula*

$$\sum_{i=1}^{m} \measuredangle(s_1 - z_i) - \sum_{i=1}^{n} \measuredangle(s_1 - p_i) = (2l + 1)\pi, \quad l = 0, \pm1, \pm2, \ldots \quad (7.26)$$

where s_1 is any point on the root locus in the neighborhood of the complex conjugate pole under consideration.

The following example demonstrates the results of Theorems 7.9 and 7.10 in the procedure of getting additional information about the root locus drawn in Figure 7.7.

Example 7.5: A magnified picture of a certain part of the root locus from Figure 7.7 is given in Figure 7.8.

If the point s_1 is chosen very close to the pole p_3, then it spans almost the same angles with the remaining poles and zeros as the pole p_3. Applying the formula (7.26) to s_1 will produce the value of the angle between p_3 and s_1, and thus the angle of departure of the root locus from the pole p_3. From (7.26) we have

$$\measuredangle(s_1 - z_1) - \measuredangle(s_1 - p_1) - \measuredangle(s_1 - p_2) - \measuredangle(s_1 - p_3) - \measuredangle(s_1 - \overline{p}_3) = \pi$$

Note that it is sufficient to take $l = 0$ in (7.26). The following angles are easily obtained

$$\measuredangle(s_1 - z_1) = \tan^{-1}\frac{1}{2} = 26.57°$$

$$\measuredangle(s_1 - p_1) = 90°, \quad \measuredangle(s_1 - p_2) = \tan^{-1}\frac{1}{3} = 18.43°$$

$$\measuredangle(s_1 - \overline{p}_3) = 90°$$

Thus, we have

$$\theta = \measuredangle(s_1 - p_3) = \tan^{-1}\frac{1}{2} - \tan^{-1}\frac{1}{3} - 2\pi = \tan^{-1}\frac{1}{2} - \tan^{-1}\frac{1}{3} = 8.13°$$

This information helps us to draw the root locus more precisely.

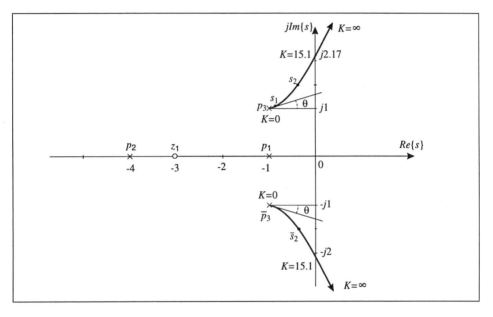

Figure 7.8: Angles of departure for Example 7.3

Now, assume that the dominant complex conjugate poles have to be located at the points s_2 and $\bar{s}_2$ in order to get the desired transient response. Using the result of Theorem 7.9 will help us to find the value of the static gain K at the point s_2. From (7.25) we have

$$K = \frac{|s_2 - p_1||s_2 - p_2||s_2 - p_3||s_2 - \bar{p}_3|}{|s_2 - z_1|} \simeq 5$$

where the corresponding distances are obtained graphically with the help of a ruler and Figure 7.8. This can also be easily done by using MATLAB and its function abs.

◇

In some cases, as K increases the pair of complex conjugate poles approaches the real axis and, at the point of intersection with the axis, two real identical poles appear. If we increase K further, the double poles break away along the real axis, producing two real distinct poles. It is also possible that two originally real poles meet each other somewhere on the real axis and then break away as the complex conjugate pair. Both phenomena are present in the next example and the corresponding Figures 7.9 and 7.10.

Example 7.6: Consider the characteristic equation of a closed-loop system in the form

$$1 + K\frac{s+2}{s(s+1)} = 0$$

This system has the finite open-loop zero at –2 and the open-loop poles at 0 and –1. Using the result of Theorem 7.5, we conclude that the parts of the real axis $[-1, 0]$ and $(-\infty, -2]$ belong to the root locus (see Figure 7.9).

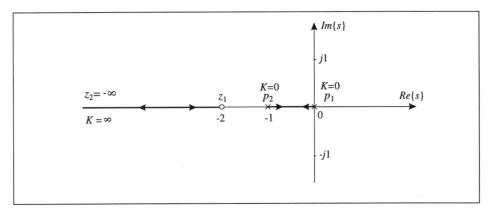

Figure 7.9: Real axis branches of the root locus from Example 7.6

According to Theorem 7.2, the root locus ends at the open-loop zeros, i.e. at $z_1 = -2$ and $z_2 = -\infty$. When K increases from zero, the real distinct poles p_1 and p_2 approach each other, and at some extremal value of K they form double real poles. Since they cannot continue their motion along the real axis,[1] they

---

[1]   Polynomial equations have unique solutions which imply that for different values of K we get different solutions so that the root locus branches cannot overlap. This fact may be used as an additional rule for plotting root locus.

must break away as a complex conjugate pair. After moving around in the space of complex conjugate pairs, they must come back to the real axis (one of the zeros is real and finite), break away, and, moving to the open-loop zeros, cover the complete part of the real axis in between -2 and $-\infty$ (Theorem 7.5). The main issue here is how to find the breakaway points. It can be observed that at the breakaway points the static gain K takes extreme values (with respect to its values on the real axis). In this example it is maximum for $-1 < s < 0$ and minimum for $-\infty < s < 2$. Thus, in order to find extremes of $K(s)$, along the real axis $s = \sigma$, we have to find $K(\sigma)$, take the derivative and equate it with zero. In this example we have

$$1 + K\frac{\sigma + 2}{\sigma(\sigma + 1)} = 0 \quad \Rightarrow \quad K(\sigma) = -\frac{\sigma(\sigma + 1)}{\sigma + 2}$$

so that

$$\frac{dK(\sigma)}{d\sigma} = 0 \quad \Rightarrow \quad -(2\sigma + 1)(\sigma + 2) + \sigma^2 + \sigma = 0$$

or

$$\sigma^2 + 4\sigma + 2 = 0 \quad \Rightarrow \quad s_{1,2} = \sigma_{1,2} = -2 \pm \sqrt{2}$$

Having obtained the breakaway points as $s_1 = -0.59$ and $s_2 = -3.41$, we are able to complete the root locus. It has the shape given in Figure 7.10.

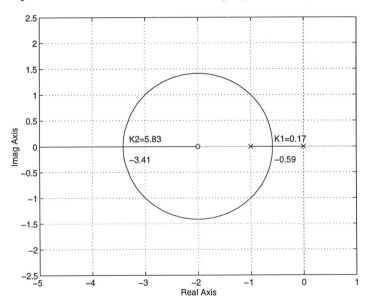

Figure 7.10: Root locus for Example 7.6

Using the result of Theorem 7.9, it is easy to find the values for the static gain at the breakaway points, namely

$$K_1 = \frac{|s_1 - p_1||s_1 - p_2|}{|s_1 - z_1|} = \frac{0.59 \times 0.41}{1.41} = 0.17$$

and

$$K_2 = \frac{|s_2 - p_1||s_2 - p_2|}{|s_2 - z_2|} = \frac{3.41 \times 2.41}{1.41} = 5.83$$

The root locus in Figure 7.10 has been drawn using the MATLAB function rlocus.

◇

In general, the breakaway points can be complex conjugate pairs. They represent the saddle points of the function

$$K(s) = -\frac{1}{G_1(s)H_1(s)}$$

which are obtained from

$$\frac{d}{ds}K(s) = 0 = \frac{\frac{d}{ds}G_1(s)H_1(s)}{[G_1(s)H_1(s)]^2}$$

or

$$\frac{d}{ds}G_1(s)H_1(s) = 0 \qquad (7.27)$$

It is well known from calculus that (7.27) represents the necessary condition only. In other words, all points satisfying (7.27) are not the breakaway points, but all the breakaway points satisfy (7.27). In summary, we can deduce the following theorem.

Theorem 7.11 *The necessary condition for the breakaway points is given by (7.27).*

From our experience about the possible shapes of root loci, we have to conclude which points satisfying (7.27) are indeed the breakaway points.

The previous eleven theorems form eleven rules for drawing the root locus. They are summarized in Table 7.1.

1	Root locus starts at the open-loop poles $(K = 0)$
2	Root locus ends at the open-loop zeros $(K = \infty)$
3	Root locus is composed of n branches
4	Root locus is symmetrical with respect to the real axis
5	If the total number of the finite open-loop poles and zeros to the right of *the point on the real axis* is odd, the point belongs to the root locus
6	Angles of the asymptotes: $$\theta_l = \frac{(2l + 1)\pi}{n - m}, \quad l = 0, 1, \ldots, n - m - 1$$
7	Intersection of the asymptotes: $$\rho = \frac{\sum_{i=1}^{n} p_i - \sum_{i=1}^{m} z_i}{n - m} = \frac{b_{m-1} - a_{n-1}}{n - m}$$
8	The Routh–Hurwitz test produces information about the points of intersection of the root locus and the imaginary axis
9	Static gain at a point s_1: $$K = \frac{\prod_{i=1}^{n} |s_1 - p_i|}{\prod_{i=1}^{m} |s_1 - z_i|}$$
10	Angles of departure and/or arrival: $$\sum_{i=1}^{m} \measuredangle(s_1 - z_i) - \sum_{i=1}^{n} \measuredangle(s_1 - p_i) = (2l + 1)\pi, \quad l = 0, \pm1, \pm2, \ldots$$ where s_1 is a point arbitrarily close to the operating point
11	The necessary condition for the breakaway points: $$\frac{d}{ds} G_1(s) H_1(s) = 0$$

Table 7.1: Root locus rules—summary

7.3 Motivation for Time Domain Controller Design

In this section, we study the impact of adding dynamic elements in the open-loop transfer function on the shape of the root loci. This will serve as motivation for controller design techniques to be developed in Section 8.5, where controllers will be added to the open-loop transfer functions in order to modify the root loci of systems such that desired locations of system poles can be achieved.

In the previous sections of this chapter we have introduced the root locus technique as one of the simplest techniques for controller design. That is, if by changing only the static gain K we can also meet the desired design specifications, then the root locus method can also be used as a controller design technique. However, choosing different values for the static gain K allows the motion of the closed-loop eigenvalues along the specified paths only—the branches of the root locus. If, according to the design specifications, the closed-loop poles of the control system under consideration do not belong to the root locus, the root locus technique alone will not produce a satisfactory answer. This is demonstrated in the next very simple example.

Example 7.7: Consider the closed-loop system with

$$1 + \frac{K}{s(s+2)} = 0$$

The root locus of the system is given in Figure 7.11.

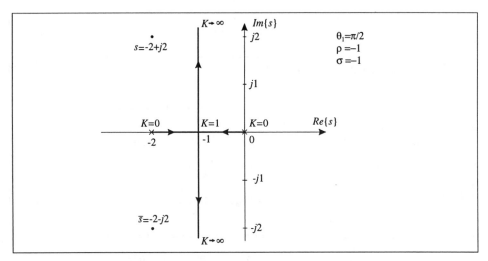

Figure 7.11: Root locus for Example 7.7

If one intends to design a control system with, for example, the natural frequency $\omega_n = \sqrt{8}$ and the damping factor $\zeta = 1/\sqrt{2}$, i.e. with the closed-loop eigenvalues located at $s_2 = -2 \pm j2$, it is obvious that the root locus technique alone will not produce the desired result since there is no way to reach the point s_2 by changing only the static gain K.

◇

We have seen that the static controller K, as given in Figure 7.1, does not solve the problem defined in Example 7.7. However, inclusion of a dynamic controller in the control system loop, as presented in Figure 7.12, gives control engineers a lot of possibilities.

Depending on the form of the dynamic controller $G_c(s)$, the root locus can be moved to almost any position in the complex plane. The proper choice of the dynamic controller answers many important control theory questions. We can use the root locus technique to design the parameters of the dynamic controller $G_c(s)$, such that all design specifications are met. In most cases, we will be able to achieve our goal by using a simple controller having only one pole and one zero, that is

$$G_c(s) = \frac{s+a}{s+b} \tag{7.28}$$

In that respect, our controller design problem demands the proper choice of the parameters a, b, and K, which produce the desired control system specifications.

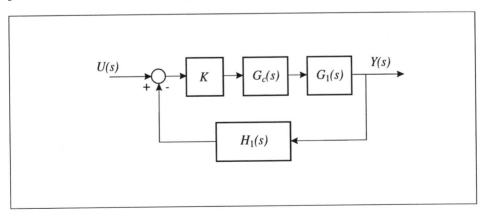

Figure 7.12: Dynamic controller configuration

The closed-loop characteristic equation that includes the controller is given by

$$1 + KG_c(s)G_1(s)H_1(s) = 0 \tag{7.29}$$

The general problem of controller design by the root locus technique will be studied in detail in Chapter 8.

Since, in general, the dynamic controller $G_c(s)$ introduces new zeros and poles, we will study in this section the effect of additional open-loop poles and zeros on the original root locus. Consider the following examples.

Example 7.8: Let the closed-loop system have the same form as in Example 7.7 with the addition of the pole at -4, that is

$$1 + \frac{K}{s(s+2)(s+4)} = 0$$

The root locus of this system is given in Figure 7.13.

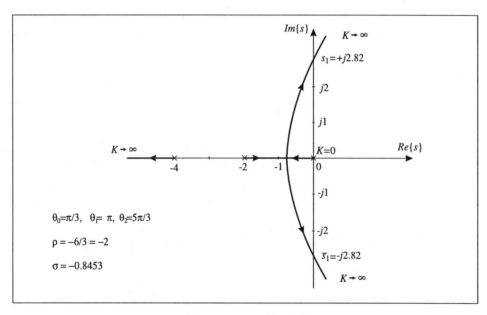

Figure 7.13: The effect of adding a pole

Comparing the root loci in Figures 7.11 and 7.13, it can be seen that the addition of a stable pole moves the root locus to the right.

◇

Example 7.9: If we add a zero at the same point where we have added the pole ($s = -4$) in Example 7.8, the root locus will move to the left. The corresponding characteristic equation is given by

$$1 + \frac{K(s+4)}{s(s+2)} = 0$$

The root locus is presented in Figure 7.14.

At the breakaway points, the static gain is given by

$$K_1 = \frac{1.17 \times 0.83}{2.83} = 0.3431, \quad K_2 = \frac{6.83 \times 4.83}{2.83} = 11.6569$$

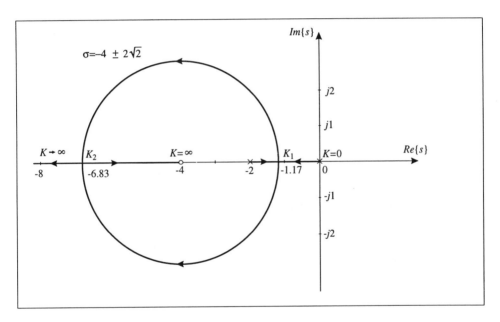

Figure 7.14: The effect of a stable zero addition

It can be observed that the addition of a stable zero causes the root locus to move to the left.

◇

The last two examples indicate that the following rule can be formulated.

Rule: *The addition of stable poles moves the root locus to the right, and the addition of stable zeros moves the root locus to the left.*

Note that this is only a rule of thumb, valid in most cases, and can thus only serve as a guideline in controller design procedure. The addition of poles and zeros can sometimes change the root locus drastically, which is demonstrated in the next example.

Example 7.10: If we add an unstable zero at 4, instead of the stable zero at −4 as was done in Example 7.9, the root locus gets a completely new shape as shown in Figure 7.15.

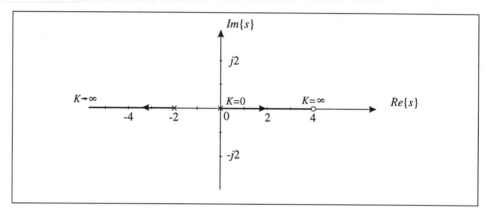

Figure 7.15: Root locus for Example 7.10

It can be seen that this system is unstable for any choice of the static gain $K > 0$, unlike the dual system represented in Figure 7.14, which is stable for all values of $K > 0$.

Going back to our original problem defined in Example 7.7, we can conclude that in order to get the desired transient response with $\omega_n = \sqrt{8}$ and $\zeta = 1/\sqrt{2}$ we have to add a stable zero which will move the root locus to the left and hopefully through the desired operating points $-2 \pm j2$. Our objective is to find the corresponding values for K and a in

$$1 + K \frac{s + a}{s(s + 2)} = 0$$

Since we have a very simple second-order system, the problem will be solved analytically. The characteristic equation of the system under consideration is

$$s^2 + (K + 2)s + Ka = 0$$

For the second-order system having poles at $-2 \pm j2$, the characteristic equation is

$$(s + 2 + j2)(s + 2 - j2) = 0 = s^2 + 4s + 8$$

Equating the coefficients in the last two equations, we get

$$K + 2 = 4 \Rightarrow K = 2$$
$$Ka = 8 \Rightarrow a = 4$$

Thus, the root locus of the closed-loop system represented by its characteristic equation

$$1 + K \frac{s + 4}{s(s + 2)}$$

passes through the desired points $-2 \pm j2$.

◇

Note that this example is used only to demonstrate the impact of addition of an open-loop stable zero to achieve the desired locations of the system poles. The result obtained has mathematical rather than engineering importance. In the general case of high-order systems an analytical solution to the controller design problem in the time domain by using the root locus technique is impossible. In that case the required controller is designed by the trial and error method with the help of the established rule on the addition of poles and zeros. This has to be done by powerful software for computer-aided control system design. In this book we will use the MATLAB package. In some cases, to be studied in Chapter 8, systematic procedures can be established for designing simple controllers containing only one pole, one zero, or one pole and one zero. The controller design technique in the time domain by the root locus method will be illustrated on several real-world examples in Sections 8.5 and 8.6.

It should be emphasized that from the root locus method it follows that the systems having unstable open-loop zeros become unstable for large values of the static gain K. Such kinds of systems are called *nonminimum phase systems*, in contrast to minimum phase systems whose definition is given below.

Definition 7.1 Systems having all open-loop zeros and poles in the closed left half of the complex plane excluding the origin are called *minimum phase systems*.

7.4 Discrete-Time Root Locus

The root locus technique for discrete-time systems is exactly the same as for continuous-time systems. This can be observed from the fact that the closed-loop discrete-time transfer function is given by

$$M(z) = \frac{C(z)}{R(z)} = \frac{G(z)}{1 + G(z)H(z)} \tag{7.30}$$

and its characteristic equation is

$$1 + G(z)H(z) = 1 + K\frac{(z - z_1)(z - z_2)\cdots(z - z_m)}{(z - p_1)(z - p_2)\cdots(z - p_n)} = 0 \tag{7.31}$$

where z_i, $i = 1, 2, ..., m$ and p_j, $j = 1, 2, ..., n$ are open-loop zeros and poles, respectively, and K is the system static gain. Expressions (7.30) and (7.31) have exactly the same forms as (7.3) and (7.8) with s replaced by z. Thus, the methodology presented in Section 7.2 for the continuous-time root locus directly extends to (7.30) and (7.31).

It is left to students, in the form of a laboratory experiment, to draw the discrete-time root locus and produce an interpretation of the corresponding rules from Table 7.1 in the context of discrete-time systems.

It is important to point out that unlike the continuous-time domain, the time delay elements in the feedback loop of discrete systems can be handled easily by using the discrete-time root locus procedure (see laboratory experiment 2, Section 7.6).

Note that exactly the same MATLAB function `rlocus` is used for drawing both continuous-time and discrete-time root loci.

7.5 MATLAB Case Studies

7.5.1 Case Study: F-8 Aircraft

Consider the model of an F-8 aircraft given in Section 5.7. The open-loop transfer functions of this aircraft are given by

$$\frac{Y_1(s)}{R(s)} = K\frac{-0.1577s^3 + 1.0747s^2 + 0.0152s}{s^4 + 1.8972s^3 + 9.9367s^2 + 0.15597s + 0.0571}$$

$$\frac{Y_2(s)}{R(s)} = K\frac{-0.433s^3 + 1.1499s^2 + 4.3643s - 34.6824}{s^4 + 1.8972s^3 + 9.9367s^2 + 0.15597s + 0.0574}$$

These expressions are obtained by using MATLAB function `ss2tf`, i.e. by `[num,den]=ss2tf(A,B,C,D)`. Note that matrix $\mathbf{D} = \mathbf{0}^{2\times1}$, and that the matrix num has two rows num1 and num2 containing the numerator coefficients of two outputs $Y_1(s)$ and $Y_2(s)$, respectively. The open-loop zeros for the first output are $z_1 = 0$, $z_2 = -0.0141$, $z_3 = 6.829$. The second output has open-loop zeros $z_{1,2} = 3.4364 \pm j2.6805$, $z_3 = -4.2171$. The open-loop system poles are $p_{1,2} = -0.9411 \pm j3.0028$, $p_{3,4} = -0.0075 \pm j0.0758$. These zeros and poles are obtained by the MATLAB function `roots`. The root locus diagrams of this system, obtained by `rlocus(num1,den)` and `rlocus(num1,den)`, are shown in Figure 7.16. Figures 7.16b and 7.16d represent enlarged parts of the root loci, i.e. those parts that are not clearly seen in Figures 7.16a and 7.16c. From Figure 7.16d it is clear that the second output of the closed-loop system is unstable even for very small values of the static gain K since it has a pair of complex conjugate poles near the imaginary axis. The closed-loop step responses obtained for two values of K are shown in Figure 7.17.

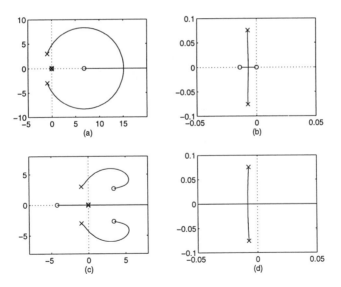

Figure 7.16: Root locus for the F-8 aircraft

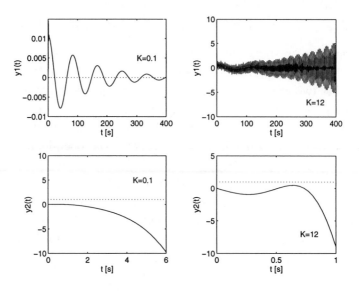

Figure 7.17: The F-8 aircraft's step responses

7.5.2 Case Study: A Synchronous Machine

The seventh-order system matrix of a synchronous machine connected to an
infinite bus is given in Problem 3.28. The remaining matrices are chosen as

$$\mathbf{B} = \begin{bmatrix} 0 & 0 & 0 & 0 & 0 & 0 & 1 \end{bmatrix}^T$$

$$\mathbf{C} = \begin{bmatrix} 0 & 0 & 0 & 0 & 0 & 0 & 1 \end{bmatrix}, \quad \mathbf{D} = \mathbf{0}$$

The open-loop poles and zeros of this system can be computed with the
following sequence of MATLAB operators

```
[num,den]=ss2tf(A,B,C,D,1);
p=roots(den);
z=roots(num);
```

where num and den stand for the numerator and denominator of the transfer
function of this single-input single-output system. Their values are given as
follows

$$\text{num} = P(s) = s^6 + 6.78s^5 + 77.1736s^4 + 330.732s^3$$
$$+ 312.4332s^2 + 47.1506s - 5.9441$$

$$\text{den} = Q(s) = s^7 + 23s^6 + 324s^5 + 2408s^4$$
$$+ 15465s^3 + 41909s^2 + 16975s + 4937$$

The open-loop zeros and poles are

$$z_{1,2} = -0.8608 \pm j7.9264, \quad z_3 = -3.8335$$

$$z_4 = -1, \quad z_5 = -0.3049, \quad z_6 = 0.08$$

$$p_{1,2} = -8.7691 \pm j8.4173, \quad p_{3,4} = -0.8499 \pm j7.8841$$

$$p_{5,6} = -0.2108 \pm j0.3077, \quad p_7 = -3.8206$$

The complete root locus of this system is shown in Figure 7.18a. Some interesting portions of the root locus are shown enlarged in Figure 7.18b. In the same figure the step response of this system is presented for $K = 1$.

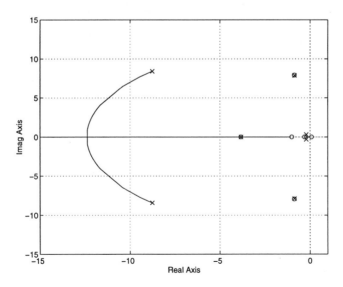

Figure 7.18a: Root locus for a synchronous machine connected to an infinite bus

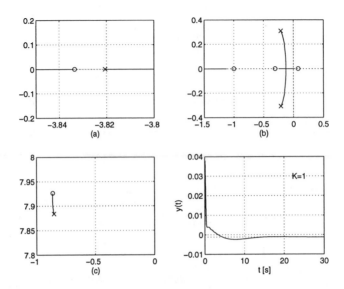

Figure 7.18b: Some parts of the root locus and the step response

7.6 MATLAB Laboratory Experiments

Two MATLAB laboratory experiments are formulated. They can be performed either as undergraduate control system laboratory assignments or as computer homework problems.

7.6.1 Hydro Power Plant Experiment

The design of a static controller for a real hydro power plant is considered in Skatarić and Gajić (1992). The hydro power plant is treated as one-unit synchronous generator connected to an infinite bus system through the transmission line. A linearized mathematical model of the synchronous generator in the dq reference frame is obtained under the assumption that both transient effects in stator windings and in damper windings are not negligible (Anderson and Fouad, 1984). The synchronous generator is assumed to be equipped with a first-order exciter. The state variables of this hydro power plant are represented by

$$\mathbf{x}^T = [\Delta\theta \quad \Delta\omega \quad \Delta u_f \quad \Delta\psi_d \quad \Delta\psi_q \quad \Delta\psi_f \quad \Delta\psi_D \quad \Delta\psi_Q]$$

where

$$\Delta\theta \;\; = \text{torque angle in rad.}$$

$$\Delta\omega \;\; = \text{rotation speed in p.u.}$$

$$\Delta u_f = \text{excitation voltage in p.u.}$$

$$\Delta\psi_d = \text{d-axis stator windings flux linkage}$$

$$\Delta\psi_q = \text{q-axis stator windings flux linkage}$$

$$\Delta\psi_f = \text{excitation flux linkage}$$

$$\Delta\psi_D = \text{d-axis damper windings flux linkage}$$

$$\Delta\psi_Q = \text{q-axis damper windings flux linkage.}$$

The control input given by $u = \Delta u_{vr}$ represents the control signal to the voltage regulation system. The system matrices for the considered nominal point are given by[2]

$$
A = \begin{bmatrix}
0 & 314.16 & 0 & 0 & 0 & 0 & 0 & 0 \\
0 & -0.286 & 0 & 0.147 & 0.528 & -0.134 & -0.04 & -0.276 \\
0 & 0 & -100 & 0 & 0 & 0 & 0 & 0 \\
255.38 & -152.49 & 0 & -13.72 & 511.14 & 8.51 & 2.556 & -135.04 \\
-182.84 & -319.5 & 0 & -534.9 & -12.24 & 137 & 41.136 & 8.389 \\
0 & 0 & 314.16 & 0.446 & 0 & -0.523 & 0.0375 & 0 \\
0 & 0 & 0 & 21.646 & 0 & 6.094 & -29.6 & 0 \\
0 & 0 & 0 & 0 & 87.236 & 0 & 0 & -97.74
\end{bmatrix}
$$

$$\mathbf{B} = \begin{bmatrix} 0 & 0 & 0.184 & 0 & 0 & 0 & 0 & 0 \end{bmatrix}^T$$

(a) Find the eigenvalues of the matrix $\mathbf{A}$ by using the MATLAB function eig. Do they represent the open-loop system poles? Justify your answer.

(b) Assume that matrix $\mathbf{D} = 0$, and take the following choices for the output matrix $\mathbf{C}$

1. $\mathbf{C} = \begin{bmatrix} 1 & 0 & 0 & 0 & 0 & 0 & 0 & 0 \end{bmatrix}$
2. $\mathbf{C} = \begin{bmatrix} 1 & 0 & 1 & 0 & 0 & 0 & 0 & 1 \end{bmatrix}$
3. $\mathbf{C} = \begin{bmatrix} 0 & 0 & 1 & 1 & 1 & 0 & 0 & 0 \end{bmatrix}$

---

[2] Note that all data (numerical values for matrices and vectors) and all MATLAB programs used in this book are available on a computer disk, which can be obtained through Math Works Inc. They may also be retrieved via anonymous FTP (see Preface).

and find the corresponding system transfer functions. Use the MATLAB functions ss2tf and/or ss2zp. Find the poles and zeros for these transfer functions and determine whether or not the corresponding systems are the minimum phase systems.

(c) Draw the root loci of the closed-loop systems for all three cases assuming that $H(s) = 1$. Use MATLAB functions rlocus(A,B,C,D) and/or rlocus(num,den). Hint: Use the MATLAB function axis to obtain detailed pictures of some parts of the root locus.

(d) Plot the unit step response of the closed-system for the choice of C given in (b.1) and comment on the system transient. Choose a value for the static gain K (obtained from the Routh–Hurwitz test) that assures the system asymptotic stability. Identify the system complex conjugate dominant poles.

(e) Examine the controllability and observability of all considered systems.

7.6.2 Discrete-Time Root Locus Experiment

Part 1. Give interpretation of the root locus rules from Table 7.1 in the context of discrete-time systems. Note that the imaginary axis from the s-plane is mapped into a unit circle in the z-plane. For stability consideration (rule number 8 in Table 7.1) Jury's test from Section 4.5.1 can be used.

Part 2. Use the MATLAB function rlocus(A,B,C,D,K) to draw the root locus for a fifth-order discrete-time model of a steam power control system considered in (Gajić and Shen, 1991) and given by

$$\mathbf{A}_d = \begin{bmatrix} 0.9150 & 0.0510 & 0.0380 & 0.0150 & 0.0380 \\ -0.0300 & 0.8890 & -0.005 & 0.0460 & 0.1110 \\ -0.0060 & 0.4680 & 0.2470 & 0.0140 & 0.0480 \\ -0.7150 & -0.0220 & -0.0211 & 0.2400 & -0.0240 \\ -0.1480 & -0.0030 & -0.0040 & 0.0900 & 0.0260 \end{bmatrix}$$

$$\mathbf{B}_d = [0.0098 \quad 0.1220 \quad 0.0360 \quad 0.5620 \quad 0.1150]^T$$

$$\mathbf{C}_d = \begin{bmatrix} 1 & 1 & 0 & 0 & 0 \\ 0 & 0 & 1 & 1 & 1 \end{bmatrix}, \quad \mathbf{D}_d = \mathbf{0}^{2 \times 1}$$

Find the step response of this system.

Part 3. Use the MATLAB function c2d in order to discretize the continuous-time F-8 aircraft model (given in Section 5.7) with the sampling period $\Delta T = 1$.

Draw the corresponding root locus. Repeat the problem with the sampling period $\Delta T = 0.1$. Does a change in the sampling rate affect the root locus?

Part 4. Draw the root locus of a discrete-time system represented by

$$1 + G(z)H(z) = 1 + \frac{K(z + 0.5)}{(z - 0.5)(z + 1)}$$

By introducing a pure time delay in the loop, the characteristic equation is changed to

$$1 + G(z)H(z) = 1 + \frac{K(z + 0.5)}{z(z - 0.5)(z + 1)}$$

Draw a modified root locus and comment on the effect of a pure time delay element on the original root locus. You may use the MATLAB function `zgrid` in order to generate a grid with lines of constant damping and constant normalized natural frequencies.

7.7 References

Anderson, P. and A. Fouad, *Power System Control and Stability,* Iowa State University Press, Ames, Iowa, 1984.

Dressler, R. and D. Tabak, "Satellite tracking by combined optimal estimation and control technique," *IEEE Transactions on Automatic Control*, vol. AC-16, 833–840, 1971.

Ewans, W., "Graphical analysis of control systems," *AIEE Transactions*, vol. 67, 547–551, 1948.

Ewans, W., "Control system synthesis by root locus method," *AIEE Transactions*, vol. 69, 1–4, 1950.

Gajić, Z. and X. Shen, "Parallel reduced-order controllers for stochastic linear singularly perturbed discrete systems," *IEEE Transactions on Automatic Control*, vol. AC-36, 87–90, 1991.

Grimble, M. and R. Patton, "The design of dynamic ship positioning control systems using stochastic optimal control theory," *Optimal Control Applications & Methods*, vol. 1, 167–202, 1980.

Jameson, A., "Design of a single-input system for specified roots using output feedback," *IEEE Transactions on Automatic Control*, vol. AC-15, 345–348, 1970.

Kadiman, K. and D. Williamson, "Discrete minimax linear quadratic regulation of continuous-time systems," *Automatica*, vol. 23, 741–747, 1987.

Kokotović, P., "Feedback design of large scale linear systems," in J. Cruz (ed.), *Feedback Systems*, McGraw-Hill, New York, 1972.

McBrinn, D. and R. Roy, "Stabilization of linear multivariable systems by output feedback," *IEEE Transactions on Automatic Control*, vol. AC-17, 243–245, 1972.

Sankaran, V., "An iterative technique to stabilize a linear constant system," *IEEE Transactions on Automatic Control*, vol. AC-20, 263–264, 1975.

Skatarić, D. and Z. Gajić, "Linear control of nearly singularly perturbed hydro power plants," *Automatica*, vol. 28, 159–163, 1992.

7.8 Problems

7.1 An open-loop control system is described by the transfer function

$$KG_1(s)H_1(s) = \frac{K(s+2)}{s(s+5)(s+10)}$$

Sketch the root locus of this system. Verify the answer obtained using the MATLAB function `rlocus`.

7.2 Examine the effect on the closed-loop eigenvalues of the compensator

$$G_c(s) = \frac{s+3}{s+5}$$

added like in Figure 7.12 to the open-loop transfer function from Problem 7.1.

7.3 Consider the inverted pendulum from Section 3.5.1 in the feedback configuration given in Figure 7.19 where $G_1(s)$ represents the pendulum open-loop transfer function. Draw the root locus of this system. Can you find the range of values for the static gain K that assures the closed-loop system stability.

7.4 Repeat Problem 7.3 for the magnetic tape control system from Problem 5.12. Assume that the output matrix is given by $C = \begin{bmatrix} 1 & 0 & 0 & 0 \end{bmatrix}$.

Use the MATLAB functions `rlocus (A, B, C, D, K)` *and* `plot (r)` *in order to find and plot the root loci for the real-world control systems given in Problems 7.5–7.11.*

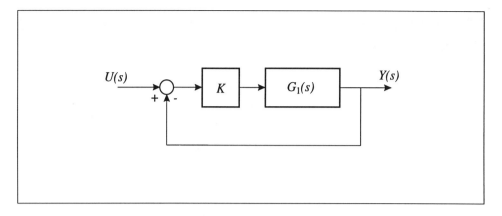

Figure 7.19: Feedback configuration

7.5 The swaying motion of a ship positioning system (Grimble and Patton, 1980; Kadiman and Williamson, 1987) has problem matrices

$$\mathbf{A} = \begin{bmatrix} -0.0546 & 0 & 0.5435 \\ 1 & 0 & 0 \\ 0 & 0 & -1.55 \end{bmatrix}, \ \mathbf{B} = \begin{bmatrix} 0 \\ 0 \\ 1.55 \end{bmatrix}, \ \mathbf{C}^T = \begin{bmatrix} 0 \\ 1 \\ 0 \end{bmatrix}, \ \mathbf{D} = 0$$

Verify the results analytically. Find a breakaway point and the range of values for the static gain K that preserves the system stability.

7.6 The lateral equations of motion of a typical medium-sized aircraft under the influence of its rudder are represented by (Jameson, 1970)

$$\mathbf{A} = \begin{bmatrix} -0.154 & 0.004 & 0.178 & 0.075 \\ -1.25 & -2.85 & 0 & 1.43 \\ 0 & 1 & 0 & 0 \\ 0.568 & -0.277 & 0 & -0.284 \end{bmatrix}, \ \mathbf{B} = \begin{bmatrix} 0.075 \\ -0.727 \\ 0 \\ -2.05 \end{bmatrix}$$

Assume the output matrices in the form

$$\mathbf{C} = \begin{bmatrix} 1 & 0 & 0 & 0 \end{bmatrix}, \quad \mathbf{D} = 0$$

7.7 Draw the root locus for a voltage regulator (Kokotović, 1972) considered in Section 6.7.

7.8   The model of the 30-ft antenna considered in Dressler and Tabak (1971) is represented by

$$
\mathbf{A} = \begin{bmatrix}
0 & 1 & 0 & 0 & 0 & 0 \\
-85.239 & -5 & 13.874 & 0 & 0 & 0 \\
0 & -352.8 & -13.3 & 1020 & 0 & 0 \\
0 & 0 & 0 & 0 & 1 & 0 \\
0 & 0 & 0 & 0 & 0 & 1 \\
0 & 0 & 0 & 0 & -0.000156 & -125
\end{bmatrix}
$$

$$
\mathbf{B} = [0 \quad 0 \quad 0 \quad 0 \quad 0 \quad 1560]^T
$$

Draw the root locus under the assumption that

$$
\mathbf{C} = [1 \quad 0 \quad 0 \quad 0 \quad 0 \quad 0], \quad \mathbf{D} = 0
$$

7.9   The 85-ft antenna tracking system is represented by the open-loop transfer function (Dressler and Tabak, 1971)

$$
\frac{\theta_a(s)}{u(s)} = \frac{K(s+1)}{s^2(s+6)(s+11.5)(s^2+8s+256)}
$$

where θ_a is the antenna's angular position, u is a control input, and K is the static gain coefficient. Assuming a unit feedback, draw the corresponding root locus.

7.10   A seven-state single-controller model of a Saturn V booster considered in McBrinn and Roy (1972) is represented by

$$
\mathbf{A} = \begin{bmatrix}
0 & 1 & 0 & 0 & 0 & 0 & 0 \\
0 & 0 & 0.2 & -0.05 & -0.002 & 2.6 & 0 \\
-0.014 & 1 & -0.041 & 0.0002 & -0.015 & -0.033 & 0 \\
0 & 0 & 0 & 0 & 1 & 0 & 0 \\
0 & 0 & 0 & -45 & -0.13 & 255 & 0 \\
0 & 0 & 0 & 0 & 0 & 0 & 1 \\
0 & 0 & 0 & 0 & 0 & -50 & -10
\end{bmatrix}
$$

$$
\mathbf{B} = [0 \quad 0 \quad 0 \quad 0 \quad 0 \quad 0 \quad 1]^T
$$

Assume that $\mathbf{C} = [1 \quad 0 \quad 0 \quad 0 \quad 0 \quad 0 \quad 0]$ and $\mathbf{D} = 0$.

7.11 A helicopter with a slung load has a linearized model as follows (Sarnkaran, 1975)

$$\mathbf{A} = \begin{bmatrix} 0 & 1 & 0 & 0 & 0 & 0 \\ -0.80435 & -0.1 & 0.80435 & 0 & 48.261 & 0 \\ 0 & 0 & 0 & 1 & 0 & 0 \\ 1.6086 & 0 & -1.6087 & 0 & 0 & 0 \\ 0 & 0 & 0 & 0 & 0 & 1 \\ 0 & 0.5 & 0 & 0 & 0 & -2 \end{bmatrix}$$

$$\mathbf{B} = \begin{bmatrix} 0 & 0 & 0 & 0 & 0 & 1 \end{bmatrix}^T$$

Assume that matrices $\mathbf{C}$ and $\mathbf{D}$ are the same as the corresponding ones in Problem 7.8. Discretize this system with the sampling period $\Delta T = 0.1$ and draw the discrete-time root locus.

Solve Problems 7.12–7.19 analytically.

7.12 A unit feedback system has the open-loop transfer function

$$G(s) = \frac{K(s+1)}{s^2(s+2)(s+4)}, \qquad K > 0$$

Sketch the root locus and determine the marginal values of K that guarantee the closed-loop system stability.

7.13 Repeat Problem 7.12 for the open-loop transfer function given by

$$G(s) = \frac{K(s+6)}{s(s+4)(s^2+4s+8)}, \qquad K > 0$$

7.14 Repeat Problem 7.12 for the open-loop transfer function given by

$$G(s) = \frac{K}{s(s^2+6s+10)}, \qquad K > 0$$

7.15 Repeat Problem 7.12 for the open-loop transfer function given by

$$G(s) = \frac{K(s-1)(s+3)}{(s+2)(s+4)(s+6)}, \qquad K > 0$$

7.16 Repeat Problem 7.12 for the open-loop transfer function given by

$$G(s) = \frac{K(s-3)}{s(s+1)(s+2)(s^2+2s+2)}, \quad K > 0$$

7.17 Repeat Problem 7.12 for the open-loop transfer function given by

$$G(s) = \frac{K(s+1)}{(s-1)(s-3)}, \quad K > 0$$

7.18 Repeat Problem 7.12 for the open-loop transfer function given by

$$G(s) = \frac{Ks^2}{(s+2)(s-4)^2}, \quad K > 0$$

7.19 Repeat Problem 7.12 for the open-loop transfer function given by

$$G(s) = \frac{Ks^2}{(s+1)(s+2)(s+3)(s+4)}, \quad K > 0$$

7.20 Sketch the root locus for the following discrete-time system

$$1 + G(z)H(z) = 1 + \frac{K(z+3)}{(z+0.5)(z-0.5)}$$

7.21 Repeat Problem 7.20 for

$$1 + G(z)H(z) = 1 + \frac{K\left(z+\frac{1}{3}\right)}{(z+0.5)(z-0.5)}$$

7.22 Repeat Problem 7.20 for

$$1 + G(z)H(z) = 1 + \frac{K(z-0.5)(z-2)}{z(z+1)(z-1)}$$

Chapter Eight

Time Domain Controller Design

8.1 Introduction

In this chapter we study the problem of controller design such that the desired system specifications are achieved. Controller design is performed in the time domain using the root locus technique. Controller design techniques in the frequency domain, based on Bode diagrams, will be presented in Chapter 9. In this book we emphasize controller design in the time domain for the following reasons: (a) with the help of MATLAB very accurate results can be obtained for both desired transient response parameters and steady state errors; (b) while designing a controller in time system stability will can be easily monitored since the root locus technique is producing information about the location of all of the system poles so that one is able to design control systems that have a specified relative degree (extent) of stability; (c) *controller design using the root locus method is simpler than the corresponding one based on Bode diagrams*; (d) root locus controller design techniques are equally applicable to both minimum and nonminimum phase systems, whereas the *corresponding techniques based on Bode diagrams are very difficult to use (if applicable at all) for nonminimum phase systems* (Kuo, 1995).

Before we actually introduce the root locus techniques for dynamic controller design (Section 8.5), in Section 8.2 we consider a class of static controllers obtained through the pole placement technique based on full state feedback. The main purpose of this controller is to stabilize the closed-loop system, and it can sometimes be used to improve the system transient response.

331

In some cases it is possible to achieve the desired system performance by changing only the static gain K. In general, *as K increases, the steady state errors decrease, but the maximum percent overshoot increases*. However, very often a static controller is not sufficient and one is faced with the problem of designing dynamic controllers.

In Section 8.3 we present common controllers used in linear system control design. Two main classes of these controllers are discussed: PI and phase-lag controllers that improve steady state errors, and PD and phase-lead controllers that improve the system transient response. Combinations of these controllers, which simultaneously improve both the system transient response and steady state errors, are also considered.

A simple class of dynamic feedback controllers can be obtained by feeding back the derivative of the output variables. These controllers, known as the rate feedback controllers, are presented in Section 8.4. It is shown that a rate feedback controller increases the damping ratio of a second-order system while keeping the natural frequency unchanged so that both the response maximum percent overshoot and the settling time are reduced.

Actual controller design in the time domain is studied in Section 8.5. Design algorithms (procedures) are outlined for several types of controllers introduced in Section 8.3. The impact of particular controllers on transient response parameters and steady state errors is examined.

Several controller design case studies involving real physical systems are presented in Section 8.6. A the end of this chapter, in Section 8.7, a MATLAB laboratory experiment is formulated, in which students are exposed to the problem of controller design for real physical control systems.

Chapter Objectives

This chapter presents systematic procedures for time domain controller design techniques based on the root locus method. Students will learn how to design different types of controllers such that the closed-loop control systems have the desired steady state errors and transient response parameters (maximum percent overshoot, settling time). In addition, the eigenvalue (pole) placement technique, which for controllable systems allows location of system eigenvalues in any desired position in the complex plane, is fully explained for the case of single-input single-output systems.

8.2 State Feedback and Pole Placement

Consider a linear dynamic system in the state space form

$$\dot{\mathbf{x}} = \mathbf{A}\mathbf{x} + \mathbf{B}\mathbf{u}$$
$$\mathbf{y} = \mathbf{C}\mathbf{x} \tag{8.1}$$

In some cases one is able to achieve the goal (e.g. stabilizing the system or improving its transient response) by using the full state feedback, which represents a linear combination of the state variables, that is

$$\mathbf{u} = -\mathbf{F}\mathbf{x} \tag{8.2}$$

so that the closed-loop system, given by

$$\dot{\mathbf{x}} = (\mathbf{A} - \mathbf{B}\mathbf{F})\mathbf{x}$$
$$\mathbf{y} = \mathbf{C}\mathbf{x} \tag{8.3}$$

has the desired specifications.

The main role of state feedback control is to stabilize a given system so that all closed-loop eigenvalues are placed in the left half of the complex plane. The following theorem gives a condition under which is possible to place system poles in the desired locations.

Theorem 8.1 *Assuming that the pair* $(\mathbf{A}, \mathbf{B})$ *is controllable, there exists a feedback matrix* $\mathbf{F}$ *such that the closed-loop system eigenvalues can be placed in arbitrary locations.*

This important theorem will be proved (justified) for *single-input single-output* systems. For the general treatment of the pole placement problem for multi-input multi-output systems, which is much more complicated, the reader is referred to Chen (1984).

If the pair $(\mathbf{A}, \mathbf{b})$ is controllable, the original system can be transformed into the phase variable canonical form, i.e. it exists a nonsingular transformation

$$\mathbf{x} = \mathbf{Q}\mathbf{z} \tag{8.4}$$

such that

$$\dot{\mathbf{z}} = \begin{bmatrix} 0 & 1 & 0 & \cdots & 0 \\ 0 & 0 & 1 & \cdots & 0 \\ \vdots & \vdots & \vdots & \ddots & \vdots \\ 0 & 0 & 0 & \cdots & 1 \\ -a_0 & -a_1 & -a_2 & \cdots & -a_{n-1} \end{bmatrix} \mathbf{z} + \begin{bmatrix} 0 \\ 0 \\ \vdots \\ 0 \\ 1 \end{bmatrix} u \tag{8.5}$$

where a_i's are coefficients of the characteristic polynomial of $\mathbf{A}$, that is

$$\det(\lambda \mathbf{I} - \mathbf{A}) = s^n + a_{n-1}s^{n-1} + a_{n-2}s^{n-2} + \cdots + a_1 s + a_0 \qquad (8.6)$$

For single-input single-output systems the state feedback is given by

$$u(\mathbf{z}) = -f_1 z_1 - f_2 z_2 - \cdots - f_n z_n = -\mathbf{f}_c \mathbf{z} \qquad (8.7)$$

After closing the feedback loop with $u(\mathbf{z})$, as given by (8.7), we get from (8.5)

$$\dot{\mathbf{z}} = \begin{bmatrix} 0 & 1 & 0 & \cdots & 0 \\ 0 & 0 & 1 & \cdots & 0 \\ \vdots & \vdots & \vdots & \ddots & \vdots \\ 0 & 0 & 0 & \cdots & 1 \\ -(a_0 + f_1) & -(a_1 + f_2) & -(a_2 + f_3) & \cdots & -(a_{n-1} + f_n) \end{bmatrix} \mathbf{z} \qquad (8.8)$$

If the desired closed-loop eigenvalues are specified by $\lambda_1^d, \lambda_2^d, ..., \lambda_n^d$, then the desired characteristic polynomial will be given by

$$\begin{aligned} \Delta^d(\lambda) &= \left(\lambda - \lambda_1^d\right)\left(\lambda - \lambda_2^d\right) \cdots \left(\lambda - \lambda_n^d\right) \\ &= s^n + a_{n-1}^d s^{n-1} + a_{n-2}^d s^{n-2} + \cdots + a_1^d s + a_0^d \end{aligned} \qquad (8.9)$$

Since the last row in (8.8) contains coefficients of the characteristic polynomial of the original system after the feedback is applied, it follows from (8.8) and (8.9) that the required feedback gains must satisfy

$$\begin{aligned} a_0 + f_1 = a_0^d &\Rightarrow f_1 = a_0^d - a_0 \\ a_1 + f_2 = a_1^d &\Rightarrow f_2 = a_1^d - a_1 \\ &\cdots \\ a_{n-1} + f_n = a_{n-1}^d &\Rightarrow f_n = a_{n-1}^d - a_{n-1} \end{aligned} \qquad (8.10)$$

The pole placement procedure using the state feedback for a system which is already in phase variable canonical form is demonstrated in the next example.

Example 8.1: Consider the following system given in phase variable canonical form

$$\dot{\mathbf{z}} = \begin{bmatrix} 0 & 1 & 0 \\ 0 & 0 & 1 \\ -2 & -5 & -10 \end{bmatrix} \mathbf{z} + \begin{bmatrix} 0 \\ 0 \\ 1 \end{bmatrix} u$$

It is required to find coefficients f_1, f_2, f_3 such that the closed-loop system has the eigenvalues located at $\lambda_{1,2}^d = -1 \pm j1$, $\lambda_3^d = -5$. The desired characteristic polynomial is obtained from (8.9) as

$$\Delta^d(\lambda) = (\lambda + 5)(\lambda + 1 + j1)(\lambda + 1 - j1) = \lambda^3 + 7\lambda^2 + 12\lambda + 10$$

so that from (8.10) we have

$$f_1 = a_0^d - a_0 = 10 - 2 = 8$$
$$f_2 = a_1^d - a_1 = 12 - 5 = 7$$
$$f_3 = a_2^d - a_2 = 7 - 10 = -3$$

◇

In general, in order to be able to apply this technique to all controllable single-input single-output systems we need to find a nonsingular transformation which transfers the original system into phase variable canonical form. This transformation can be obtained by using the linearly independent columns of the system controllability matrix

$$\mathcal{C} = \left[\mathbf{b} \vdots \mathbf{Ab} \vdots \mathbf{A^2b} \vdots \cdots \vdots \mathbf{A^{n-1}b} \right]$$

It can be shown (Chen, 1984) that the required transformation is given by

$$\mathbf{Q} = [\mathbf{q}_1 \quad \mathbf{q}_2 \quad \cdots \quad \mathbf{q}_n] \tag{8.11}$$

where

$$\mathbf{q}_n = \mathbf{b}$$
$$\mathbf{q}_{n-1} = \mathbf{Aq}_n + a_0\mathbf{q}_n = \mathbf{Ab} + a_0\mathbf{b}$$
$$\mathbf{q}_{n-2} = \mathbf{Aq}_{n-1} + a_1\mathbf{q}_n = \mathbf{A^2b} + a_0\mathbf{Ab} + a_1\mathbf{b} \tag{8.12}$$
$$\cdots$$
$$\mathbf{q}_1 = \mathbf{Aq}_2 + a_{n-2}\mathbf{q}_n = \mathbf{A^{n-1}b} + a_0\mathbf{A^{n-2}b} + \cdots + a_{n-2}\mathbf{b}$$

where a_i's are coefficients of the characteristic polynomial of matrix $\mathbf{A}$. After the feedback gain has been found for phase variable canonical form, $\mathbf{f}_c$, in the original coordinates it is obtained as (similarity transformation)

$$\mathbf{f} = \mathbf{f}_c\mathbf{Q}^{-1} \tag{8.13}$$

Example 8.2: Consider the following linear system given by

$$\mathbf{A} = \begin{bmatrix} -1 & 2 & 0 \\ 1 & -3 & 4 \\ -1 & 1 & -9 \end{bmatrix}, \quad \mathbf{b} = \begin{bmatrix} 1 \\ 2 \\ -1 \end{bmatrix}, \quad \mathbf{c} = \begin{bmatrix} 1 & 0 & 1 \end{bmatrix}$$

The characteristic polynomial of this system is

$$\det(\lambda \mathbf{I} - \mathbf{A}) = s^3 + 13s^2 + 33s + 13$$

Its phase variable canonical form can be obtained either from its transfer function (see Section 3.1.2) or by using the nonsingular (similarity) transformation (8.11). The system transfer function is given by

$$\mathbf{c}(s\mathbf{I} - \mathbf{A})^{-1}\mathbf{b} = \frac{46s + 13}{s^3 + 13s^2 + 33s + 13}$$

Using results from Section 3.1.2 we are able to write new matrices in phase variable canonical form representation as

$$\mathbf{A}_c = \begin{bmatrix} 0 & 1 & 0 \\ 0 & 0 & 1 \\ -13 & -33 & -13 \end{bmatrix}, \quad \mathbf{b}_c = \begin{bmatrix} 0 \\ 0 \\ 1 \end{bmatrix}, \quad \mathbf{c}_c = \begin{bmatrix} 13 & 46 & 0 \end{bmatrix}$$

The same matrices could have been obtained by using the similarity transformation with

$$\mathbf{A}_c = \mathbf{Q}^{-1}\mathbf{A}\mathbf{Q}, \quad \mathbf{b}_c = \mathbf{Q}^{-1}\mathbf{b}, \quad \mathbf{c}_c = \mathbf{c}\mathbf{Q} \tag{8.14}$$

with $\mathbf{Q}$ obtained from (8.11) and (8.12) as

$$\mathbf{Q} = \begin{bmatrix} 51 & 16 & 1 \\ 19 & 17 & 0 \\ -5 & -3 & 1 \end{bmatrix}$$

Assume that we intend to find the feedback gain for the original system such that its closed-loop eigenvalues are located at $-1, -2, -3$, then

$$\Delta^d(\lambda) = (\lambda + 1)(\lambda + 2)(\lambda + 3) = \lambda^3 + 6\lambda^2 + 11\lambda + 6$$

From equation (8.10) we get expressions for the feedback gains for the system in phase variable canonical form as

$$f_1 = a_0^d - a_0 = 6 - 13 = -7$$

$$f_2 = a_1^d - a_1 = 11 - 33 = -22$$

$$f_3 = a_2^d - a_2 = 6 - 13 = -7$$

In the original coordinates the feedback gain is obtained from (8.13)

$$\mathbf{f} = \mathbf{f}_c \mathbf{Q}^{-1} = [0.8149 \quad -1.0540 \quad 5.7069]$$

Using this gain in order to close the state feedback around the system we get

$$\dot{\mathbf{x}} = \begin{bmatrix} -1.8149 & 3.0540 & -5.7069 \\ -0.6298 & -0.8920 & -7.4139 \\ -0.1851 & -0.0540 & -3.2931 \end{bmatrix} \mathbf{x}$$

It is easy to check by MATLAB that the eigenvalues of this systems are located at $-1, -2, -3$.

◇

Comment: Exactly the same procedure as the one given in this section can be used for placing the observer poles in the desired locations. The observers have been considered in Section 5.6. Choosing the observer gain $\mathbf{K}$ such that the closed-loop observer matrix $\mathbf{A} - \mathbf{KC}$ has the desired poles corresponds to the problem of choosing the feedback gain $\mathbf{F}$ such that the closed-loop system matrix $\mathbf{A}^T - \mathbf{F}^T \mathbf{B}^T$ has the same poles. Thus, for the observer pole placement problem, matrix $\mathbf{A}$ should be replaced by $\mathbf{A}^T$, $\mathbf{B}$ replaced by $\mathbf{C}^T$ and $\mathbf{F}$ replaced by $\mathbf{K}^T$. In addition, it is known from Chapter 5 that the observability of the pair $(\mathbf{A}, \mathbf{C})$ is equal to the controllability of the pair $(\mathbf{A}^T, \mathbf{C}^T)$, and hence the controllability condition stated in Theorem 8.1—the pair $(\mathbf{A}, \mathbf{B})$ is controllable, which for observer pole placement requires that the pair $(\mathbf{A}^T, \mathbf{C}^T)$ be controllable—is satisfied by assuming that the pair $(\mathbf{A}, \mathbf{C})$ is observable.

8.3 Common Dynamic Controllers

Several common dynamic controllers appear very often in practice. They are known as PD, PI, PID, phase-lag, phase-lead, and phase-lag-lead controllers. In this section we introduce their structures and indicate their main properties. In the follow-up sections procedures for designing these controllers by using the root locus technique such that the given systems have the desired specifications are presented. In the most cases these controllers are placed in the forward path at the front of the plant (system) as presented in Figure 8.1.

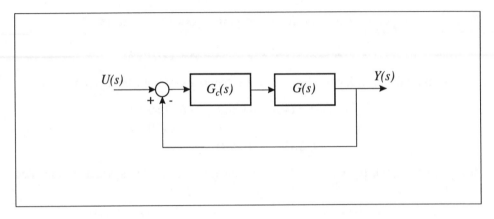

Figure 8.1: A common controller-plant configuration

8.3.1 PD Controller

PD stands for a proportional and derivative controller. The output signal of this controller is equal to the sum of two signals: the signal obtained by multiplying the input signal by a constant gain K_p and the signal obtained by differentiating and multiplying the input signal by K_d, i.e. its transfer function is given by

$$G_c(s) = K_p + K_d s \tag{8.15}$$

This controller is used to improve the system transient response.

8.3.2 PI Controller

Similarly to the PD controller, the PI controller produces as its output a weighted sum of the input signal and its integral. Its transfer function is

$$G_c(s) = K_p + K_i \frac{1}{s} = \frac{K_p s + K_i}{s} \tag{8.16}$$

In practical applications the PI controller zero is placed very close to its pole located at the origin so that the angular contribution of this "dipole" to the root locus is almost zero. *A PI controller is used to improve the system response steady state errors* since it increases the control system type by one (see Definition 6.1).

8.3.3 PID Controller

The PID controller is a combination of PD and PI controllers; hence its transfer function is given by

$$G_c(s) = K_p + K_d s + K_i \frac{1}{s} = \frac{K_i + K_p s + K_d s^2}{s} \qquad (8.17)$$

The PID controller can be used to improve both the system transient response and steady state errors. This controller is very popular for industrial applications.

8.3.4 Phase-Lag Controller

The phase-lag controller belongs to the same class as the PI controller. The phase-lag controller can be regarded as a generalization of the PI controller. It introduces a negative phase into the feedback loop, which justifies its name. It has a zero and pole with the pole being closer to the imaginary axis, that is

$$G_c(s) = \left(\frac{p_1}{z_1}\right) \frac{s + z_1}{s + p_1}, \qquad z_1 > p_1 > 0$$

$$(8.18)$$

$$\arg G_c(s) = \arg(s + z_1) - \arg(s + p_1) = \theta_{z_1} - \theta_{p_1} < 0$$

where p_1/z_1 is known as the lag ratio. The corresponding angles θ_{z_1} and θ_{p_1} are given in Figure 8.2a. *The phase-lag controller is used to improve steady state errors.*

8.3.5 Phase-Lead Controller

The phase-lead controller is designed such that its phase contribution to the feedback loop is positive. It is represented by

$$G_c(s) = \frac{s + z_2}{s + p_2}, \qquad p_2 > z_2 > 0$$

$$(8.19)$$

$$G_c(s) = \arg(s + z_2) - \arg(s + p_2) = \theta_{z_2} - \theta_{p_2} > 0$$

where θ_{z_2} and θ_{p_2} are given in Figure 8.2b. This controller introduces a positive phase shift in the loop (phase lead). *It is used to improve the system response transient behavior.*

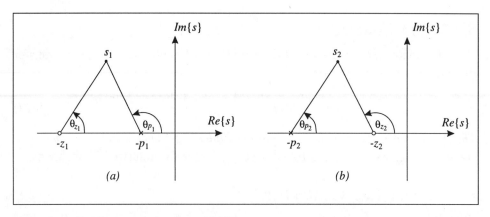

Figure 8.2: Poles and zeros of phase-lag (a) and phase-lead (b) controllers

8.3.6 Phase-Lag-Lead Controller

The phase-lag-lead controller is obtained as a combination of phase-lead and phase-lag controllers. Its transfer function is given by

$$G_c(s) = \frac{(s + z_1)(s + z_2)}{(s + p_1)(s + p_2)}, \quad p_2 > z_2 > z_1 > p_1 > 0, \quad z_1 z_2 = p_1 p_2 \quad (8.20)$$

It has features of both phase-lag and phase-lead controllers, i.e. *it can be used to improve simultaneously both the system transient response and steady state errors.* However, it is harder to design phase-lag-lead controllers than either phase-lag or phase-lead controllers.

Note that all controllers presented in this section can be realized by using active networks composed of operational amplifiers (see, for example, Dorf, 1992; Nise, 1992; Kuo, 1995).

8.4 Rate Feedback Control

The controllers considered in the previous section have simple forms and in most cases they are placed in the forward loop in the front of the system to be controlled. Another simple controller that is always used in the feedback loop is known as the rate feedback controller. The rate feedback controller is obtained by feeding back the derivative of the output of a second-order system (or a system which can be approximated by a second-order system, i.e. a system with dominant complex conjugate poles) according to the block diagram given in Figure 8.3.

The rate feedback control helps to increase the system damping. This follows from the fact that the closed-loop transfer function for this configuration is given by

$$\frac{Y(s)}{U(s)} = \frac{\omega_n^2}{s^2 + 2\left(\zeta + \frac{1}{2}K_t\omega_n\right)\omega_n s + \omega_n^2} = \frac{\omega_n^2}{s^2 + 2\zeta_c\omega_n s + \omega_n^2}, \quad K_t > 0 \quad (8.21)$$

Compared with the closed-loop transfer function of the second-order system without control (6.4), we see that the damping factor is now increased to

$$\zeta_c = \zeta + \frac{1}{2}K_t\omega_n \quad (8.22)$$

Since the natural frequency is unchanged, this controller decreases the response settling time (see (6.20)). The system response maximum percent overshoot is also decreased (see Problem 8.3).

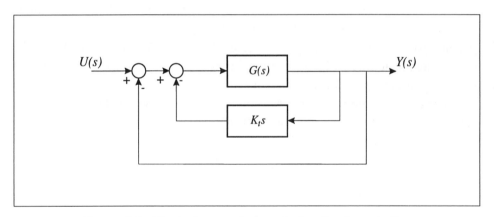

Figure 8.3: Block diagram for a rate feedback controller

Example 8.3: Design a rate feedback controller such that the damping ratio of the second-order system considered in Example 6.1 is increased to $\zeta_c = 0.75$ and determine the system response settling time and maximum percent overshoot.

From Example 6.1 we know

$$\omega_n = 2\,\text{rad/s}, \quad \zeta = 0.5, \quad t_s \approx 3\,\text{s}, \quad MPOS = 16.3\,\%$$

The gain for the rate controller is obtained from (8.22) as

$$K_t = \frac{1}{\omega_n}2(\zeta_c - \zeta) = \frac{1}{2}2(0.75 - 0.5) = 0.25$$

The new values for the response settling time and maximum percent overshoot are given by

$$t_{sc} \approx \frac{3}{\zeta_c\omega_n} = \frac{3}{0.75 \times 2} = 2\,\text{s}, \quad MPOS = e^{-\frac{\zeta_c\pi}{\sqrt{1-\zeta_c}}}100(\%) = 2.83\,\%$$

It can be seen that both the system response settling time and maximum percent overshoot are reduced.

$$\diamond$$

8.5 Compensator Design by the Root Locus Method

Sometimes one is able to improve control system specifications by changing the static gain K only. It can be observed that *as K increases, the steady state errors decrease (assuming system's asymptotic stability), but the maximum percent overshoot increases*. However, using large values for K may damage system stability. Even more, in most cases the desired operating points for the system dominant poles, which satisfy the transient response requirements, do not lie on the original root locus. Thus, in order to solve the transient response and steady state errors improvement problem, one has to design dynamic controllers, considered in Section 8.3, and put them in series with the plant (system) to be controlled (see Figure 8.1).

In the following we present dynamic controller design techniques in three categories: improvement of steady state errors (PI and phase-lag controllers), improvement of system transient response (PD and phase-lead controllers), and improvement of both steady state errors and transient response (PID and phase-lag-lead controllers). Note that transient response specifications are obtained under the assumption that a given system has a pair of dominant complex conjugate closed-loop poles; hence this assumption has to be checked after a controller is added to the system. This can be easily done using the root locus technique.

8.5.1 Improvement of Steady State Errors

It has been seen in Chapter 6 that the steady state errors can be improved by increasing the type of feedback control system, in other words, by adding a pole at the origin to the open-loop system transfer function. The simplest way to achieve this goal is to add in series with the system a PI controller as defined in (8.16), i.e. to get

$$G_c(s)G(s) = \frac{K_p s + K_i}{s} G(s)$$

Since this controller also introduces a zero at $-K_i/K_p$, *the zero should be placed as close as possible to the pole.* In that case the pole at $p = 0$ and the zero at $z \approx p$ act as a dipole, and so their mutual contribution to the root locus is almost negligible. Since the root locus is practically unchanged, the system transient response remains the same and the effect due to the PI controller is to increase the type of the control system by one, which produces improved steady state errors. The effect of a dipole on the system response is studied in the next example.

Example 8.4: Consider the open-loop transfer functions

$$G_1(s) = \frac{(s+2)}{(s+1)(s+3)}$$

and

$$G_2(s) = \frac{(s+2)(s+5)}{(s+1)(s+3)(s+5.1)}$$

Note that the second transfer function has a dipole with a stable pole at -5.1. The corresponding step responses are given in Figure 8.4. It can be seen from this figure that the system with a stable dipole and the system without a stable dipole have almost identical responses. These responses have been obtained by the following sequence of MATLAB instructions.

```
num1=[1  2];
den1=[1  4  3];
num2=[1  7  10];
d1=[1  1];
d2=[1  3];
d3=[1  5.1];
d12=conv(d1,d2);
```

```
den2=conv(d12,d3);
[cnum1,cden1]=cloop(num1,den1,-1);
[cnum2,cden2]=cloop(num2,den2,-1);
t=0:0.1:2;
step(cnum1,den1,t)
step(cnum1,den1,t)
```

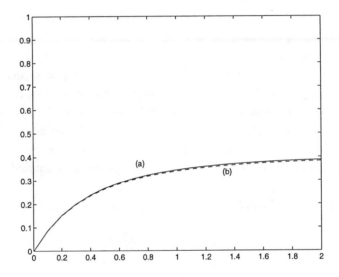

Figure 8.4: Step responses of a system without
a stable dipole (a) and with a stable dipole (b)

It is important to point out that in the case of an *unstable dipole* the effect
of a dipole is completely different. Consider, for example, the open-loop transfer
function given by

$$G_3(s) = \frac{(s+2)(s-5)}{(s+1)(s+3)(s-5.1)}$$

Its step response is presented in Figure 8.5b and compared with the corresponding
step response after a dipole is eliminated (Figure 8.5a). In fact, the system without
a dipole is stable and the system with a dipole is unstable; hence their responses
are drastically different. Thus, we can conclude that *it is not correct to cancel an*

unstable dipole since it has a big impact on the system response.

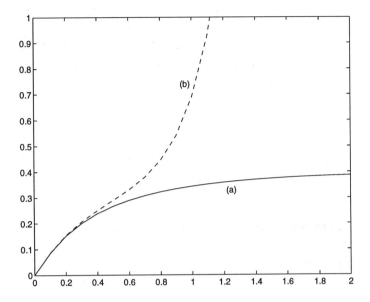

Figure 8.5: Step responses of a system without an
unstable dipole (a) and with an unstable dipole (b)

◇

Both the PI and phase-lag controller use this *"stable dipole effect"*. They do
not change the system transient response, but they do have an important impact
on the steady state errors.

PI Controller Design

As we have already indicated, the PI controller represents a stable dipole with a
pole located at the origin and a stable zero placed near the pole. Its impact on the
transient response is negligible since it introduces neither significant phase shift
nor gain change (see root locus rules 9 and 10 in Table 7.1). Thus, the transient
response parameters with the PI controller are almost the same as those for the
original system, but the steady state errors are drastically improved due to the
fact that the feedback control system type is increased by one.

The PI controller is represented, in general, by

$$G_c(s) = K_p \frac{s + \frac{K_i}{K_p}}{s}, \qquad K_i \ll K_p \tag{8.23}$$

where K_p represents its static gain and K_i/K_p is a stable zero near the origin. Very often it is implemented as

$$G_c(s) = \frac{s + z_c}{s} \tag{8.24}$$

This implementation is sufficient to justify its main purpose. The design algorithm for this controller is extremely simple.

Design Algorithm 8.1:

1. Set the PI controller's pole at the origin and locate its zero arbitrarily close to the pole, say $z_c = 0.1$ or $z_c = 0.01$.
2. If necessary, adjust for the static loop gain to compensate for the case when K_p is different from one. Hint: Use $K_p = 1$, and avoid gain adjustment problem.

Comment: Note that while drawing the root locus of a system with a PI controller (compensator), the stable open-loop zero of the compensator will attract the compensator's pole located at the origin as the static gain increases from 0 to $+\infty$ so that there is no danger that the closed-loop system may become unstable due to addition of a PI compensator (controller).

The following example demonstrates the use of a PI controller in order to reduce the steady state errors.

Example 8.5: Consider the following open-loop transfer function

$$G(s) = \frac{K(s + 6)}{(s + 10)(s^2 + 2s + 2)}$$

Let the choice of the static gain $K = 10$ produce a pair of dominant poles on the root locus, which guarantees the desired transient specifications. The corresponding position constant and the steady state unit step error are given by

$$K_p = \frac{10 \times 6}{10 \times 2} = 3 \ \Rightarrow \ e_{ss} = \frac{1}{1 + K_p} = 0.25$$

Using a PI controller in the form of (8.24) with the zero at -0.1 ($z_c = 0.1$), we obtain the improved values as $K_p = \infty$ and $e_{ss} = 0$. The step responses of the original system and the compensated system, now given by

$$G_c(s)G(s) = \frac{10(s + 0.1)(s + 6)}{s(s + 10)(s^2 + 2s + 2)}$$

are presented in Figure 8.6.

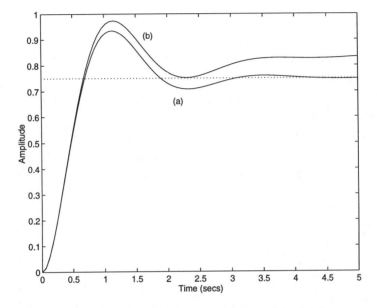

Figure 8.6: Step responses of the original (a)
and compensated (b) systems for Example 8.5

The closed-loop poles of the original system are given by

$$\lambda_1 = -9.5216, \qquad \lambda_{2,3} = -1.2392 \pm j2.6204$$

For the compensated system they are

$$\lambda_{1c} = -9.5265, \qquad \lambda_{2c,3c} = -1.1986 \pm j2.6109$$

Having obtained the closed-loop system poles, it is easy to check that the dominant system poles are preserved for the compensated system and that the damping ratio and natural frequency are only slightly changed. Using information about the dominant system poles and relationships obtained from Figure 6.2, we get

$$\zeta\omega_n = 1.2392, \quad \omega_n^2 = (1.2392)^2 + (2.6204)^2 \implies \omega_n^2 = 2.9019, \quad \zeta = 0.4270$$

and

$$\zeta_c\omega_{nc} = 1.1986, \quad \omega_{nc}^2 = (1.1986)^2 + (2.6109)^2$$
$$\implies \omega_{nc}^2 = 2.8901, \quad \zeta_c = 0.4147$$

In Figure 8.7 we draw the step response of the compensated system over a long period of time in order to show that the steady state error of this system is theoretically and practically equal to zero.

Figures 8.6 and 8.7 are obtained by using the same MATLAB functions as those used in Example 8.4.

The root loci of the original and compensated systems are presented in Figures 8.8 and 8.9. It can be seen from these figures that the root loci are almost identical, with the exception of a tiny dipole branch near the origin.

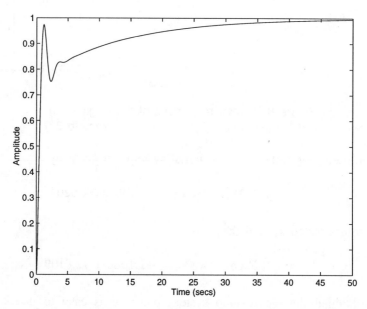

Figure 8.7: Step response of the compensated system for Example 8.5

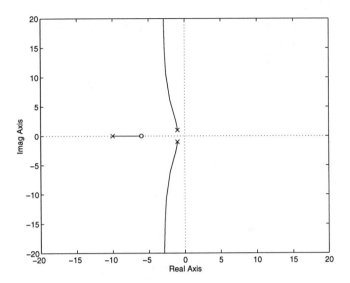

Figure 8.8: Root locus of the original system for Example 8.5

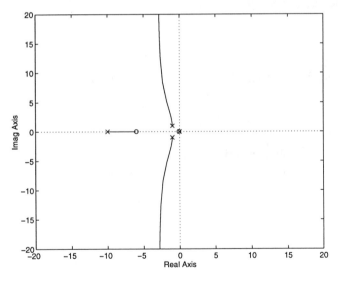

Figure 8.9: Root locus of the compensated system for Example 8.5

◇

Phase-Lag Controller Design

The phase-lag controller, in the context of root locus design methodology, is also implemented as a dipole that has no significant influence on the root locus, and thus on the transient response, but increases the steady state constants and reduces the corresponding steady state errors. Since it is implemented as a dipole, its zero and pole have to be placed very close to each other.

The lag controller's impact on the steady state errors can be obtained from the expressions for the corresponding steady state constants. Namely, from (6.31), (6.33), and (6.35) we know that

$$K_p = \lim_{s \to 0} \{H(s)G(s)\}, \quad K_v = \lim_{s \to 0} \{sH(s)G(s)\}, \quad K_a = \lim_{s \to 0} \{s^2 H(s)G(s)\}$$

and from (6.30), (6.32), and (6.34) we have

$$e_{ss_{step}} = \frac{1}{1 + K_p}, \quad e_{ss_{ramp}} = \frac{1}{K_v}, \quad e_{ss_{parabolic}} = \frac{2}{K_a}$$

For control systems of type zero, one, and two, respectively, the constants K_p, K_v, and K_a are all given by the same expression, that is

$$K_l = K \frac{z_1 z_2 \cdots}{p_1 p_2 \cdots}, \quad l = p, v, a \tag{8.25}$$

Consider, first, a phase-lag compensator of the form

$$G_c(s) = \frac{s + z_c}{s + p_c}, \quad z_c > p_c > 0 \tag{8.26}$$

If we put this controller in series with the system, the corresponding steady state constants of the compensated system will be given by

$$K_{lc} = K \frac{z_1 z_2 \cdots z_c}{p_1 p_2 \cdots p_c} = K_l \frac{z_c}{p_c}, \quad lc, l = p, v, a \tag{8.27}$$

In order to increase these constants and reduce the steady state errors, the ratio of z_c/p_c should be as large as possible. Since at the same time z_c must be close to p_c (they form a dipole), a large value for the ratio z_c/p_c can be achieved if both of them are placed close to zero. For example, the choice of $z_c = 0.1$ and $p_c = 0.01$ increases the constants $K_l, l = p, v, a$, ten times and reduces the corresponding steady state errors ten times.

Now consider a phase-lag controller defined by (8.18), that is

$$G_c(s) = \left(\frac{p_c}{z_c}\right)\frac{s + z_c}{s + p_c}, \quad z_c > p_c > 0$$

This controller will change the value of the static gain K by a factor of p_c/z_c, which will produce a movement of the desired operating point along the root locus in the direction of smaller static gains. Thus, the plant static gain has to be adjusted to a higher value in order to preserve the same operating point. The consequence of using a phase-lag controller as defined in (8.18) is that *the same (desired) operating point is obtained with higher static gain*. We already know that by increasing the static gain, the steady state errors are reduced. In this case, the static gain adjustment has to be done by choosing a new static gain $\widetilde{K} = Kz_c/p_c$. Note that the effects of both phase-lag controllers (8.18) and (8.26) are exactly the same, since the gain adjustment in the case of controller (8.18) in fact cancels its lag ratio p_c/z_c.

The following simple algorithm is used for phase-lag controller design.

Design Algorithm 8.2:

1. Choose a point that has the desired transient specifications on the root locus branch with dominant system poles. Read from the root locus the value for the static gain K at the chosen point, and determine the corresponding steady state errors.
2. Set both the phase-lag controller's pole and zero near the origin with the ratio z_c/p_c obtained from (8.27) such that the desired steady state error requirement is satisfied.
3. In the case of controller (8.18), adjust for the static loop gain, i.e. take a new static gain as $\widetilde{K} = Kz_c/p_c$.

The next example demonstrates the controller design procedure with a phase-lag compensator according to the steps outlined in Design Algorithm 8.2.

Example 8.6: The steady state errors of the system considered in Example 8.5 can be improved by using a phase-lag controller of the form

$$G_c(s) = \frac{s + 0.1}{s + 0.01}$$

Since $z_c/p_c = 10$, the position constant is increased ten times, that is

$$K_{pc} = K_p\frac{z_c}{p_c} = 3 \times 10 = 30$$

so that the steady state error due to a unit step input is reduced to

$$e_{ss_{step}} = \frac{1}{1 + K_{pc}} = \frac{1}{31} = 0.03226$$

It can be easily checked that the transient response is almost unchanged; in fact, the dominant system poles with this phase-lag compensator are $-1.2026 \pm j2.6119$, which is very close to the dominant poles of the original system (see Example 8.5).

◇

Example 8.7: Consider the following open-loop transfer function

$$G(s)H(s) = \frac{K(s + 15)}{s(s + 20)(s^2 + 4s + 8)}$$

Let the choice of the static gain $K = 20$ produce a pair of dominant poles on the root locus that guarantees the desired transient specifications. The system closed-loop poles for $K = 20$ are given by

$$\lambda_{1,2} = -0.5327 \pm j2.2024, \quad \lambda_3 = -2.9194, \quad \lambda_4 = -20.0153$$

so that for this value of the static gain K the dominant poles exist, i.e. the absolute value of the real part of the dominant poles (0.5327) is about six times smaller than the absolute value of the real part of the next pole (2.9194), which is in practice sufficient to guarantee poles' dominance. Since we have a type one feedback control system, the steady state error due to a unit step is zero. The velocity constant and the steady state unit ramp error are obtained as

$$K_v = \frac{20 \times 15}{20 \times 8} = \frac{15}{8} \quad \Rightarrow \quad e_{ss_{ramp}} = \frac{1}{K_v} = 0.53$$

Using the phase-lag controller with a zero at -0.1 ($z_c = 0.1$) and a pole at -0.01 ($p_c = 0.01$), we get

$$K_{vc} = K_v \frac{z_c}{p_c} = \frac{150}{8} \quad \Rightarrow \quad e_{ss_{c_{ramp}}} = 0.053$$

It can be easily shown by using MATLAB that the ramp responses of the original and the compensated systems are very close to each other. The same holds for the root loci. Note that even smaller steady state errors can be obtained if we increase the ratio z_c/p_c, e.g. to $z_c/p_c = 100$.

◇

8.5.2 Improvement of Transient Response

The transient response can be improved by using either the PD or phase-lead controllers. In the following, we consider these two controllers independently. However, both of them have the common feature of introducing a positive phase shift, and both of them can be implemented in a similar manner.

PD Controller Design

The PD controller is represented by

$$G_c(s) = s + z_c, \quad z_c > 0 \tag{8.28}$$

which indicates that the compensated system open-loop transfer function will have one additional zero. The effect of this zero is to introduce a positive phase shift. The phase shift and position of the compensator's zero can be determined by using simple geometry. That is, for the chosen dominant complex conjugate poles that produce the desired transient response we apply the root locus angle rule given in formula (7.10) and presented in Table 7.1 as rule number 10. This rule basically says that for a chosen point, s_d, on the root locus the difference of the sum of the angles between the point s_d and the open-loop zeros, and the sum of the angles between the point s_d and the open-loop poles must be $180°$. Applying the root locus angle rule to the compensated system, we get

$$\angle G_c(s_d)G(s_d) = \angle(s_d + z_c) + \sum_{i=1}^{m} \angle G(s_d + z_i) - \sum_{i=1}^{n} \angle G(s_d + p_i) = 180°$$
$$\tag{8.29}$$

which implies

$$\angle(s_d + z_c) = 180° - \sum_{i=1}^{m} \angle G(s_d + z_i) + \sum_{i=1}^{n} \angle G(s_d + p_i) = \alpha_c \tag{8.30}$$

From the obtained angle $\angle(s_d + z_c)$ the location of the compensator's zero is obtained by playing simple geometry as demonstrated in Figure 8.10. Using this figure it can be easily shown that the value of z_c is given by

$$z_c = \frac{\omega_n}{\tan \alpha} \left(\zeta \tan \alpha + \sqrt{1 - \zeta^2} \right) \tag{8.31}$$

An algorithm for the PD controller design can be formulated as follows.

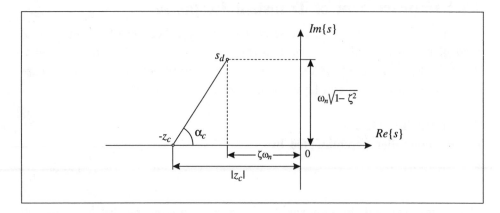

Figure 8:10 Determination of a PD controller's zero location

Design Algorithm 8.3:

1. Choose a pair of complex conjugate dominant poles in the complex plane that produces the desired transient response (damping ratio and natural frequency). Figure 6.2 helps to accomplish this goal.
2. Find the required phase contribution of a PD regulator by using formula (8.30).
3. Find the absolute value of a PD controller's zero by using formula (8.31); see also Figure 8.10.
4. Check that the compensated system has a pair of dominant complex conjugate closed-loop poles.

Example 8.8: Let the design specifications be set such that the desired maximum percent overshoot is less than 20% and the 5%-settling time is 1.5 s. Then, the formula for the maximum percent overshoot given by (6.16) implies

$$-\frac{\zeta\pi}{\sqrt{1-\zeta^2}} = \ln\{OS\} \ \Rightarrow \ \zeta = \sqrt{\frac{\ln^2\{OS\}}{\pi^2 + \ln^2\{OS\}}} = 0.456$$

We take $\zeta = 0.46$ so that the expected maximum percent overshoot is less than 20%. In order to have the 5%-settling time of 1.5 s, the natural frequency should satisfy

$$t_s \approx \frac{3}{\zeta\omega_n} \ \Rightarrow \ \omega_n \approx \frac{3}{\zeta t_s} = 4.348\,\text{rad/s}$$

The desired dominant poles are given by

$$s_d = \lambda_d = -\zeta\omega_n \pm j\omega_n\sqrt{1-\zeta^2} = -2.00 \pm j3.86$$

Consider now the open-loop control system

$$G(s) = \frac{K(s+10)}{(s+1)(s+2)(s+12)}$$

The root locus of this system is represented in Figure 8.11a.

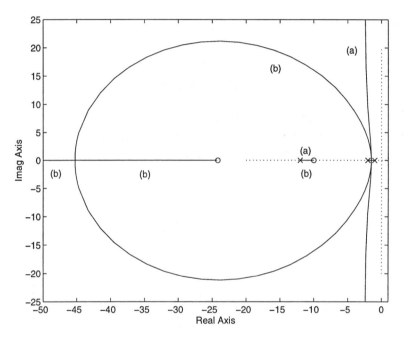

Figure 8.11: Root loci of the original (a) and compensated (b) systems

It is obvious from the above figure that the desired dominant poles do not belong
to the original root locus since the breakaway point is almost in the middle
of the open-loop poles located at -1 and -2. In order to move the original
root locus to the left such that it passes through s_d, we design a PD controller
by following Design Algorithm 8.3. Step 1 has been already completed in the
previous paragraph. Since we have determined the desired operating point, s_d, we
now use formula (8.30) to determine the phase contribution of a PD controller. By

MATLAB function `angle` (or just using a calculator), we can find the following angles

$$\angle(s_d + z_1) = 0.4495\,\text{rad}, \quad \angle(s_d + p_1) = 1.8243\,\text{rad}$$
$$\angle(s_d + p_2) = 1.5708\,\text{rad}, \quad \angle(s_d + p_3) = 0.3684\,\text{rad}$$

Note that MATLAB function `angle` produces results in radians. Using formula (8.30), we get

$$\angle(s_d + z_c) = \pi - 0.4495 + 1.8243 + 1.5708 + 0.3684$$
$$= 0.1723\,\text{rad} = 9.8734° = \alpha$$

Having obtained the angle α, the formula (8.31) produces the location of the controller's zero, i.e. $z_c = 24.1815$, so that the required PD controller is given by

$$G_c(s) = s + 24.1815$$

The root locus of the compensated system is presented in Figures 8.11b and 8.12b. It can be seen from Figure 8.12 that the point $s_d = -2 \pm j3.86$ lies on the root locus of the compensated system.

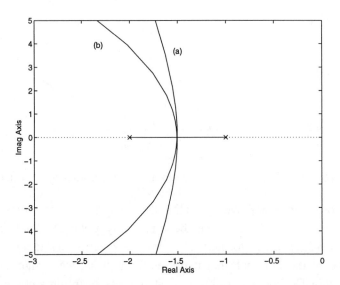

Figure 8.12: Enlarged portion of the root loci in the neighborhood of the desired operating point of the original (a) and compensated (b) systems

At the desired point, s_d, the static gain K, obtained by applying the root locus rule number 9 from Table 7.1, is given by $K = 0.825$. This value can be obtained either by using a calculator or the MATLAB function abs as follows:

```
d1=abs(sd+p1);
d2=abs(sd+p2);
d3=abs(sd+p3);
d4=abs(sd+z1);
d5=abs(sd+zc);
K=(d1*d2*d3)/(d4*d5)
```

For this value of the static gain K, the steady state errors for the original and compensated systems are given by $e_{ss} = 0.7442$, $e_{ssc} = 0.1074$. Note that in the case when $z_c > 1$, this controller can also improve the steady state errors. In addition, since the controller's zero will attract one of the system poles for large values of K, it is not advisable to choose small values for z_c since it may damage the transient response dominance by the pair of complex conjugate poles closest to the imaginary axis.

The closed-loop step response for this value of the static gain is presented in Figure 8.13. It can be observed that both the maximum percent overshoot and the settling time are within the specified limits.

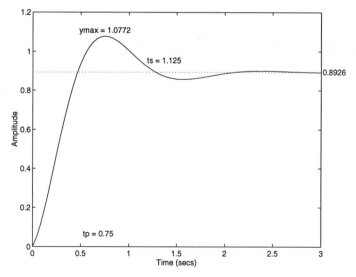

Figure 8.13: Step response of the compensated system for Example 8.8

The values for the overshoot, peak time, and settling time are obtained by the following MATLAB routine:

```
[yc,xc,t]=step[cnumc,cdenc];
% t is a time vector of length i=73;
% cnumc = closed-loop compensated numerator
% cdenc = closed-loop compensated denominator
plot(t,yc);
[ymax,imax]=max(yc);
% ymax is the function maximum;
% imax = time index where maximum occurs;
tp=t(imax)
essc=0.1074;
yss=1-essc;
os=ymax-yss
% procedure for finding the settling time;
delt5=0.05*yss;
i=73;
while (yc(i)-yss)<delt5;
i=i-1;
end;
ts=t(i)
```

Using this program, we have found that $t_s = 1.125\,\mathrm{s}$ and $MPOS = 20.68\%$. Our starting assumptions have been based on a model of the second-order system. Since the second-order systems are only approximations for higher-order systems that have dominant poles, the obtained results are satisfactory.

Finally, we have to check that the system response is dominated by a pair of complex conjugate poles. Finding the closed-loop eigenvalues we get $\lambda_1 = -11.8251$, $\lambda_{2,3} = -2.000 \pm j3.8600$, which indicates that the presented controller design results are correct since the transient response is dominated by the eigenvalues $\lambda_{2,3}$.

◇

Phase-Lead Controller Design

The phase-lead controller works on the same principle as the PD controller. It uses the argument rule, formula (7.10), of the root locus method, which indicates the phase shift that needs to be introduced by the phase-lead controller such that the desired dominant poles (having the specified transient response characteristics) belong to the root locus.

The general form of this controller is given by (8.19), that is

$$G_c(s) = \frac{s + z_c}{s + p_c}, \quad p_c > z_c > 0$$

By choosing a point s_d for a dominant pole that has the required transient response specifications, the design of a phase-lead controller can be done in similar fashion to that of a PD controller. First, find the angle contributed by a controller such that the point s_d belongs the root locus, which can be obtained from

$$\angle G_c(s_d) = 180° - \angle G(s_d) \tag{8.32}$$

that is

$$\theta_c = \angle(s_d + z_c) - \angle(s_d + p_c) = 180° - \sum_{i=1}^{m} \angle(s_d + z_i) + \sum_{i=1}^{n} \angle(s_d + p_i) \tag{8.33}$$

Second, find locations of controller's pole and zero. This can be done in many ways as demonstrated in Figure 8.14.

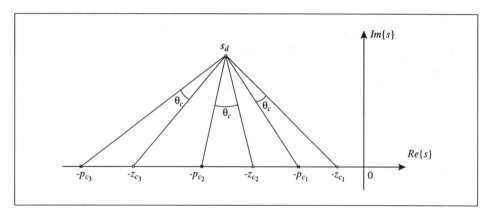

Figure 8.14: Possible locations for poles and zeros of phase-lead controllers that have the same angular contribution

All these controllers introduce the same phase shift and have the same impact on the transient response. However, the impact on the steady state errors is different since it depends on the ratio of z_c/p_c. Since this ratio for a phase-lead controller is less than one, we conclude from formula (8.27) that the corresponding steady state constant is reduced and the steady state error is increased.

Note that if the location of a phase-lead controller zero is chosen, then simple geometry, similar to that used to derive formula (8.31), can be used to find the location of the controller's pole. For example, let $-z_{c3}$ be the required zero, then using Figure 8.14 the pole $-p_{c3}$ is obtained as

$$p_{c3} = \zeta\omega_n + \omega_n\sqrt{1-\zeta^2}\tan\left(\theta_c - \varphi + \pi/2\right) \tag{8.34}$$

where $\varphi = \angle(s_d + z_{c3})$. Note that $\varphi > \theta_c$.

An algorithm for the phase-lead controller design can be formulated as follows.

Design Algorithm 8.4:

1. Choose a pair of complex conjugate poles in the complex plane that produces the desired transient response (damping ratio and natural frequency). Figure 6.2 helps to accomplish this goal.
2. Find the required phase contribution of a phase-lead controller by using formula (8.33).
3. Choose values for the controller's pole and zero by placing them arbitrarily such that the controller will not damage the response dominance of a pair of complex conjugate poles. Some authors (e.g. Van de Verte, 1994) suggest placing the controller zero at $-\zeta\omega_n$.
4. Find the controller's pole by using formula (8.34).
5. Check that the compensated system has a pair of dominant complex conjugate closed-loop poles.

Example 8.9: Consider the following control system represented by its open-loop transfer function

$$G(s) = \frac{K(s+6)}{(s+10)(s^2 + 2s + 2)}$$

It is desired that the closed-loop system have a settling time of $1.5\,\mathrm{s}$ and a maximum percent overshoot of less than 20%. From Example 8.8 we know that the system operating point should be at $s_d = -2 \pm j3.86$. A controller's phase contribution, obtained from formula (8.33) is

$$\theta_c = \pi - 0.7676 + 0.4495 + 1.9072 + 1.7737$$
$$= 6.5044\,\text{rad} = 0.2213\,\text{rad} = 12.6769°$$

Let us locate a zero at -15 ($z_c = 15$), then by (8.34) the compensator's pole is at $-p_c = -59.2025$. The root loci of the original and compensated systems are given in Figure 8.15, and the corresponding step responses in Figure 8.16.

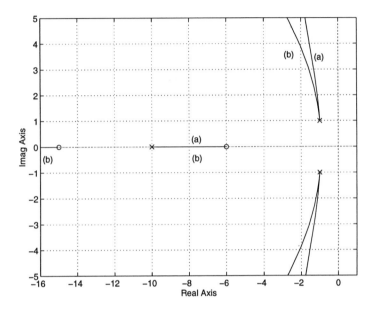

Figure 8.15: Root loci for the original (a) and compensated (b) systems

It can be seen that the root locus indeed passes through the point $-2 \pm j3.86$. For this operating point the static gain is obtained as $K = 101.56$; hence the steady state constants of the original and compensated systems are given by $K_p = 30.468$ and $K_{pc} = K_p(z_c/p_c) = 7.7196$, and the steady state errors are $e_{ss} = 0.0317, e_{ssc} = 0.1147$. Figure 8.16 reveals that for the compensated system both the maximum percent overshoot and settling time are reduced. However, the steady state unit step error is increased, as previously noted analytically.

Consider now another phase-lead compensator with a zero set at -9. From (8.34) we get $p_c = 15.291$. The root locus of the compensated system with a new controller is given in Figure 8.17.

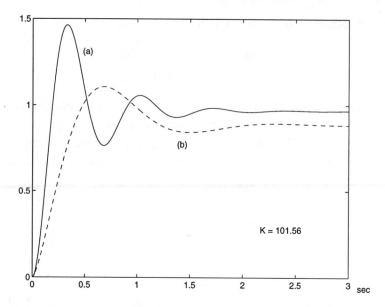

Figure 8.16: Step responses of the original (a) and compensated (b) systems

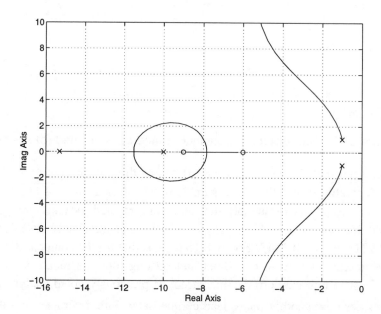

Figure 8.17: Root locus for the compensated
system with the second controller for Example 8.9

The static gain at the desired operating point $-2 \pm j3.86$ is $K = 41.587$, and hence the steady state errors are $e_{ss} = 0.0742$, $e_{ssc} = 0.11986$. The step responses of the original and compensated systems, for $K = 41.587$, are presented in Figure 8.18.

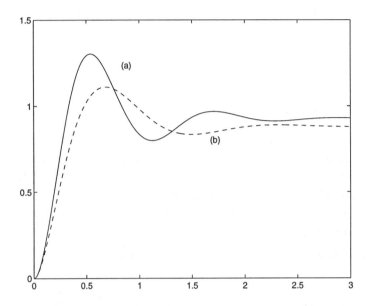

Figure 8.18: Step responses of the original (a) and compensated (b) systems with the second controller for Example 8.9

It can be seen that this controller also reduces both the overshoot and settling time, while the steady state error is slightly increased.

We can conclude that both controllers produce similar transient characteristics and similar steady state errors, but the second one is preferred since the smaller value for the static gain of the compensated system has to be used. The eigenvalues of the closed-loop system for $K = 41.587$ are given by

$$\lambda_{1c} = -12.4165, \quad \lambda_{2c} = -10.8725, \quad \lambda_{2c,3c} = -2.000 \pm j3.8600$$

which indicates that the response of this system is still dominated by a pair of complex conjugate poles.

◇

Remark: In some applications for a chosen desired point, s_d, the required phase increase, θ_c, may be very high. In such cases one can use a *multiple phase-lead controller* having the form

$$G_{lead}^n(s) = \left(\frac{s + z_c}{s + p_c}\right)^n, \qquad p_c > z_c > 0 \tag{8.35}$$

so that each single phase-lead controller has to introduce a phase increase of θ_c/n.

8.5.3 PID and Phase-Lag-Lead Controller Designs

It can be observed from the previous design algorithms that implementation of a PI (phase-lag) controller does not interfere with implementation of a PD (phase-lead) controller. Since these two groups of controllers are used for different purposes—one to improve the transient response and the other to improve the steady state errors—implementing them jointly and independently will take care of both controller design requirements.

Consider first a PID controller. It is represented as

$$G_{PID}(s) = K_p + K_d s + \frac{K_i}{s} = K_d \frac{s^2 + \frac{K_p}{K_d}s + \frac{K_i}{K_d}}{s}$$

$$= K_d(s + z_{c_1})\frac{(s + z_{c_2})}{s} = G_{PD}(s)G_{PI}(s) \tag{8.36}$$

which indicates that the transfer function of a PID controller is the product of transfer functions of PD and PI controllers. Since in Design Algorithms 8.1 and 8.3 there are no conflicting steps, the design algorithm for a PID controller is obtained by combining the design algorithms for PD and PI controllers.

Design Algorithm 8.5: PID Controller

1. Check the transient response and steady state characteristics of the original system.
2. Design a PD controller to meet the transient response requirements.
3. Design a PI controller to satisfy the steady state error requirements.
4. Check that the compensated system has the desired specifications.

Example 8.10: Consider the problem of designing a PID controller for the open-loop control system studied in Example 8.8, that is

$$G(s) = \frac{K(s+10)}{(s+1)(s+2)(s+12)}$$

In fact, in that example, we have designed a PD controller of the form

$$G_{PD}(s) = s + 24.1815$$

such that the transient response has the desired specifications. Now we add a PI controller in order to reduce the steady state error. The corresponding steady state error of the PD compensated system in Example 8.8 is $e_{ssc} = 0.1074$. Since a PI controller is a dipole that has its pole at the origin, we propose the following PI controller

$$G_{PI}(s) = \frac{s + 0.1}{s}$$

In comparison to (8.36), we are in fact using a PID controller with $K_d = 1$, $z_{c1} = 24.1815$, $z_{c2} = 0.1$. The corresponding root locus of this system compensated by a PID controller is represented in Figure 8.19.

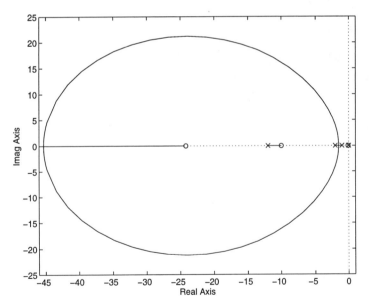

Figure 8.19: Root locus for the system from
Example 8.8 compensated by the PID controller

It can be seen that the PI controller does not affect the root locus, and hence Figures 8.11b and 8.19 are almost identical except for a dipole branch.

On the other hand, the step responses of the system compensated by the PD controller and by the PID controller (see Figures 8.13 and 8.20) differ in the steady state parts. In Figure 8.13 the steady state step response tends to $y_{ss} = 0.8926$, and the response from Figure 8.20 tends to 1 since due to the presence of an open-loop pole at the origin, the steady state error is reduced to zero. Thus, we can conclude that the transient response is the same one as that obtained by the PD controller in Example 8.8, but the steady state error is improved due to the presence of the PI controller.

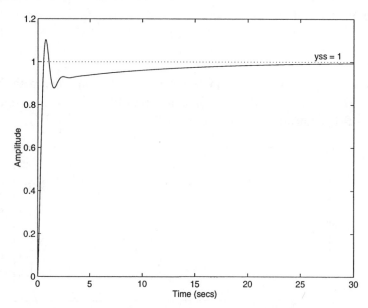

Figure 8.20: Step response of the system from
Example 8.8 compensated by the PID controller

◇

Similarly to the PID controller, the design for the phase-lag-lead controller combines Design Algorithms 8.2 and 8.4. Looking at the expression for a phase-lag-lead controller given in formula (8.20), it is easy to conclude that

$$G_{lag/lead}(s) = G_{lag}(s)G_{lead}(s) \tag{8.37}$$

The phase-lag-lead controller design can be implemented by the following algorithm.

Design Algorithm 8.6: Phase-Lag-Lead Controller

1. Check the transient response and steady state characteristics of the original system.
2. Design a phase-lead controller to meet the transient response requirements.
3. Design a phase-lag controller to satisfy the steady state error requirements.
4. Check that the compensated system has the desired specifications.

Example 8.11: In this example we design a phase-lag-lead controller for a control system from Example 8.9, that is

$$G(s) = \frac{K(s+6)}{(s+10)(s^2+2s+2)}$$

such that both the system transient response and steady state errors are improved. We have seen in Example 8.9 that a phase-lead controller of the form

$$G_{lead}(s) = \frac{s+9}{s+15.291}$$

improves the transient response to the desired one. Now we add in series with the phase-lead controller another phase-lag controller, which is in fact a dipole near the origin. For this example we use the following phase-lag controller

$$G_{lag}(s) = \frac{s+0.1}{s+0.01}$$

so that the compensated system becomes

$$G(s) = G(s)G_c(s) = \frac{K(s+6)}{(s+10)(s^2+2s+2)} \frac{(s+9)}{(s+15.291)} \frac{(s+0.1)}{(s+0.01)}$$

The corresponding root locus of the compensated system and its closed-loop step response are represented in Figures 8.21 and 8.22. We can see that the addition of the phase-lag controller does not change the transient response, i.e. the root loci in Figures 8.17 and 8.21 are almost identical. However, the phase-lag controller reduces the steady state error from $e_{ss,lead} = 0.11986$ to $e_{ss,lag/lead} = 0.01344$ since the position constant is increased to

$$K_{p,lag/lead} = K_{p,lead}\frac{0.1}{0.01} = \frac{41.587 \times 9 \times 0.1}{10 \times 2 \times 15 \times 0.01} = 73.432$$

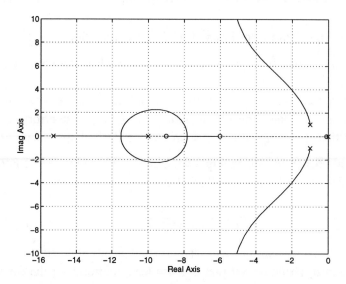

Figure 8.21: Root locus for the system from Example
8.9 compensated by the phase-lag-lead controller

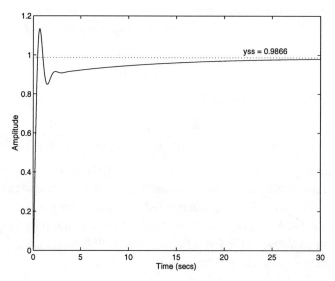

Figure 8.22: Step response of the system from Example
8.9 compensated by the phase-lag-lead controller

so that

$$e_{ss,lag/lead} = \frac{1}{1 + K_{p,lag/lead}} = 0.01344$$

$\diamond$

8.6 MATLAB Case Studies

In this section we consider the compensator design for two real control systems: a PD controller designed to stabilize a ship, and a PID controller used to improve the transient response and steady state errors of a voltage regulator control system.

8.6.1 Ship Stabilization by a PD Controller

Consider a ship positioning control system defined in the state space form in Problem 7.5. The open-loop transfer function of this control system is

$$G(s) = \frac{0.8424}{s(s + 0.0546)(s + 1.55)}$$

The root locus of the original system is presented in Figure 8.23a.

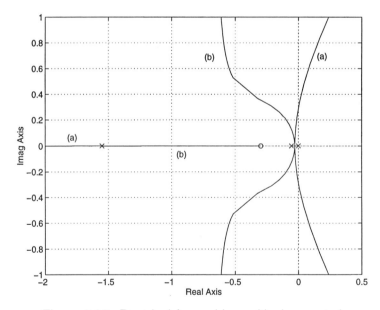

Figure 8.23: Root loci for a ship positioning control problem: (a) original system, (b) compensated system

It can be seen that this system is unstable even for very small values of the static gain. Thus, the system transient response blows up very quickly due to the system's instability. Our goal is to design a PD controller in order to stabilize the system and improve its transient response. Let the desired operating point be located at $s_d = -0.2 \pm j0.3$, which implies $w_n = 0.3606\,\text{rad/s}$ and $\zeta = 0.5547$. We find from (8.30) that the required phase shift is $\alpha_c = 72.0768°$, and from (8.31) the location of the compensator zero is obtained at -0.297. Thus, the PD compensator sought is of the form

$$G_c(s) = s + 0.297$$

It can be seen from Figure 8.23 that the root locus of the compensated system indeed passes through the point $s_d = -0.2 \pm j0.3$ and that the compensated system is stable for all values of the static gain. The static gain at the desired operating point is given by $K_{s_d} = 0.6258$ and the corresponding closed-loop eigenvalues at this operating point are $\lambda_{1c} = -1.2046$, $\lambda_{2c,3c} = -0.2 \pm j0.3$. In Figure 8.24 the unit step response of the compensated system is presented.

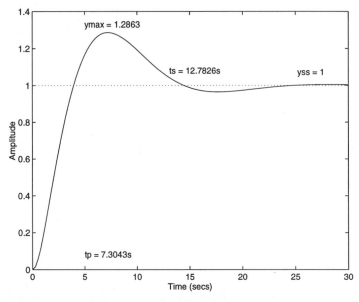

Figure 8.24: Step response of a ship positioning compensated control system

It is found that $y_{max} = 1.2863$, $t_p = 7.3043\,\text{s}$, and $t_s = 12.7826\,\text{s}$. From the same figure we observe that the steady state error for this system is zero, which also follows from the fact that the system open-loop transfer function has one pole at the origin.

8.6.2 PID Controller for a Voltage Regulator Control System

The mathematical model of a voltage regulator control system (Kokotović, 1972) is given in Section 6.7. The open-loop transfer function of this system is

$$G(s) = \frac{154280}{(s+0.2)(s+0.5)(s+10)(s+14.28)(s+25)}$$

The corresponding root locus is presented in Figure 8.25. Since one of the branches goes quite quickly into the instability region, our design goal is to move this branch to the left so that it passes through the operating point selected as $s_d = -1 \pm j1$. For this operating point, we have $w_n = \sqrt{2}$ rad/s and $\zeta = 0.7071$ so that the expected maximum percent overshoot and the 5%-settling time of the compensated system are $MPOS = 4.3214\%$, $t_s = 3$ s. In addition, the design objective is to reduce the steady state error due to a unit step to zero.

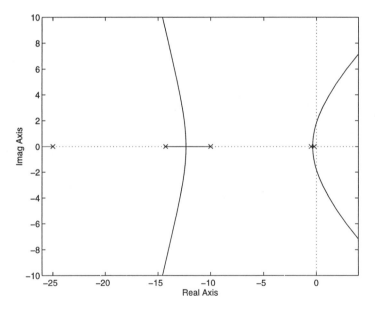

Figure 8.25: Root locus for a voltage regulator system

We use a PID controller to solve the controller design problem defined above. The required phase improvement for the selected operating point is found by using

(8.30) as $\alpha_c = 1.3658\,\text{rad} = 78.2573°$. From formula (8.31) the location of the compensator's zero is obtained as $-z_c = -1.2079$, so that the PD part of a PID compensator is

$$G_{PD}(s) = s + 1.2079$$

The branches of the root loci in the neighborhood of the desired operating point of the original and PD compensated systems are presented in Figure 8.26. It can be seen that the compensated root locus indeed passes through the point $s_d = -1 \pm j1$.

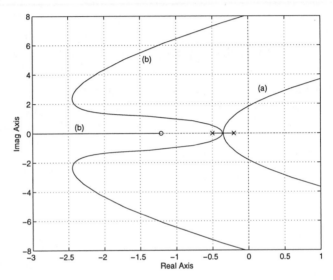

Figure 8:26 Root loci of the original (a) and PD (b) compensated systems

The closed-loop unit step response of the system compensated by the PD controller is represented in Figure 8.27. Using the MATLAB programs given in Example 8.8, gives $MPOS = 6.08\%$, $t_p = 2.1\,\text{s}$, and $t_s = 3.5\,\text{s}$, which is quite satisfactory. However, the steady state unit step error is $e_{ssPD} = 0.0808$. Note that the static gain at the operating point, obtained by applying the root locus rule number 9 from Table 7.1, is $K_{s_d} = 4060.8$. The closed-loop eigenvalues at the operating point are

$$\lambda_{1_{PD}} = -23.7027, \quad \lambda_{2_{PD}} = -18.1675, \quad \lambda_{3_{PD}} = -6.1105$$
$$\lambda_{4,5_{PD}} = -0.997 \pm j1.0011$$

which indicates that the system has preserved a pair of dominant complex conjugate poles.

In order to reduce this steady state error to zero we use a PI controller of the form

$$G_{PI}(s) = \frac{s + 0.1}{s}$$

Since the compensated system open-loop transfer function now has a pole at the origin, we conclude that the steady state error is reduced to zero, which can also be observed from Figure 8.27.

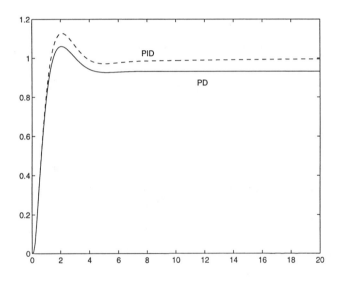

Figure 8.27: Step responses of PD and PID compensated systems

The transient response specifications for the system compensated by the proposed PID controller are $MPOS = 11.277\%$, $t_p = 2.1\,\text{s}$, and $t_s = 3.1\,\text{s}$. Thus, the proposed PI controller has slightly worsened the transient response characteristics. It is left to students, in the form of a MATLAB laboratory experiment, to check that the transient response specifications of the compensated system obtained by using PI controllers that have zeros located at -0.01 and -0.001 are improved.

8.7 Comments on Discrete-Time Controller Design

Similarly to the continuous-time controller design, the root locus method can be used for design of controllers (compensators) in the discrete-time domain. In Section 7.4 we have indicated that the discrete-time root locus method is identical to its continuous-time counterpart. Thus, the results presented in this chapter can be easily extended and used for the controller design of discrete-time systems. For more details, the reader is referred to Kuo (1992) for theoretical aspects, and to Shahian and Hassul (1993) for MATLAB discrete-time controller design.

8.8 MATLAB Laboratory Experiment

Part 1. Consider the control system given in Example 8.8, that is

$$G(s) = \frac{K(s+10)}{(s+1)(s+2)(s+12)}$$

with the transient response requirements determined by a desired operating point located at $s_d = -2 + j3$.

(a) Design a PD controller such that the transient response requirements are met.
(b) Design a PID controller such that the steady state error is reduced to zero.
(c) Design a phase-lead controller for the design problem defined in (a).
(d) Add a phase-lag controller in series with the controller obtained in (c) in order to reduce the steady state error by 50%.

Plot the root loci and unit step responses for (b) and (d) of both the original and compensated systems and compare the results obtained with PID and phase-lag-lead controllers.

Part 2. Use MATLAB to design a phase-lag-lead controller for the control system of Example 8.9 such that $t_s < 2\,\mathrm{s}$, $MPOS < 10\%$, and $e_{ssc} < 0.01$.

Use the MATLAB program from Example 8.8 to find the actual response overshoot and settling time. Find the eigenvalues of the compensated system and check whether the system response is dominated by a pair of complex conjugate poles.

Part 3. For the voltage regulator system considered in Section 8.6.2, use the same PD controller as in Section 8.6.2, but take for a PI controller the following forms:

(a) $G_{PI}(s) = (s + 0.01)/s$.

(b) $G_{PI}(s) = (s + 0.001)/s$.

For both cases, find the unit step responses of the PID compensated systems and determine the transient response parameters. Compare the results obtained with the results from Section 8.6.2.

8.9 References

Arnautović, D. and D. Skatarić, "Suboptimal design of hydroturbine governors," *IEEE Transactions on Energy Conversion*, vol. 6, 438–444, 1991.

Chen, C., *Linear System Theory Design*, CBS College Publishing, New York, 1984.

Dorf, D., *Modern Control Systems*, Addison-Wesley, Reading, Massachusetts, 1992.

Kokotović, P., "Feedback design of large scale linear systems," in *Feedback Systems*, J. Cruz (ed.), McGraw-Hill, New York, 1972.

Kuo, B., *Digital Control Systems*, Saunders College Publishing, New York, 1992.

Kuo, B., *Automatic Control Systems*, Prentice Hall, Englewood Cliffs, New Jersey, 1995.

Nise, N., *Control System Engineering*, Benjamin Cummings, Redwood City, California, 1992.

Shahian, B. and M. Hassul, *Control System Design Using MATLAB*, Prentice Hall, Englewood Cliffs, New Jersey, 1993.

Van de Verte, J., *Feedback Control Systems*, Prentice Hall, Englewood Cliffs, New Jersey, 1994.

8.10 Problems

8.1 Consider a single-input single-output system whose open-loop transfer function is given by

$$G(s) = \frac{10}{s^4 + 3s^3 + 4s^2 + 1}$$

Find the phase variable canonical form for this system and design the full state feedback static controller such that the system closed-loop poles are located at $\lambda_{1,2} = -1 \pm j2, \lambda_3 = -3, \lambda_4 = -10$.

8.2   Find the feedback gain **f** such that the system

$$\dot{\mathbf{x}} = \begin{bmatrix} -1 & 2 & 3 \\ 0 & -2 & 1 \\ 1 & -1 & -5 \end{bmatrix} \mathbf{x} + \begin{bmatrix} 2 \\ 0 \\ 1 \end{bmatrix} u$$

with $u = -\mathbf{f}\mathbf{x}$ has a pair of dominant complex conjugate closed-loop poles at $-1 \pm j2$ and a closed-loop pole at -10.

8.3   Plot the function

$$e^{-\frac{\zeta\pi}{\sqrt{1-\zeta^2}}}$$

and comment on the maximum percent overshoot dependence on the damping ratio.

8.4   Design a rate feedback controller such that for the second-order system represented by the closed-loop transfer function

$$\frac{9}{s^2 + 3s + 9}$$

the maximum percent overshoot is 10%. Find the settling time of the compensated system.

8.5   Consider the second-order system from Example 6.6. Its steady state error is $e_{ss} = 0.4$. Design a PI controller to reduce the steady state error to zero. Use MATLAB to plot the root loci and unit step responses of the original and compensated systems.

8.6   For the open-loop control system

$$G(s) = \frac{K(s+10)}{s(s+20)(s^2 + 2s + 10)}$$

draw the root locus. Check that the static gain $K = 3$ produces a pair of complex conjugate dominant poles. Find the corresponding steady state errors and transient response parameters. Design a PI controller such that the steady state error is reduced to zero while the transient response characteristics are preserved.

8.7 Consider the controller design problem for the hydroturbine governors of a power system (Arnautović and Skatarić, 1991), represented by

$$A = \begin{bmatrix} -0.71 & 0 & 0 & 0 & 0 \\ 0 & -2 & 0 & 0 & 0 \\ 0.61 & 1.28 & -1.46 & 0.566 & 0 \\ -0.18 & -0.37 & 0.56 & -0.594 & -0.23 \\ 0 & 0 & 0 & 314.16 & 0 \end{bmatrix}, \quad B = \begin{bmatrix} 0.71 \\ 2 \\ 0 \\ 0 \\ 0 \end{bmatrix}$$

Assume that the output matrices are given by

$$C = \begin{bmatrix} 0 & 0 & 1 & 0 & 1 \end{bmatrix}, \quad D = 0$$

Using MATLAB, perform the following:

(a) Find its open-loop transfer function and the steady state error.
(b) Suggest a phase-lag controller to reduce its steady state error ten times.
(c) Locate the system operating point on the root locus approximately at $-1.5 + j1$ and find the static gain at that point.

8.8 Design a phase-lag controller for the system represented by

$$G(s) = \frac{K}{(s+1)(s+5)(s+10)}$$

which produces a steady state unit step error of less than 0.01. Take $K = 200$.

8.9 Consider the synchronous machine from Section 7.5.2. Design a PI controller to reduce the steady state unit step error by 50%. Choose an operating point on the root locus and find the corresponding static gain. Use MATLAB to check that the closed-loop system is asymptotically stable.

8.10 For the second-order system from Example 6.1, design a PD controller such that the compensated system has $t_s \approx 2\,\mathrm{s}$ and $MPOS < 15\%$.

8.11 Repeat Problem 8.10 with the following requirements $t_s \approx 3\,\mathrm{s}$ and $MPOS < 10\%$.

8.12 Solve Problem 8.11 using a phase-lead controller.

8.13 Consider the problem of a PD controller design. Assume that a controller's zero and desired operating point satisfy $\angle(s_d + z_c) > 90°$. Derive a formula corresponding to (8.31).

8.14 Consider the phase-lag controller design for the system given in Example 8.9. Use different values for the controller's poles and zeros and examine their impact on the steady state errors. Suggest at least five different controllers.

8.15 Derive formula (8.34) for finding the location of the pole of a phase-lead controller assuming that the location of its zero is chosen.

8.16 Design a phase lead-controller such that

$$G(s) = \frac{K(s+10)}{(s+20)(s^2+2s+2)}$$

has $t_s \leq 1.5\,\mathrm{s}$ and $MPOS \leq 20\%$.

8.17 Consider the voltage regulator control system from Section 8.6.2. Use a double PD controller of the form

$$G_{PD}(s) = (s+1.2079)^2$$

to compensate this system.

(a) Draw the root locus of this system compensated with a double PD controller and compare it with the corresponding one from Section 8.6.2 obtained using a single PD controller.

(b) Choose the operating point for a pair of complex conjugate poles such that the damping ratio is the same as in Section 8.6.2, i.e. $\zeta = 0.7071$. Find the static gain at that point and the closed-loop eigenvalues. Does the compensated system preserve the response dominance of a pair of complex conjugate poles?

8.18 Design a phase-lag-lead controller for the voltage regulator system from Section 8.6.2 such that $MPOS < 5\%$, $t_s < 3\,\mathrm{s}$, and $e_{ss} < 1\%$.

8.19 Consider the ship position control system from Section 8.6.1. Design a phase-lead controller such that $MPOS < 10\%$ and $t_s < 10\,\mathrm{s}$.

8.20 Use MATLAB to find and plot the ramp responses of the original and compensated systems studied in Example 8.7.

8.21 Consider the F-15 aircraft under supersonic flight conditions with its state space matrices given in Example 1.4. Note that this aircraft has one input and four outputs, where the outputs represent the state space variables. Thus, we are able to get four transfer functions of the form

$$G_i(s) = \mathbf{C}_i(s\mathbf{I} - \mathbf{A})^{-1}\mathbf{B}, \qquad i = 1, 2, 3, 4$$

(a) Using MATLAB, find all four transfer functions of the F-15 aircraft.
(b) Plot the root loci for $G_i(s)$, $i = 1, 2, 3, 4$. Comment on the stability properties of this aircraft with respect to the values of the static feedback gains K_i, $i = 1, 2, 3, 4$.
(c) Find the closed-loop transfer functions with unit feedback and examine their stability.
(d) Propose controllers which will stabilize all outputs of this aircraft.
(e) For the proposed controllers that assure stabilization, find the closed-loop step responses and determine the response steady state errors and transient parameters.
(f) If necessary, design dynamic controllers based on the root locus technique, as discussed in this chapter, such that steady state errors and transient response parameters are improved.

8.22 Repeat Problem 8.21 for the F-15 aircraft under subsonic flight conditions with the state space matrices given in Example 1.4.[1]

---

[1] Problems 8.21 and 8.22 can be assigned as either term papers or final projects.

Chapter Nine

Frequency Domain Controller Design

9.1 Introduction

Frequency domain techniques together with the root locus method have been very popular classical methods for both analysis and design of control systems. As they are still used extensively in industry for solving controller design problems for many industrial processes and systems, they have to be included in academic curricula and modern textbooks on control systems. This chapter is organized as follows.

In Section 9.2 we study the open- and closed-loop frequency transfer functions and identify the frequency response parameters such as system frequency bandwidth, peak resonance, and resonant frequency.

In Section 9.3 we show how to read the phase and gain stability margins, and the values of the steady state error parameters K_p, K_v, K_a from the corresponding frequency diagrams, known as Bode diagrams. Bode diagrams represent the magnitude and phase plots of the open-loop transfer function with respect to the angular frequency ω. They can be obtained either analytically or experimentally. In this chapter we present only the analytical study of Bode diagrams, though it should be mentioned that Bode diagrams can be obtained from experimental measurements performed on a real physical system driven by a sinusoidal input with a broad range of frequencies.

The controller design technique based on Bode diagrams is considered in Section 9.4. It is shown how to use the phase-lag, phase-lead, and phase-lag-lead controllers such that the compensated systems have the desired phase and gain stability margins and steady state errors. It is also possible in some cases to improve the transient response since the damping ratio is proportional to the phase margin, and the response rise time is inversely proportional to the system bandwidth.

Section 9.5 contains a case study for a ship positioning control system, and Section 9.7 represents a laboratory experiment. In Section 9.6 we comment on discrete-time controller design.

Chapter Objectives

The main objective of this chapter is to show how to use Bode diagrams as a tool for controller design such that compensated systems have, first of all, the desired phase and gain stability margins and steady state errors. Controllers based on Bode diagrams can also be used to improve some of the transient response parameters, but their design is far more complicated and far less accurate than the design of the corresponding controllers based on the root locus method.

9.2 Frequency Response Characteristics

The open- and closed-loop system transfer functions are defined in Chapter 2. For a feedback control system these transfer functions are respectively given by

$$\text{Open–loop: } G(s)H(s), \quad \text{Closed–loop}: \frac{G(s)}{1 + G(s)H(s)} = M(s) \quad (9.1)$$

The frequency transfer functions are defined for sinusoidal inputs having all possible frequencies $\omega \in [0, +\infty)$. They are obtained from (9.1) by simply setting $s = j\omega$, that is

$$\text{Open–loop: } G(j\omega)H(j\omega), \quad \text{Closed–loop:} \frac{G(j\omega)}{1 + G(j\omega)H(j\omega)} = M(j\omega)$$

$$(9.2)$$

Using the frequency transfer functions in the system analysis gives complete information about the system's steady state behavior, but not about the system's transient response. That is why controller design techniques based on frequency transfer functions improve primarily the frequency domain specifications such as phase and gain relative stability margins and steady state errors. However, some frequency domain specifications, to be defined soon, can be related to certain time domain specifications. For example, the wider the system bandwidth, the faster system response, which implies a shorter response rise time.

Typical diagrams for the magnitude and phase of the open-loop frequency transfer function are presented in Figure 9.1. From this figure one is able to read directly the phase and gain stability margins and the corresponding phase and gain crossover frequencies. In Section 9.3, where we present the Bode diagrams, which also represent the magnitude and phase plots of the open-loop frequency transfer function with respect to frequency, with magnitude being calculated in decibels (dB), we will show how to read the values for K_p, K_v, K_a.

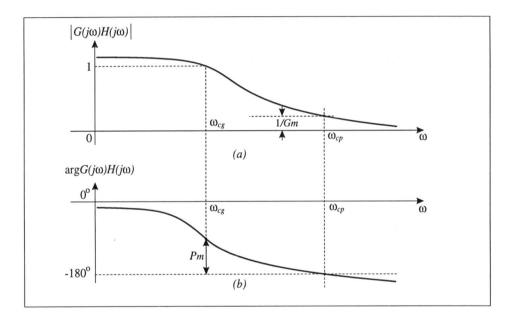

Figure 9.1: Magnitude (a) and phase (b) of the open-loop transfer function

In addition to the phase and gain margins, obtained from the *open-loop* frequency transfer function, other frequency response parameters can be obtained from the frequency plot of the magnitude of the *closed-loop* frequency transfer function. This plot is given in Figure 9.2. The main *closed-loop* frequency response parameters are: system bandwidth, peak resonance, and resonant frequency. They are formally defined below.

System Bandwidth: This represents the frequency range in which the magnitude of the closed-loop frequency transfer function drops no more than 3 dB (decibels) from its zero-frequency value. The system bandwidth can be obtained from the next equality, which indicates the attenuation of 3 dB, as

$$|M(j\omega_{BW})| = \frac{1}{\sqrt{2}}|M(0)| \Rightarrow \omega_{BW} \qquad (9.3)$$

It happens to be computationally very involved to solve equation (9.3) for higher-order systems, and hence the system bandwidth is mostly determined experimentally. For second-order systems the frequency bandwidth can be found analytically (see Problem 9.2).

Peak Resonance: This is obtained by finding the maximum of the function $|M(j\omega)|$ with respect to frequency ω. It is interesting to point out that the systems having large maximum overshoot have also large peak resonance. This is analytically justified for a second-order system in Problem 9.1.

Resonant Frequency: This is the frequency at which the peak resonance occurs. It can be obtained from

$$\frac{d}{d\omega}|M(j\omega)| = 0 \quad \Rightarrow \quad \omega_r$$

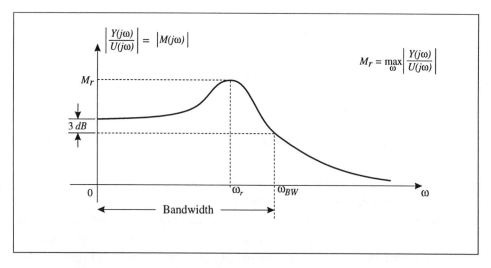

Figure 9.2: Magnitude of the closed-loop transfer function

9.3 Bode Diagrams

Bode diagrams are the main tool for frequency domain controller design, a topic that will be presented in detail in Section 9.4. In this section we show how to plot these diagrams and how to read from them certain control system characteristics, such as the phase and gain stability margins and the constants required to determine steady state errors.

Bode diagrams represent the frequency plots of the magnitude and phase of the open-loop frequency transfer function $G(j\omega)H(j\omega)$. *The magnitude is plotted in* dB *(decibels) on the* $\log \omega$ *scale.* In general, the open-loop frequency transfer function contains elementary frequency transfer functions representing a constant term (static gain) and dynamic elements like system real poles and zeros and complex conjugate poles and zeros. We first study independently the magnitude and frequency plots of each of these elementary frequency transfer functions. Since the open-loop frequency transfer function $G(j\omega)H(j\omega)$ is given in terms of products and ratios of elementary transfer functions, it is easy to see that *the phase of* $G(j\omega)H(j\omega)$ *is obtained by summing and subtracting phases of the elementary transfer functions.* Also, by expressing the magnitude of the open-loop transfer function in decibels, *the magnitude* $|G(j\omega)H(j\omega)|_{dB}$ *is obtained by adding the magnitudes of the elementary frequency transfer functions.* For example

$$|G(j\omega)H(j\omega)|_{dB} = 20\log_{10}\left|\frac{K(j\omega + z_1)(j\omega + z_2)}{(j\omega)(j\omega + p_2)(j\omega + p_3)}\right|$$

$$= 20\log_{10}|K| + 20\log_{10}|j\omega + z_1| + 20\log_{10}|j\omega + z_2|$$

$$+ 20\log_{10}\left|\frac{1}{j\omega}\right| + 20\log_{10}\left|\frac{1}{j\omega + p_2}\right| + 20\log_{10}\left|\frac{1}{j\omega + p_3}\right|$$

and

$$\arg\left\{G(j\omega)H(j\omega)\right\} = \arg\left\{K\right\} + \arg\left\{j\omega + z_1\right\} + \arg\left\{j\omega + z_2\right\}$$

$$- \arg\left\{j\omega\right\} - \arg\left\{j\omega + p_2\right\} - \arg\left\{j\omega + p_3\right\}$$

In the following we show how to draw Bode diagrams for elementary frequency transfer functions.

Constant Term: Since

$$K_{dB} = 20\log_{10} K = \begin{cases} \text{positive number} & K > 1 \\ \text{negative number} & K < 1 \end{cases}$$

(9.4)

$$\arg K = \begin{cases} 0°, & K > 0 \\ -180°, & K < 0 \end{cases}$$

the magnitude and phase of this element are easily drawn and are presented in Figure 9.3.

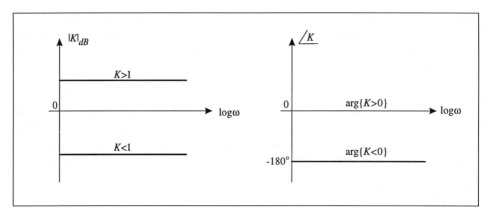

Figure 9.3: Magnitude and phase diagrams for a constant

Pure Integrator: The transfer function of a pure integrator, given by

$$G(j\omega) = \frac{1}{j\omega}$$

(9.5)

has the following magnitude and phase

$$|G(j\omega)|_{dB} = 20\log_{10}\frac{1}{\omega} = -20\log_{10}\omega, \quad \arg G(j\omega) = -90°$$

(9.6)

It can be observed that the phase for a pure integrator is constant, whereas the magnitude is represented by a straight line intersecting the frequency axis at $\omega = 1$ and having the slope of $-20\,\text{dB/decade}$. Both diagrams are represented in Figure 9.4. Thus, *a pure integrator introduces a phase shift of* $-90°$ *and a gain attenuation of* $-20\,\text{dB/decade}$.

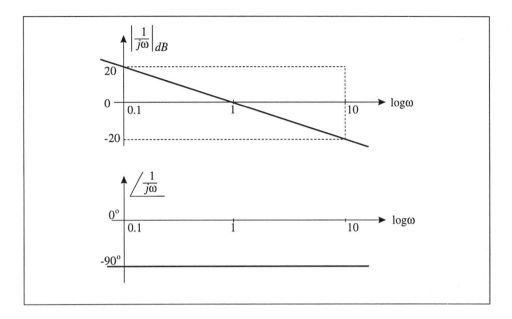

Figure 9.4: Magnitude and phase diagrams for a pure integrator

Pure Differentiator: The transfer function of a pure differentiator is given by

$$G(j\omega) = j\omega \tag{9.7}$$

Its magnitude and phase are easily obtained as

$$|G(j\omega)|_{dB} = 20\log_{10}\omega, \quad \arg G(j\omega) = 90° \tag{9.8}$$

The corresponding frequency diagrams are presented in Figure 9.5. *It can be concluded that a pure differentiator introduces a positive phase shift of* $90°$ *and an amplification of* $20\,\mathrm{dB/decade}$.

Real Pole: The transfer function of a real pole, given by

$$G(j\omega) = \frac{p}{p + j\omega} = \frac{1}{1 + j\frac{\omega}{p}} \tag{9.9}$$

has the following magnitude and phase

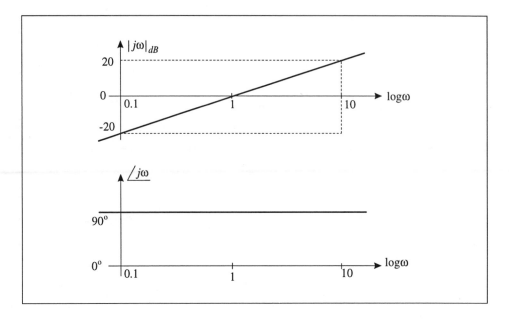

Figure 9.5: Magnitude and phase diagrams for a pure differentiator

$$|G(j\omega)|_{dB} = -20\log_{10}\left[1 + \left(\frac{\omega}{p}\right)^2\right]^{1/2}, \quad \arg G(j\omega) = -\tan^{-1}\left(\frac{\omega}{p}\right) \tag{9.10}$$

The phase diagram for a real pole can be plotted directly from (9.10). It can be seen that for large values of ω, $\omega \gg p$, the phase contribution is $-90°$. For ω small, $\omega \ll p$, the phase is close to zero, and for $\omega = p$ the phase contribution is $-45°$. This information is sufficient to sketch $\arg G(j\omega)$ as given in Figure 9.6.

For the magnitude, we see from (9.10) that for small ω the magnitude is very close to zero. For large values of ω we can neglect 1 compared to ω/p so that we have a similar result as for a pure integrator, i.e. we obtain an attenuation of $20\,\mathrm{dB/decade}$. For small and large frequencies we have straight-line approximations. These straight lines intersect at $\omega = p$, which is also known as a *corner frequency*. The actual magnitude curve is below the straightline approximations. It has the biggest deviation from the asymptotes at the corner frequency (see Figure 9.6).

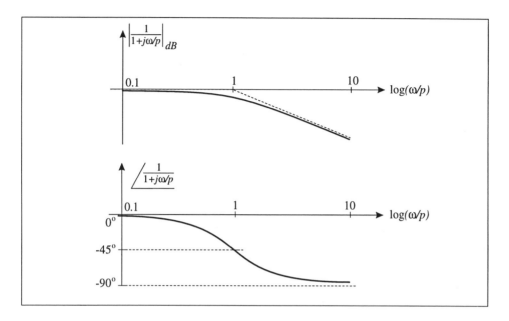

Figure 9.6: Magnitude and phase diagrams for a real pole

Real Zero: The transfer function of an element representing a real zero is given by

$$G(j\omega) = \frac{1}{z}(z + j\omega) = 1 + j\left(\frac{\omega}{z}\right) \qquad (9.11)$$

Its magnitude and phase are

$$|G(j\omega)|_{dB} = 20\log_{10}\left[1 + \left(\frac{\omega}{z}\right)^2\right]^{1/2}, \qquad \arg G(j\omega) = \tan^{-1}\left(\frac{\omega}{z}\right) \qquad (9.12)$$

Using analysis similar to that performed for a real pole, we can conclude that for small frequencies an asymptote for the magnitude is equal to zero and for large frequencies the magnitude asymptote has a slope of 20 dB/decade and intersects the real axis at $\omega = z$ (the corner frequency). The phase diagram for small frequencies also has an asymptote equal to zero and for large frequencies an asymptote of 90°. The magnitude and phase Bode diagrams for a real-zero element are represented in Figure 9.7.

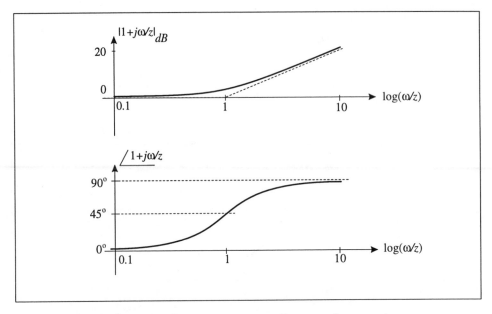

Figure 9.7: Magnitude and phase diagrams for a real zero

Complex Conjugate Poles: The transfer function of an element representing a pair of complex conjugate poles is in fact a transfer function of a second-order system, which has the form

$$G(j\omega) = \frac{w_n^2}{(j\omega)^2 + 2\zeta\omega_n(j\omega) + \omega_n^2} = \frac{1}{\left(1 - \frac{\omega^2}{\omega_n^2}\right) + j2\zeta\frac{\omega}{\omega_n}} \tag{9.13}$$

The magnitude and phase of this second-order system are given by

$$|G(j\omega)|_{dB} = -20\log_{10}\left[\left(\frac{2\zeta\omega}{\omega_n}\right)^2 + \left(1 - \frac{\omega^2}{\omega_n^2}\right)^2\right]^{1/2} \tag{9.14}$$

$$\arg G(j\omega) = -\tan^{-1}\left(\frac{2\zeta\omega_n\omega}{\omega_n^2 - \omega^2}\right)$$

For large values of ω the corresponding approximations of (9.14) are

$$|G(j\omega)|_{dB} \approx -20\log_{10}\left(\frac{\omega^2}{\omega_n^2}\right) = -40\log_{10}\left(\frac{\omega}{\omega_n}\right)$$

$$\arg\left\{G(j\omega)\right\} \approx -\tan^{-1}\left(\frac{2\zeta\omega_n}{-\omega}\right) \rightarrow -\tan^{-1}\left(0^-\right) = -180°$$

At low frequencies the approximations can be obtained directly from (9.13), that is

$$G(j\omega) \approx \frac{\omega_n^2}{\omega_n^2} = 1 \Rightarrow |G(j\omega)|_{dB} = 0, \quad \arg\left\{G(j\omega)\right\} = 0°$$

Thus, the corresponding asymptotes for small and large frequencies are, respectively, zero and $-40\,\text{dB/decade}$ (with the corner frequency at $\omega = \omega_n$) for the magnitude, and zero and $-180°$ for the phase. At the corner frequency ω_n the phase is equal to $-90°$. The corresponding Bode diagrams are represented in Figure 9.8. Note that the actual plot in the neighborhood of the corner frequency depends on the values of the damping ratio ζ. Several curves are shown for $0.1 \leq \zeta \leq 1$. It can be seen from Figure 9.8 that the smaller ζ, the higher peak of the magnitude plot.

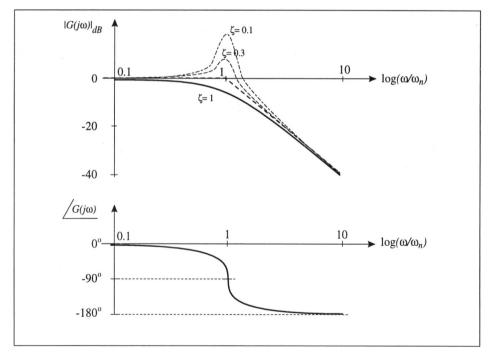

Figure 9.8: Magnitude and phase diagrams for complex conjugate poles

Complex Conjugate Zeros: An element that has complex conjugate zeros can be represented in the form

$$G(j\omega) = 1 + 2\zeta j\left(\frac{\omega}{\omega_n}\right) + \left(\frac{j\omega}{\omega_n}\right)^2 = 1 - \left(\frac{\omega}{\omega_n}\right)^2 + j2\zeta\left(\frac{\omega}{\omega_n}\right) \qquad (9.15)$$

so that the corresponding Bode diagrams will be the mirror images of the Bode diagrams obtained for the complex conjugate poles represented by (9.13). In the case of complex conjugate zeros, the asymptotes for small frequencies are equal to zero for both the magnitude and phase plots; for high frequencies the magnitude asymptote has a slope of 40 dB/decade and starts at the corner frequency of $\omega = \omega_n$, and the phase plot asymptote is $180°$.

9.3.1 Phase and Gain Stability Margins from Bode Diagrams

It has been already indicated in Figure 9.1 how to read the phase and gain stability margins from the frequency magnitude and phase plots of the open-loop feedback transfer function. In the case of Bode diagrams the magnitude plot is expressed in dB (decibels) so that the gain crossover frequency is obtained at the point of intersection of the Bode magnitude plot and the frequency axis. Bearing in mind the definition of the phase and gain stability margins given in (4.54) and (4.55), and the corresponding phase and gain crossover frequencies defined in (4.56) and (4.57), it is easy to conclude that these margins can be found from Bode diagrams as indicated in Figure 9.9.

Example 9.1: In this example we use MATLAB to plot Bode diagrams for the following open-loop frequency transfer function

$$G(j\omega)H(j\omega) = \frac{(j\omega + 1)}{j\omega(j\omega + 2)\left[(j\omega)^2 + 2(j\omega) + 2\right]}$$

Bode diagrams are obtained by using the MATLAB function bode(num, den). The phase and gain stability margins and the phase and gain crossover frequencies can be obtained by using [Gm, Pm, wcp, wcg]=margin(num, den). Note that the open-loop frequency transfer function has to be specified in terms of polynomials num (numerator) and den (denominator). The MATLAB function conv helps to multiply polynomials as explained below in the program written to plot Bode diagrams and find the phase and gain margins for Example 9.1.

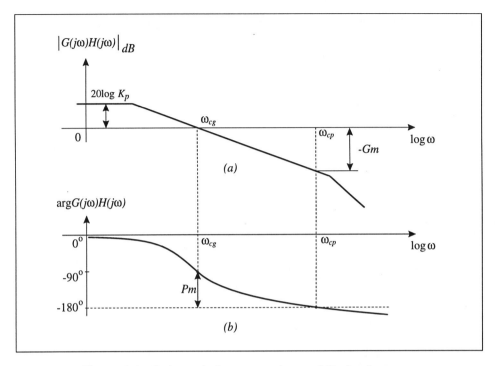

Figure 9.9: Gain and phase margins and Bode diagrams

```
num=[1  1];
d1=[1  0];
d2=[1  2];
d3=[1  2  2];
den1=conv(d1,d2);
den=conv(den1,d3);
bode(num,den);
[Gm,Pm,wcp,wcg]=margin(num,den);
```

The corresponding Bode diagrams are presented in Figure 9.10. The phase and gain stability margins and the corresponding crossover frequencies are obtained as

$$Gm = 8.9443\,\text{dB}, \; Pm = 82.2462°, \; \omega_{cp} = 1.7989\,\text{rad/s}, \; \omega_{cg} = 0.2558\,\text{rad/s}$$

◇

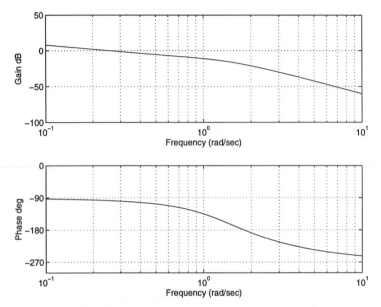

Figure 9.10: Bode diagrams for Example 9.1

Example 9.2: Consider the following open-loop frequency transfer function

$$G(j\omega)H(j\omega) = \frac{10(j\omega + 4)}{(j\omega + 1)(j\omega + 2)(j\omega + 3)}$$

The corresponding Bode diagrams obtained by following the same MATLAB instructions as in Example 9.1 are given in Figure 9.11. The phase and gain margins and the phase and gain crossover frequencies are obtained as

$$Gm = \infty, \quad Pm = 43.1488°, \quad \omega_{cp} = \infty, \quad \omega_{cg} = 3.0576\,\text{rad/s}$$

◇

9.3.2 Steady State Errors and Bode Diagrams

Steady state errors can be indirectly determined from Bode diagrams by reading the values for constants K_p, K_v, K_a from them. Knowing these constants, the corresponding errors are easily found by using formulas (6.30), (6.32), and (6.34). The steady state errors and corresponding constants K_p, K_v, K_a are first of all determined by the system type, which represents the multiplicity of the pole at the origin of the open-loop feedback transfer function, in general, represented by

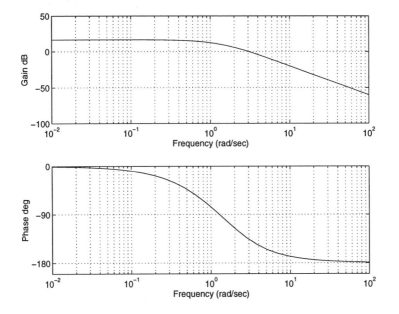

Figure 9.11: Bode diagrams for Example 9.2

$$G(j\omega)H(j\omega) = \frac{K(j\omega + z_1)(j\omega + z_2) \cdots}{(j\omega)^r(j\omega + p_1)(j\omega + p_2) \cdots} \qquad (9.16)$$

This can be rewritten as

$$G(j\omega)H(j\omega) = \frac{Kz_1z_2 \cdots \left(1 + \frac{j\omega}{z_1}\right)\left(1 + \frac{j\omega}{z_2}\right) \cdots}{p_1p_2 \cdots (j\omega)^r\left(1 + \frac{j\omega}{p_1}\right)\left(1 + \frac{j\omega}{p_2}\right) \cdots}$$

$$\qquad (9.17)$$

$$= \frac{K_B\left(1 + \frac{j\omega}{z_1}\right)\left(1 + \frac{j\omega}{z_2}\right) \cdots}{(j\omega)^r\left(1 + \frac{j\omega}{p_1}\right)\left(1 + \frac{j\omega}{p_2}\right) \cdots}$$

where

$$K_B = \frac{Kz_1z_2 \cdots}{p_1p_2 \cdots} \qquad (9.18)$$

is known as *Bode's gain*, and r is the type of feedback control system.

For control systems of type $r = 0$, the position constant according to formula (6.31) is obtained from (9.17) as

$$K_p = \frac{K_B\left(1 + \frac{j\omega}{z_1}\right)\left(1 + \frac{j\omega}{z_2}\right) \cdots}{(j\omega)^0 \left(1 + \frac{j\omega}{p_1}\right)\left(1 + \frac{j\omega}{p_2}\right) \cdots} \Big|_{j\omega=0} = K_B \qquad (9.19)$$

It follows from (9.17)–(9.19) that the corresponding magnitude Bode diagram of type zero control systems for small values of ω is flat (has a slope of 0 dB) and the value of $20 \log K_B = 20 \log K_p$. This is graphically represented in Figure 9.12.

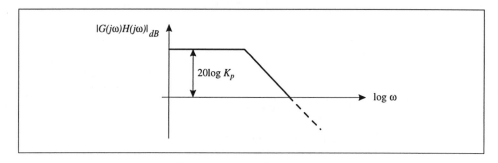

Figure 9.12: Magnitude Bode diagram of type
zero control systems at small frequencies

For control systems of type $r = 1$, the open-loop frequency transfer function is approximated at low frequencies by

$$\frac{K_B\left(1 + \frac{j\omega}{z_1}\right)\left(1 + \frac{j\omega}{z_2}\right) \cdots}{(j\omega)^1 \left(1 + \frac{j\omega}{p_1}\right)\left(1 + \frac{j\omega}{p_2}\right) \cdots} \approx \frac{K_B}{(j\omega)^1} \qquad (9.20)$$

It follows that the corresponding magnitude Bode diagram of type one control systems for small values of ω has a slope of -20 dB/decade and the values of

$$20 \log \left|\frac{K_B}{jw}\right| = 20 \log |K_B| - 20 \log |\omega| \qquad (9.21)$$

From (9.20) and (6.33) it is easy to conclude that for type one control systems the velocity constant is $K_v = K_B$. Using this fact and the frequency plot of

(9.21), we conclude that K_v is equal to the frequency ω^* at which the line (9.21) intersects the frequency axis, that is

$$0 = 20\log|K_B| - 20\log|\omega^*| \quad \Rightarrow \quad K_B = \omega^* = K_v \qquad (9.22)$$

This is graphically represented in Figure 9.13.

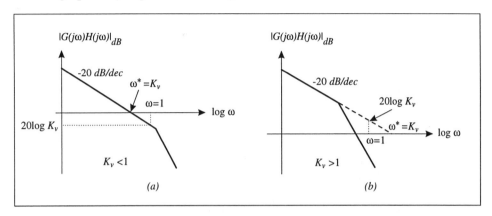

Figure 9.13: Magnitude Bode diagram of type
one control systems at small frequencies

Note that if $K_v = \omega^* > 1$, the corresponding frequency ω^* is obtained at the point where the extended initial curve, which has a slope of $-20\,\mathrm{dB/decade}$, intersects the frequency axis (see Figure 9.13b).

Similarly, for type two control systems, $r = 2$, we have at low frequencies

$$\frac{K_B\left(1+\frac{j\omega}{z_1}\right)\left(1+\frac{j\omega}{z_2}\right)\cdots}{(j\omega)^2\left(1+\frac{j\omega}{p_1}\right)\left(1+\frac{j\omega}{p_2}\right)\cdots} \approx \frac{K_B}{(j\omega)^2} \qquad (9.23)$$

which indicates an initial slope of $-40\,\mathrm{dB/decade}$ and a frequency approximation of

$$20\log\left|\frac{K_B}{(jw)^2}\right| = 20\log|K_B| - 20\log|\omega^2| = 20\log|K_B| - 40\log\omega \qquad (9.24)$$

From (9.23) and (6.35) it is easy to conclude that for type two control systems the acceleration constant is $K_a = K_B$. From the frequency plot of the straight line (9.24), it follows that $K_B = (\omega^{**})^2$, where ω^{**} represents the intersection of the initial magnitude Bode plot with the frequency axis as represented in Figure 9.14.

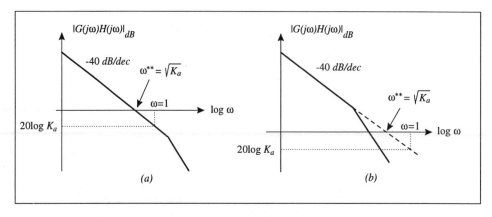

Figure 9.14: Magnitude Bode diagram of type
two control systems at small frequencies

It can be seen from Figures 9.12–9.14 that by increasing the values for the magnitude Bode diagrams at low frequencies (i.e. by increasing K_B), the constants K_p, K_v, and K_a are increased. According to the formulas for steady state errors, given in (6.30), (6.32), and (6.34) as

$$e_{ss_{step}} = \frac{1}{1 + K_p}, \quad e_{ss_{ramp}} = \frac{1}{K_v}, \quad e_{ss_{parabolic}} = \frac{1}{K_a}$$

we conclude that in this case the steady state errors are decreased. Thus, *the bigger K_B, the smaller the steady state errors.*

Example 9.3: Consider Bode diagrams obtained in Examples 9.1 and 9.2. The Bode diagram in Figure 9.10 has an initial slope of $-20\,\mathrm{dB/decade}$ which intersects the frequency axis at roughly $\omega^* = 0.2\,\mathrm{rad/s}$. Thus, we have for the Bode diagram in Figure 9.10

$$K_p = \infty, \quad K_v \approx 0.2, \quad K_a = 0$$

Using the exact formula for K_v, given by (6.33), we get

$$K_v = \lim_{s \to 0} \left\{ s \frac{(s+1)}{s(s+2)(s^2 + 2s + 2)} \right\} = 0.25$$

In Figure 9.11 the initial slope is 0 dB, and hence we have from this diagram

$$20 \log K_p \approx 15 \quad \Rightarrow \quad K_p \approx 5.62, \quad K_v = 0, \quad K_a = 0$$

Using the exact formula for K_p as given by (6.31) produces

$$K_p = \lim_{s \to 0} \left\{ \frac{10(s+4)}{(s+1)(s+2)(s+3)} \right\} = 6.67$$

Note that the accurate results about steady state error constants are obtained easily by using the corresponding formulas; hence the Bode diagrams are used only for quick and rough estimates of these constants.

◇

9.4 Compensator Design Using Bode Diagrams

In this section we show how to use Bode diagrams in order to design controllers such that the closed-loop system has the desired specifications. Three main types of controllers—phase-lead, phase-lag, and phase-lag-lead controllers—have been introduced in the time domain in Chapter 8. Here we give their interpretation in the frequency domain.

We present the design procedure for the general controllers mentioned above. Similar and simpler procedures can be developed for PD, PI, and PID controllers. After mastering the design with phase-lead, phase-lag, and phase-lag-lead controllers, students will be able to propose their own algorithms for PD, PI, and PID controllers.

Controller design techniques in the frequency domain will be governed by the following facts:

   (a) *Steady state errors are improved by increasing Bode's gain K_B.*
   (b) *System stability is improved by increasing phase and gain margins.*
   (c) *Overshoot is reduced by increasing the phase stability margin.*
   (d) *Rise time is reduced by increasing the system's bandwidth.*

However, very often it is not possible to satisfy all of these requirements at the same time, and control engineers have to compromise between several contradicting requirements.

The first two items, (a) and (b), have been already clarified. In order to justify item (c), we consider the open-loop transfer function of a second-order system given by

$$G(j\omega)H(j\omega) = \frac{\omega_n^2}{(j\omega)(jw + 2\zeta\omega_n)}$$

(9.25)

whose gain crossover frequency can be easily found from

$$|G(j\omega_{cg})H(j\omega_{cg})| = \frac{\omega_n^2}{\omega\sqrt{\omega^2 + 4\zeta^2\omega_n^2}} = 1$$

(9.26)

leading to

$$\omega_{cg} = \omega_n\sqrt{\sqrt{1 + 4\zeta^2} - 2\zeta^2}$$

(9.27)

The phase of (9.25) at the gain crossover frequency is

$$\arg\{G(j\omega_{cg})H(j\omega_{cg})\} = -90° - \tan^{-1}\frac{\omega_{cg}}{2\zeta\omega_n}$$

(9.28)

so that the corresponding phase margin becomes

$$Pm = \tan^{-1}\frac{2\zeta}{\sqrt{\sqrt{1 + 4\zeta^2} - 2\zeta^2}} = Pm(\zeta)$$

(9.29)

Plotting the function $Pm(\zeta)$, it can be shown that it is a monotonically increasing function with respect to ζ; we therefore conclude that *the higher phase margin, the larger the damping ratio, which implies the smaller the overshoot.*

Item (d) cannot be analytically justified since we do not have an analytical expression for the response rise time. However, it is very well known from undergraduate courses on linear systems and signals that rapidly changing signals have a wide bandwidth. Thus, *systems that are able to accommodate fast signals must have a wide bandwidth.*

In the remainder of this section, we first present standard controllers (phase-lag, phase-lead, and phase-lag-lead) in the frequency domain, and then show how to use these in order to achieve the desired system specifications. Each design technique will be given in an algorithmic form, and each will be demonstrated by an example.

9.4.1 Phase-Lag Controller

The transfer function of a phase-lag controller is given by

$$G_{lag}(j\omega) = \left(\frac{p_1}{z_1}\right)\frac{z_1 + j\omega}{p_1 + j\omega} = \frac{1 + j\frac{\omega}{z_1}}{1 + j\frac{\omega}{p_1}}, \quad z_1 > p_1 \qquad (9.30)$$

The corresponding magnitude and phase frequency diagrams for a phase-lag controller are presented in Figure 9.15.

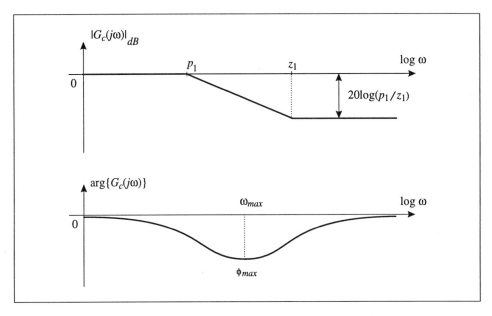

Figure 9.15: Magnitude approximation and exact phase of a phase-lag controller

Note that for the magnitude diagram it is sufficient to use only the straightline approximations for a complete understanding of the role of this controller. In general, straightline approximations can be used for almost all Bode magnitude diagrams in controller design problems. However, phase Bode diagrams are very sensitive to changes in frequency in the neighborhood of the corner frequencies, and so should be drawn as accurately as possible.

Due to attenuation of the phase-lag controller at high frequencies, the frequency bandwidth of the compensated system (controller and system in series) is reduced. Thus, *the phase-lag controllers are used in order to decrease the system bandwidth (to slow down the system response).* In addition, they can be used *to improve the stability margins (phase and gain) while keeping the steady state errors constant.*

Expressions for ω_{max} and ϕ_{max} of a phase-lag controller will be derived in the next subsection in the context of the study of a phase-lead controller. As a matter of fact, both types of controllers have the same expressions for these two important design quantities.

9.4.2 Phase-Lead Controller

The transfer function of a phase-lead controller is

$$G_{lead}(j\omega) = \left(\frac{p_2}{z_2}\right)\frac{z_2 + j\omega}{p_2 + j\omega} = \frac{1 + j\frac{\omega}{z_2}}{1 + j\frac{\omega}{p_2}}, \qquad p_2 > z_2 \qquad (9.31)$$

and the corresponding magnitude and phase Bode diagrams are shown in Figure 9.16.

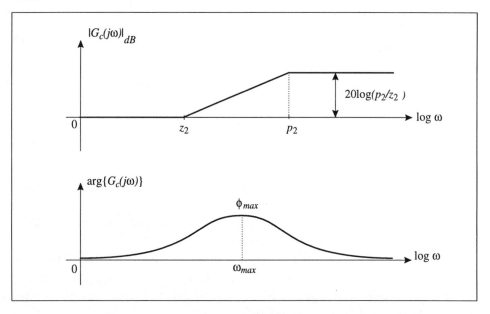

Figure 9.16: Magnitude approximation and exact phase of a phase-lead controller

Due to phase-lead controller (compensator) amplification at higher frequencies, it increases the bandwidth of the compensated system. *The phase-lead controllers are used to improve the gain and phase stability margins and to increase the system bandwidth (decrease the system response rise time).*

It follows from (9.31) that the phase of a phase-lead controller is given by

$$\arg\left\{G_{lead}(j\omega)\right\} = \tan^{-1}\left(\frac{\omega}{z_2}\right) - \tan^{-1}\left(\frac{\omega}{p_2}\right) \tag{9.32}$$

so that

$$\frac{d}{d\omega}\arg\left\{G_{lead}(j\omega)\right\} = 0 \;\Rightarrow\; \omega_{max} = \sqrt{z_2 p_2} \tag{9.33}$$

Assume that

$$p_2 = az_2, \quad a > 1 \quad \Rightarrow \quad \omega_{max} = \frac{p_2}{\sqrt{a}} \tag{9.34}$$

Substituting ω_{max} in (9.32) implies

$$\tan\phi_{max} = \frac{a-1}{2\sqrt{a}} \tag{9.35}$$

It is left as an exercise for students to give detailed derivations of formula (9.35)—see Problem 9.3.

It is easy to find, from (9.35), that the value for parameter a in terms of ϕ_{max} is given by

$$a = \frac{1 + \sin\phi_{max}}{1 - \sin\phi_{max}} \tag{9.36}$$

Note that the same formulas for ω_{max}, (9.33), and the parameter a, (9.36), hold for a *phase-lag controller* with p_1, z_1 replacing p_2, z_2 and with $p_1 = az_1$, $a < 1$.

9.4.3 Phase-Lag-Lead Controller

The phase-lag-lead controller has the features of both phase-lag and phase-lead controllers and can be used to improve both the transient response and steady state errors. However, its design is more complicated than the design of either phase-lag or phase-lead controllers. The frequency transfer function of the phase-lag-lead controller is given by

$$G_c(j\omega) = \frac{(j\omega + z_1)(j\omega + z_2)}{(j\omega + p_1)(j\omega + p_2)} = \frac{z_1 z_2}{p_1 p_2} \frac{\left(1 + j\frac{\omega}{z_1}\right)\left(1 + j\frac{\omega}{z_2}\right)}{\left(1 + j\frac{\omega}{p_1}\right)\left(1 + j\frac{\omega}{p_2}\right)}$$

$$= \frac{\left(1 + j\frac{\omega}{z_1}\right)\left(1 + j\frac{\omega}{z_2}\right)}{\left(1 + j\frac{\omega}{p_1}\right)\left(1 + j\frac{\omega}{p_2}\right)}, \qquad z_1 z_2 = p_1 p_2, \qquad p_2 > z_2 > z_1 > p_1$$

(9.37)

The Bode diagrams of this controller are shown in Figure 9.17.

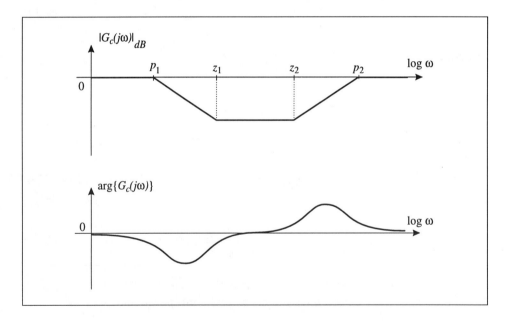

Figure 9.17: Bode diagrams of a phase-lag-lead controller

9.4.4 Compensator Design with Phase-Lead Controller

The following algorithm can be used to design a controller (compensator) with a phase-lead network.

Algorithm 9.1:

1. Determine the value of the Bode gain K_B given by (9.18) as

$$K_B = \frac{K z_1 z_2 \cdots}{p_1 p_2 \cdots}$$

such that the steady state error requirement is satisfied.

2. Find the phase and gain margins of the original system with K_B determined in step 1.

3. Find the phase difference, $\Delta\phi$, between the actual and desired phase margins and take ϕ_{max} to be $5°$–$10°$ greater than this difference. Only in rare cases should this be greater than $10°$. This is due to the fact that we have to give an estimate of a new gain crossover frequency, which can not be determined very accurately (see step 5).

4. Calculate the value for parameter a from formula (9.36), i.e. by using

$$a = \frac{1 + \sin \phi_{max}}{1 - \sin \phi_{max}} > 1$$

5. Estimate a value for a compensator's pole such that ω_{max} is roughly located at the new gain crossover frequency, $\omega_{max} \approx \omega_{cgnew}$. As a rule of thumb, add the gain of $\Delta G = 20\log(a)[\text{dB}]$ at high frequencies to the uncompensated system and estimate the intersection of the magnitude diagram with the frequency axis, say ω_1. The new gain crossover frequency is somewhere in between the old ω_{cg} and ω_1. Some authors (Kuo, 1991) suggest fixing the new gain crossover frequency at the point where the magnitude Bode diagram has the value of $-0.5\Delta G[\text{dB}]$. Using the value for parameter a obtained in step 4 find the value for the compensator pole from (9.34) as $-p_c = -\omega_{max}\sqrt{a}$ and the value for compensator's zero as $-z_c = -p_c/a$. Note that one can also guess a value for p_c and then evaluate z_c and ω_{max}. The phase-lead compensator now can be represented by

$$G_c(s) = \frac{as + p_c}{s + p_c}$$

6. Draw the Bode diagram of the given system with controller and check the values for the gain and phase margins. If they are satisfactory, the controller design is done, otherwise repeat steps 1–5.

The next example illustrates controller design using Algorithm 9.1.

Example 9.4: Consider the following open-loop frequency transfer function

$$G(j\omega)H(j\omega) = \frac{K(j\omega + 6)}{(j\omega + 1)(j\omega + 2)(j\omega + 3)}$$

Step 1. Let the design requirements be set such that the steady state error due to a unit step is less than 2% and the phase margin is at least 45°. Since

$$e_{ss} = \frac{1}{1 + K_p} = \frac{1}{1 + K_B}, \quad K_B = \frac{K \times 6}{1 \times 2 \times 3} = K$$

we conclude that $K \geq 50$ will satisfy the steady state error requirement of being less than 2%. We know from the root locus technique that high static gains can damage system stability, and so for the rest of this design problem we take $K = 50$.

Step 2. We draw Bode diagrams of the uncompensated system with the Bode gain obtained in step 1 and determine the phase and gain margins and the crossover frequencies. This can be done by using the following sequence of MATLAB functions.

```
[den]=input('enter denominator');
% for this example [den]=[1 6 11 6];
[num]=input('enter numerator');
% for this example [num]=[50 300];
[Gm,Pm,wcp,wcg]=margin(num,den);
bode(num,den)
```

The corresponding Bode diagrams are presented in Figure 9.18a. The phase and gain margins are obtained as $Gm = \infty$, $Pm = 5.59°$ and the crossover frequencies are $\omega_{cg} = 7.5423\,\text{rad/s}$, $\omega_{cp} = \infty$.

Step 3. Since the desired phase is well above the actual one, the phase-lead controller must make up for $45° - 5.59° = 39.41°$. We add 10°, for the reason explained in step 3 of Algorithm 9.1, so that $\phi_{max} = 49.41°$. The above operations can be achieved by using MATLAB as follows

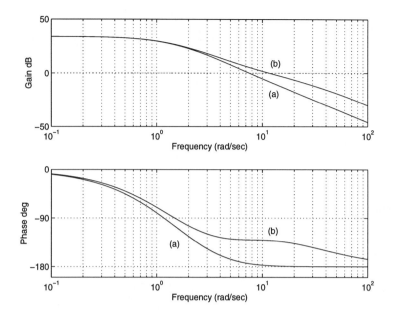

Figure 9.18: Bode diagrams for the original system
(a) and compensated system (b) of Example 9.4

```
% estimate Phimax with Pmd = desired phase margin;
Pmd=input('enter desired value for phase margin');
Phimax=Pmd-Pm+10;
% convert Phimax in radians;
Phirad=(Phimax/180)*pi;
```

Step 4. Here we evaluate the parameter a according to the formula (9.36) and get $a = 7.3144$. This can be done in MATLAB by

```
a=(1+sin(Phirad))/(1-sin(Phirad));
```

Step 5. In order to obtain an estimate for the new gain crossover frequency we first find the controller amplification at high frequencies, which is equal to $20 \log (a) = 17.2836 \, \text{dB} = \Delta G_{dB}$. The magnitude Bode diagram increased by ΔG_{dB} at high frequencies intersects the frequency axis at $\omega_1 \approx 10.5 \, \text{rad/s}$. We guess (estimate) the value for p_c as $p_c = 25$, which is roughly equal to $\omega_1 \sqrt{a}$. By using $p_c = 25$ and forming the corresponding compensator, we get for the compensated system $Pmc = 48.2891°$ at $\omega_{cgnew} = 13.8519 \, \text{rad/s}$, which is satisfactory. This step can be performed by MATLAB as follows.

```
% Find amplification at high frequencies, DG;
DG=20*log(a);
% estimate value for pole —pc from Step 5;
pc=input('enter estimated value for pole pc');
% form compensator's numerator;
nc=[a pc];
% form compensator's denominator;
dc=[1 pc];
% find the compensated system transfer function;
numc=conv(num,nc);
denc=conv(den,dc);
[Gmc,Pmc,wcp,wcg]=margin(numc,denc);
bode(numc,denc)
```

The phase-lead compensator obtained is given by

$$G_c(s) = \frac{7.3144s + 25}{s + 25} = \frac{as + p_2}{s + p_2}$$

Step 6. The Bode diagrams of the compensated control system are presented in Figure 9.18b. Both requirements are satisfied, and therefore the controller design procedure is successfully completed.

It is interesting to compare the transient response characteristics of the compensated and uncompensated systems. This cannot be easily done analytically since the orders of both systems are greater than two, but it can be simply performed by using MATLAB. Note that num, den, numc, denc represent, respectively, the numerators and denominators of the open-loop transfer functions of the original and compensated systems. In order to find the corresponding closed-loop transfer functions, we use the MATLAB function cloop, that is

```
[cnum,cden]=cloop(num,den,-1);
% —1 indicates a negative unit feedback
[cnumc,cdenc]=cloop(numc,denc,-1);
```

The closed-loop step responses are obtained by

```
[y,x]=step(cnum,cden);
[yc,xc]=step(cnumc,cdenc);
```

and are represented in Figure 9.19. It can be seen from this figure that both the maximum percent overshoot and the settling time are drastically reduced. In addition, the rise time of the compensated system is shortened since the phase-lead controller increases the frequency bandwidth of the system.

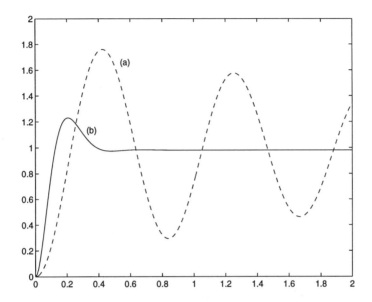

Figure 9.19: Step responses for the original (a) and compensated (b) systems

The highly oscillatory behavior of the step response of the original system can be fully understood from its root locus, which is given in Figure 9.20. It can be seen that the real part of a pair of complex conjugate poles is very small for almost all values of the static gain, which causes high-frequency oscillations and very slow convergence to the response steady state value.

The closed-loop eigenvalues of the original and compensated systems, obtained by MATLAB functions `roots(cden)` and `roots(cdenc)`, are given by

$$\lambda_1 = -5.3288, \quad \lambda_{2,3} = -0.3356 \pm j7.5704$$

$$\lambda_{1,2c} = -9.6736 \pm j13.3989, \quad \lambda_{3c} = -8.2628, \quad \lambda_{4c} = -3.3900$$

which indicates a big difference in the real parts of complex conjugate poles for the original and compensated systems.

◇

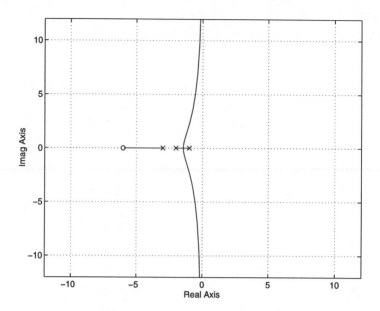

Figure 9.20: Root locus of the original system

9.4.5 Compensator Design with Phase-Lag Controller

Compensator design using phase-lag controllers is based on the compensator's attenuation at high frequencies, which causes a shift of the gain crossover frequency to the lower frequency region where the phase margin is high. The phase-lag compensator can be designed by the following algorithm.

Algorithm 9.2:

1. Determine the value of the Bode gain K_B that satisfies the steady state error requirement.

2. Find on the phase Bode plot the frequency which has the phase margin equal to the desired phase margin increased by 5° to 10°. This frequency represents the new gain crossover frequency, ω_{cgnew}.

3. Read the required attenuation at the new gain crossover frequency, i.e. $|\Delta G(j\omega_{cgnew})|_{dB}$, and find the parameter a from

$$-20\log\left(\frac{p_1}{z_1}\right) = -20\log\left(a\right) = |\Delta G(j\omega_{cgnew})|_{dB}$$

which implies

$$a = 10^{-\frac{1}{20}|\Delta G(j\omega_{cgnew})|_{dB}} = \frac{1}{|\Delta G(j\omega_{cgnew})|}$$

Note that

$$|\Delta G(j\omega_{cgnew})| = \frac{|K||j\omega_{cgnew} + z_1||j\omega_{cgnew} + z_2|\cdots}{|j\omega_{cgnew} + p_1||j\omega_{cgnew} + p_2|\cdots}$$

4. Place the controller zero one decade to the left of the new gain crossover frequency, that is

$$z_c = \frac{\omega_{cgnew}}{10}$$

Find the pole location from $p_c = az_c = a\omega_{cgnew}/10$. The required compensator has the form

$$G_c(s) = \frac{as + p_c}{s + p_c}$$

5. Redraw the Bode diagram of the given system with the controller and check the values for the gain and phase margins. If they are satisfactory, the controller design is done, otherwise repeat steps 1–5.

Example 9.5: Consider a control system represented by

$$G(s) = \frac{K}{s(s + 2)(s + 30)}$$

Design a phase-lag compensator such that the following specifications are met: $e_{ss_{ramp}} \leq 0.05$, $Pm \geq 45°$. The minimum value for the static gain that produces the required steady state error is equal to $K = 1200$. The original system with this static gain has phase and gain margins given by $Pm = 6.6449°$, $Gm = 4.0824\,dB$ and crossover frequencies of $\omega_{cg} = 6.1031\,rad/s$, $\omega_{cp} = 7.746\,rad/s$.

The new gain crossover frequency can be estimated as $\omega_{cgnew} = 1.4\,rad/s$ since for that frequency the phase margin of the original system is approximately $50°$. At $\omega_{cgnew} = 1.4\,rad/s$ the required gain attenuation is obtained by MATLAB as

```
wcgnew=1.4;
d1=1200;
g1=abs(j*1.4);
g2=abs(j*1.4+2);
g3=abs(j*1.4+30);
dG=d1/(g1*g2*g3);
```

which produces $|\Delta G(j1.4)| = 11.6906$ and $a = 1/|\Delta G(j1.4)| = 0.0855$. The compensator's pole and zero are obtained as $-z_c = -\omega_{cgnew}/10 = -0.14$ and $-p_c = -a\omega_{cgnew}/10 = -0.0120$ (see step 4 of Algorithm 9.2). The transfer function of the phase-lag compensator is

$$G_c(s) = \frac{0.0855s + 0.0120}{s + 0.0120}$$

The Bode diagrams of the original and compensated systems are given in Figure 9.21.

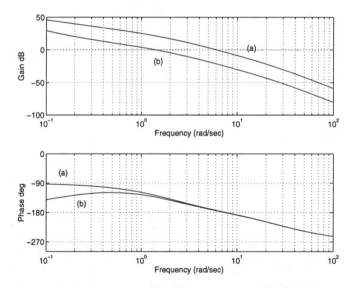

Figure 9.21: Bode diagrams for the original system (a) and compensated system (b) of Example 9.5

The new phase and gain margins and the actual crossover frequencies are $Pmc = 47.03°$, $Gmc = 24.82\,\text{dB}$, $\omega_{cgnew} = 1.405\,\text{rad/s}$, $\omega_{cpnew} = 7.477\,\text{rad/s}$ and so the design requirements are satisfied. The step responses of the original and compensated systems are presented in Figure 9.22.

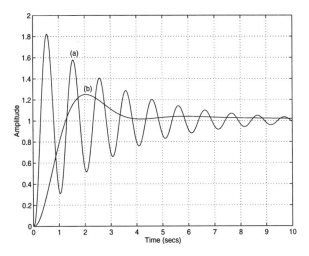

Figure 9.22: Step responses for the original system
(a) and compensated system (b) of Example 9.5

It can be seen from this figure that the overshoot is reduced from roughly 0.83 to
0.3. In addition, it can be observed that the settling time is also reduced. Note
that the phase-lag controller reduces the system bandwidth ($\omega_{cgnew} < \omega_{cg}$) so
that the rise time of the compensated system is increased.

◇

9.4.6 Compensator Design with Phase-Lag-Lead Controller

Compensator design using a phase-lag-lead controller can be performed according
to the algorithm given below, in which we first form a phase-lead compensator
and then a phase-lag compensator. Finally, we connect them together in series.
Note that several different algorithms for the phase-lag-lead controller design
can be found in the control literature.

Algorithm 9.3:

1. Set a value for the static gain K_B such that the steady state error requirement
    is satisfied.
2. Draw Bode diagrams with K_B obtained in step 1 and find the corresponding
    phase and gain margins.

3. Find the difference between the actual and desired phase margins, $\Delta\phi = Pmd - Pm$, and take ϕ_{max} to be a little bit greater than $\Delta\phi$. Calculate the parameter a_2 of a phase-lead controller by using formula (9.36), that is

$$a_2 = \frac{1 + \sin\phi_{max}}{1 - \sin\phi_{max}}$$

4. Locate the new gain crossover frequency at the point where

$$20\log|G(j\omega_{cgnew})| = -10\log a_2 \qquad (9.38)$$

5. Compute the values for the phase-lead compensator's pole and zero from

$$p_{c2} = \omega_{cgnew}\sqrt{a_2}, \qquad z_{c2} = p_{c2}/a_2 \qquad (9.39)$$

6. Select the phase-lag compensator's zero and pole according to

$$z_{c1} = 0.1z_{c2}, \qquad p_{c1} = z_{c1}/a_2 \qquad (9.40)$$

7. Form the transfer function of the phase-lag-lead compensator as

$$G_c(s) = G_{lag}(s) \times G_{lead}(s) = \frac{s + z_{c1}}{s + p_{c1}} \times \frac{s + z_{c2}}{s + p_{c2}}$$

8. Plot Bode diagrams of the compensated system and check whether the design specifications are met. If not, repeat some of the steps of the proposed algorithm—in most cases go back to steps 3 or 4.

The phase-lead part of this compensator helps to increase the phase margin (increases the damping ratio, which reduces the maximum percent overshoot and settling time) and broaden the system's bandwidth (reduces the rise time). The phase-lag part, on the other hand, helps to improve the steady state errors.

Example 9.6: Consider a control system that has the open-loop transfer function

$$G(s) = \frac{K(s + 10)}{(s^2 + 2s + 2)(s + 20)}$$

For this system we design a phase-lag-lead controller by following Algorithm 9.3 such that the compensated system has a steady state error of less than 4% and a phase margin greater than 50°. In the first step, we choose a value for the static gain K that produces the desired steady state error. It is easy to check that $K = 100 \Rightarrow e_{ss} = 3.85\%$, and therefore in the following we stick with this value for the static gain. Bode diagrams of the original system with $K = 100$ are presented in Figure 9.23.

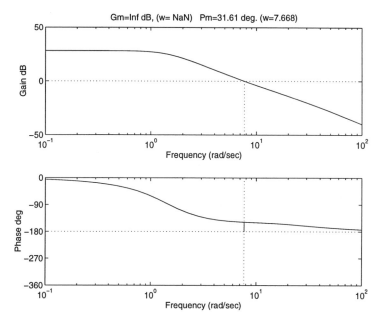

Figure 9.23: Bode diagrams of the original system

It can be seen from these diagrams—and with help of MATLAB determined accurately—that the phase and gain margins and the corresponding crossover frequencies are given by $Pm = 31.61°, Gm = \infty$, and $\omega_{cg} = 7.668 \, \text{rad/s}$, $\omega_{cp} = \infty$. According to step 3 of Algorithm 9.3, a controller has to introduce a phase lead of 18.39°. We take $\phi_{max} = 25°$ and find the required parameter $a_2 = 2.4639$. Taking $\omega_{cgnew} = 20 \, \text{rad/s}$ in step 4 and completing the design steps 5–8 we find that $Pm = 39.94°$, which is not satisfactory. We go back to step 3 and take $\phi_{max} = 30° = 0.5236 \, \text{rad}$, which implies $a_2 = 3$.

Step 4 of Algorithm 9.3 can be executed efficiently by MATLAB by performing the following search. Since $-10 \log 3 = -10.9861$ dB we search the magnitude diagram for the frequency where the attenuation is approximately equal to -11 dB. We start search at $\omega = 20 \, \text{rad/s}$ since at that point, according to Figure 9.23, the attenuation is obviously smaller than -11 dB. The following

MATLAB program is used to find the new gain crossover frequency, i.e. to solve approximately equation (9.38)

```
w=20;
while 20*log(100*abs(j*w+10)/
abs(((j*w)^2+2*j*w+2)*(j*w+20)))<-11;
w=w-1;
end
```

This program produces $\omega_{cgnew} = 10\,\text{rad/s}$. In steps 5 and 6 the phase-lag-lead controller zeros and poles are obtained as $-p_{c2} = -17.3205$, $-z_{c2} = -5.7735$ for the phase-lead part and $-p_{c1} = -0.1925$, $-z_{c1} = -0.5774$ for the phase-lag part; hence the phase-lag-lead controller has the form

$$G_c(s) = \frac{s + 0.5774}{s + 0.1925} \times \frac{s + 17.3205}{s + 5.7735}$$

The Bode diagrams of the compensated system are given in Figure 9.24.

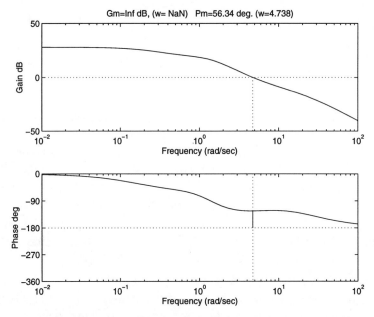

Figure 9.24 Bode diagrams of the compensated system

It can be seen that the phase margin obtained of 56.34° meets the design requirement and that the actual gain crossover frequency, 4.738 rad/s, is considerably smaller than the one predicted. This contributes to the generally accepted inaccuracy of frequency methods for controller design based on Bode diagrams.

The step responses of the original and compensated systems are compared in Figure 9.25. The transient response of the compensated system is improved since the maximum percent overshoot is considerably reduced. However, the system rise time is increased due to the fact that the system bandwidth is shortened ($\omega_{cgnew} = 4.738\,\text{rad/s} < \omega_{cg} = 7.668\,\text{rad/s}$).

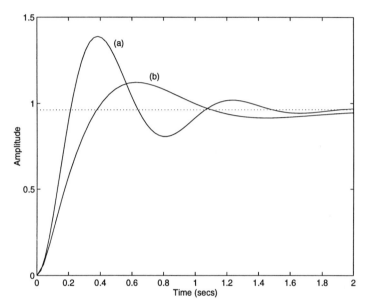

Figure 9.25 Step responses of the original (a) and compensated (b) systems

◇

9.5 MATLAB Case Study

Consider the problem of finding a controller for the ship positioning control system given in Problem 7.5. The goal is to increase stability phase margin above 30°. The problem matrices are given by

$$\mathbf{A} = \begin{bmatrix} -0.0546 & 0 & 0.5435 \\ 1 & 0 & 0 \\ 0 & 0 & -1.55 \end{bmatrix}, \quad \mathbf{B} = \begin{bmatrix} 0 \\ 0 \\ 1.55 \end{bmatrix}, \quad \mathbf{C} = \begin{bmatrix} 0 & 1 & 0 \end{bmatrix}, \quad \mathbf{D} = 0$$

The transfer function of the ship positioning system is obtained by the MATLAB instruction [num,den]=ss2tf(A,B,C,D) and is given by

$$G(s) = \frac{0.8424}{s(s + 1.55)(s + 0.0546)}$$

The phase and gain stability margins of this system are $Pm = -19.94°$ and $Gm = -15.86$ dB, with the crossover frequencies $\omega_{cp} = 0.2909\,\text{rad/s}$ and $\omega_{cg} = 0.7025\,\text{rad/s}$ (see the Bode diagrams in Figure 9.26). From known values for the phase and gain margins, we can conclude that this system has very poor stability properties.

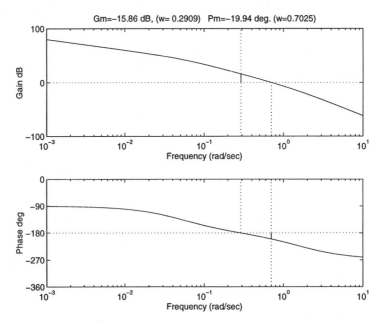

Figure 9.26: Bode diagrams of a ship positioning control system

Since the phase margin is well below the desired one, we need a controller which will make up for almost a 50° increase in phase. In general, it is hard to stabilize systems that have large negative phase and gain stability margins. In the following we will design phase-lead, phase-lag, and phase-lag-lead controllers to solve this problem and compare the results obtained.

Phase-Lead Controller: By using Algorithm 9.1 with $\phi_{max} = 50° + 10° = 60°$ we get a phase margin of only 23.536°, which is not satisfactory. It is necessary to make up for $\phi_{max} = 50° + 27° = 87°$. In the latter case the compensator has the transfer function

$$G_c(s) = 76.31 \frac{s + 0.2038}{s + 15.55}$$

Figure 9.27 shows Bode diagrams of both the original (a) and compensated (b) systems.

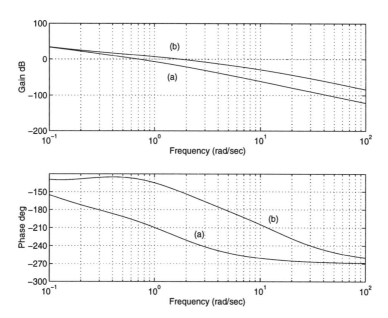

Figure 9.27: Bode diagrams for a ship positioning system:
(a) original system, (b) phase-lead compensated system

The gain and phase stability margins of the compensated system are found from the above Bode diagrams as $Gmc = 15.1603$ dB, $Pmc = 30.0538°$, and the crossover frequencies are $\omega_{cpc} = 1.7618$ rad/s, $\omega_{cgc} = 4.6419$ rad/s. The step response of the compensated system exhibits an overshoot of 45.47% (see Figure 9.28).

Phase-Lag-Lead Controller: By using Algorithm 9.3 we find the compensator transfer function as

$$G_c(s) = 7.524 \frac{s + 0.1599}{s + 1.203} \times 0.1329 \frac{s + 0.016}{s + 0.002125}$$

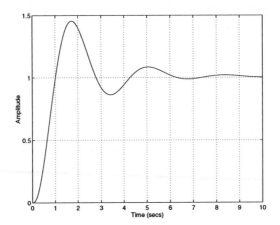

Figure 9.28: Step response of the compensated
system with a phase-lead controller

The Bode diagrams of the original and compensated systems are shown in Figure 9.29.

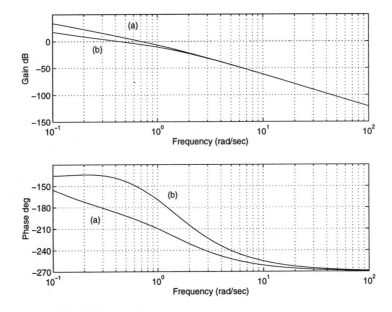

Figure 9.29: Bode diagrams for a ship positioning control system:
(a) original system, (b) phase-lag-lead compensated system

The phase and gain margins of the compensated system are given by $Pmc = 39.6694°$, $Gmc = 14$ dB and the crossover frequencies are $\omega_{cgc} = 0.4332\,\text{rad/s}$, $\omega_{cpc} = 1.2401\,\text{rad/s}$.

From the step response of the compensated system (see Figure 9.30), we can observe that this compensated system has a smaller overshoot and a larger rise time than the system compensated only by the phase-lead controller.

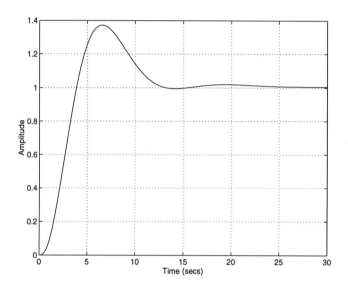

Figure 9.30: Step response of the compensated
system with a phase-lag-lead controller

Phase-Lag Controller: If we choose a new gain crossover frequency at $\omega_{cgnew} = 0.03\,\text{rad/s}$, the phase margin at that point will clearly be above 50°. Proceeding with a phase-lag compensator design, according to Algorithm 9.2, we get $|\Delta G(j0.03)| = 290.7390$ and $a = 0.034$, which implies $z_c = 0.003$ and $p_c = 1.0319 \times 10^{-5}$. Using the corresponding phase-lag compensator produces very good stability margins for the compensated system, i.e. $Gm = 32.91$ dB and $Pm = 54.33°$. The maximum percent overshoot obtained is much better than with the previously used compensators and is equal to $MPOS = 18\%$. However, the closed-loop step response reveals that the obtained system is too sluggish since the response peak time is $t_p = 95.2381\,\text{s}$ (note that in the previous two cases the peak time is only a few seconds).

One may try to get better agreement by designing a phase-lag compensator, which will reduce the phase margin of the compensated system to just above $30°$. In order to do this we write a MATLAB program, which searches the phase Bode diagram and finds the frequency corresponding to the prespecified value of the phase. That frequency is used as a new gain crossover frequency. Let $Pm = 35° = 0.6109$ rad. The MATLAB program is

```
w=0.1;
while pi+
angle(1/((j*w)*(j*w+1.55)*(j*w+0.0546)))<0.6109;
w=w-0.01;
end
dG=abs(1/((j*w)*(j*w+1.55)*(j*w+0.0546)));
```

This program produces $\omega_{cgnew} = 0.07$ rad/s and $|\Delta G(j0.07)| = 87.3677$. From step 5 of Algorithm 9.2 we obtain the phase-lag controller of the form

$$G_c(s) = \frac{s + 0.009}{s + 0.000081}$$

The Bode diagrams of the compensated system are given in Figure 9.31.

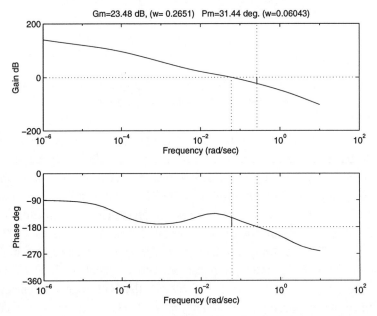

Figure 9.31: Bode diagram of the phase-lag compensated system

It can be seen that the phase and gain margins are satisfactory and given by $Pm = 31.44°$ and $Gm = 23.48$ dB. The actual gain crossover frequencies are $w_{cgnew} = 0.06043 \, \text{rad/s}$ and $\omega_{cpnew} = 0.2651 \, \text{rad/s}$.

The closed-loop step response of the phase-lag compensated system, given in Figure 9.32, shows that the peak time is reduced to $t_p = 50.15 \, \text{s}$—which is still fairly big—and that the maximum percent overshoot is increased to $MPOS = 45.82\%$, which is comparable to the phase-lead and phase-lag-lead compensation.

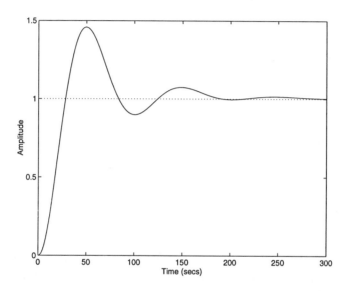

Figure 9.32: Step response of the phase-lag compensated system

Comparing all three controllers and their performances, we can conclude that, for this particular problem, the phase-lag compensation produces the worst result, and therefore either the phase-lead or phase-lag-lead controller should be used.

◇

9.6 Comments on Discrete-Time Controller Design

Bode diagrams were originally introduced for studying continuous-time systems (Bode, 1940). However, discrete-time systems can be studied using the same diagrams. The bilinear transformation, already used in this book in the stability study of discrete-time systems, that has the form

$$s = \frac{z-1}{z+1}, \qquad z = \frac{1+s}{1-s} \tag{9.41}$$

maps the imaginary axis from the s-plane into the unit circle in the z-plane and vice versa. Since on the unit circle $z = e^{j\omega_z T}$, it is easy to establish the relationship between angular frequencies in the s and z domains. It is left as an exercise for students to show that

$$w_s = \tan \frac{\omega_z T}{2}, \qquad \omega_z = \frac{2}{T} \tan^{-1} w_s \tag{9.42}$$

The above transformation allows one to map the discrete-time open-loop transfer function into the continuous-time open-loop transfer function, that is

$$G(z)|_{z=\frac{1+s}{1-s}} = G(s) \tag{9.43}$$

and to perform controller design in the continuous-time domain. The results obtained have to be mapped back into the discrete-time domain by using (9.42), that is

$$G_c(s)|_{s=\frac{z-1}{z+1}} = G_c(z) \tag{9.44}$$

Note that several bilinear transformations, which are just scaled versions of (9.41), can be found in the control literature. For more details the reader is referred, for example, to Franklin *et al.* (1990), DiStefano *et al.* (1990), and Phillips and Nagle (1995). MATLAB discrete-time controller design problems can be found in Shahian and Hassul (1993).

9.7 MATLAB Laboratory Experiment

Part 1. Consider the closed-loop system represented in Figure 9.33. This system has a transport lag-element, e^{-Ts}, which represents a time delay of T time units. The transport lag-element can be approximated for small values of time delay T by

$$\text{(i)} \quad e^{-Ts} \approx \frac{1}{1+Ts}, \qquad \text{(ii)} \quad e^{-Ts} \approx 1 - Ts$$

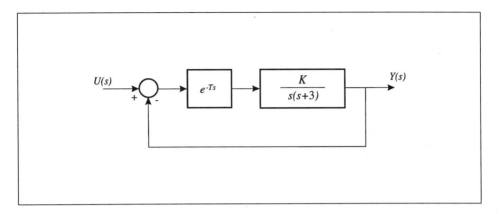

Figure 9.33: Block diagram of a control system

Using the following values for the time delay $T = 0$, 0.01, 0.1, design phase-lead, phase-lag, and phase-lag-lead compensators to meet the following closed-loop design requirements: steady state error $e_{ss_{ramp}} \leq 0.02$ and $Pm \geq 50°$. Consider both approximations (i) and (ii) and compare the results obtained.

Part 2. Draw the exact phase Bode diagrams of the systems compensated in Part 1 including the exact contribution from the time delay element. Note that the time delay element does not affect the magnitude Bode diagram, but modifies the phase Bode diagrams by a factor of $\angle e^{-j\omega T}$. Compare the approximated and exact phase Bode diagrams. Draw conclusions about the impact of time delay elements on phase and gain stability margins.

Part 3. Consider the controller design problem for a system represented by its open-loop transfer function

$$G(s) = \frac{K(s + 6)}{(s + 10)(s^2 + 2s + 2)}$$

This system has been studied in Example 8.9 and in the MATLAB laboratory experiment for the root locus controller design in Section 8.8.

(a) Design a phase-lag-lead controller using Bode diagrams such that the compensated system has the same specifications as those in Section 8.8, i.e. the steady state error is less than 1% and the phase margin is such that $t_s < 2\,\mathrm{s}$ and $MPOS < 10\%$. Note that the maximum percent overshoot and settling time are inversely proportional to the phase margin. Experiment with several values for the phase margin and take the one that satisfies both transient response requirements.

(b) Compare the results obtained with those from Section 8.8 and comment on the differences between root locus and Bode diagram phase-lag-lead controller design. Which one is easier to design? Which one is more accurate?

9.8 References

Bode, H., "Relations between attenuation and phase in feedback amplifier design," *Bell System Technical Journal*, vol. 19, 421–454, 1940.

DiStefano, J., A. Stubberud, and I. Williams, *Feedback and Control Systems*, McGraw-Hill, New York, 1990.

Franklin, G., J. Powel, and M. Workman, *Digital Control of Dynamic Systems*, Addison-Wesley, Reading, Massachusetts, 1990.

Kuo, B., *Automatic Control Systems*, Prentice Hall, Englewood Cliffs, New Jersey, 1991.

Phillips, C. and H. Nagle, *Digital Control System Analysis and Design*, Prentice Hall, Englewood Cliffs, New Jersey, 1995.

Shahian, B. and M. Hassul, *Control System Design with MATLAB*, Prentice Hall, Englewood Cliffs, New Jersey, 1993.

9.9 Problems

9.1 Show that for a second-order closed-loop system

$$M(j\omega) = \frac{\omega_n^2}{(j\omega)^2 + 2\zeta\omega_n(j\omega) + \omega_n^2}$$

the resonant frequency is given by

$$\omega_r = \omega_n\sqrt{1 - \zeta^2}$$

and the peak resonance is

$$M_p = \frac{1}{2\zeta\sqrt{1-\zeta^2}}$$

9.2 Show that the frequency bandwidth for a second-order closed-loop system given in Problem 9.1 is

$$\omega_{BW} = \omega_n\sqrt{(1-2\zeta^2) + \sqrt{4\zeta^4 - 4\zeta^2 + 2}}$$

Since the 5%-settling time is given by formula (6.20) as

$$t_s \approx \frac{3}{\zeta\omega_n}$$

conclude that the settling time is inversely proportional to the system bandwidth, in other words, *the wider the system bandwidth, the shorter the settling time.*

9.3 Derive formula (9.35) for the maximum phase of a phase-lead controller.

9.4 Using MATLAB, draw Bode diagrams for a magnetic tape control system considered in Problem 5.12. Matrices $\mathbf{A}$ and $\mathbf{B}$ are given in Problem 5.12. The output matrices are

$$\mathbf{C} = [1 \quad 0 \quad 1 \quad 0], \qquad \mathbf{D} = 0$$

Find the phase and gain stability margins for this system.

9.5 Based on Algorithms 9.1–9.3, propose algorithms for controller design with

(a) a PD controller;
(b) a PI controller;
(c) a PID controller.

9.6 Solve the controller design problem defined in Example 9.5 by using both phase-lead and phase-lag-lead controllers.

9.7 Design a phase-lag-lead network for the system

$$G(s) = \frac{K}{s(s+3)}$$

such that $Pm \geq 45°$ and $e_{ss_{ramp}} \leq 0.03$.

9.8 The block diagram of a servo control system is shown in Figure 9.34.

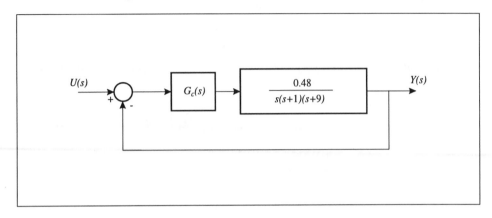

Figure 9.34: Block diagram for Problem 9.8

Design phase-lag, phase-lead, and phase-lag-lead controllers such that the phase margin is greater than $60°$.

9.9 A unit feedback system has the transfer function

$$G(s) = \frac{40K}{s(s+4)(s+10)}$$

(a) Construct Bode diagrams for $K = 1$.

(b) Using the MATLAB function `margin`, determine the phase and gain margins of the system.

(c) Determine the value for the static gain K such that the system has a gain margin of $Gm = 10$ dB. Find the corresponding steady state errors.

(d) From Bode diagrams determine the value for the static gain K such that the phase margin is $45°$. Determine the damping ratio and the natural frequency for the obtained value of K.

9.10 A zero type plant has the transfer function

$$G(s) = \frac{10}{(s+5)(s+20)}$$

(a) Determine the damping ratio and the natural frequency of the corresponding closed-loop system.

(b) Find the steady state errors and the system overshoot.

(c) Determine the phase and gain margins.

(d) Design a controller such that the compensated system has a phase margin of at least 50° and a steady state error less than 0.02.

9.11 A unit feedback system with transport delay has the transfer function

$$G(s) = \frac{10Ke^{-0.5s}}{(s+4)(s+10)}$$

Approximate time delay as $e^{-Ts} \approx 1/(1+Ts)$.

(a) Plot on the same figure the Bode diagrams for the system with and without transport delay for $K = 1$. Comment on system stability.

(b) Plot the unit step response of the system for both cases and compare the results (steady state errors, transient response parameters).

(c) Determine the value of the static gain K that gives a steady state unit step error of 0.02 for both the approximated system and the original system without time delay.

(d) Design phase-lag, phase-lead, and phase-lag-lead controllers to achieve a steady state error of $e_{ss} \leq 0.05$ and a phase margin $Pm \geq 50°$ for the approximated system.

(e) Draw the exact phase Bode diagrams of the compensated systems, including the time delay, and check the values obtained for the phase margins.

9.12 Determine a passive cascade compensator for a unit feedback system

$$G(s) = \frac{4}{s(s^2 + 4s + 8)}$$

such that the compensated system has a phase margin of 45° and a steady state ramp error of less than 2%.

9.13 Derive formulas (9.42).

9.14 Solve the controller design problem defined in Example 9.4 by using a phase-lag-lead controller. What is the advantage of the controller's phase-lag part?

9.15 Consider a control system that has the open-loop transfer function

$$G(s) = \frac{K(s+20)}{(s+1)(s+3)(s+10)}$$

(a) Use MATLAB to design any controller by using Bode diagrams such that the compensated system has the best possible transient response and steady state specifications.

(b) Solve the same problem using the root locus technique for controller design.

(c) Compare the obtained results and comment on the simplicity (or complexity) of the root locus and Bode diagram methods for controller design.

9.16 Repeat Problem 9.15 for the open-loop control system defined in Problem 9.9.

9.17 In order to relate the phase margin (gain margin) and the real parts of the system eigenvalues as quantities for relative stability measure, perform the following experiment. Consider the system

$$G(s) = \frac{K(s+6)}{(s+10)(s^2+2s+2)}$$

(a) Vary the value for static gain K from zero to 100 in increments of 10, and for each value of K find the phase and gain margins. Plot both phase and gain margins as functions of K.

(b) For each value of K find the closed-loop eigenvalues and plot the magnitude of the eigenvalue real part with respect to K.

(c) Compare diagrams obtained in (a) and (b) and draw the corresponding conclusion.

9.18 Repeat Problem 9.17 for the open-loop control system defined in Problem 9.9.

9.19 Consider a unit feedback control system that has the open-loop transfer function

$$G(s) = \frac{(2+s)e^{-Ts}}{(1+s)(s+10)}$$

Note that the term e^{-Ts} represents a time delay.

(a) Assuming that the time delay is negligible, draw the corresponding Bode diagrams and determine the phase and gain stability margins.

(b) Since the time delay affects only the phase diagram, draw the corrected phase Bode diagram for $T = 0.1$. Determine the phase and gain stability margins and compare them to the corresponding quantities found in (a).

(c) Repeat part (b) for $T = 0.2$ and $T = 0.5$. Comment on the impact of the time delay element on the phase and gain stability margins.

Chapter Ten

Control System Theory Overview

In this book we have presented results mostly for continuous-time, time-invariant, deterministic control systems. We have also, to some extent, given the corresponding results for discrete-time, time-invariant, deterministic control systems. However, in control theory and its applications several other types of system appear. If the coefficients (matrices $\mathbf{A}, \mathbf{B}, \mathbf{C}, \mathbf{D}$) of a linear control system change in time, one is faced with *time-varying control systems*. If a system has some parameters or variables of a random nature, such a system is classified as a stochastic system. Systems containing variables delayed in time are known as *systems with time delays*.

In applying control theory results to real-world systems, it is very important to minimize both the amount of energy to be spent while controlling a system and the difference (error) between the actual and desired system trajectories. Sometimes a control action has to be performed as fast as possible, i.e. in a minimal time interval. These problems are addressed in modern *optimal control theory*. The most recent approach to optimal control theory emerged in the early eighties. This approach is called the H_∞ *optimal control theory*, and deals simultaneously with the optimization of certain performance criteria and minimization of the norm of the system transfer function(s) from undesired quantities in the system (disturbances, modeling errors) to the system's outputs.

Obtaining mathematical models of real physical systems can be done either by applying known physical laws and using the corresponding mathematical equations, or through an experimental technique known as *system identification*. In the latter case, a system is subjected to a set of standard known input functions and by measuring the system outputs, under certain conditions, it is possible to obtain a mathematical model of the system under consideration.

In some applications, systems change their structures so that one has first to perform on-line estimation of system parameters and then to design a control law that will produce the desired characteristics for the system. These systems are known as *adaptive control systems*. Even though the original system may be linear, by using the closed-loop adaptive control scheme one is faced, in general, with a nonlinear control system problem.

Nonlinear control systems are described by nonlinear differential equations. One way to control such systems is to use the linearization procedure described in Section 1.6. In that case one has to know the system nominal trajectories and inputs. Furthermore, we have seen that the linearization procedure is valid only if deviations from nominal trajectories and inputs are small. In the general case, one has to be able to solve nonlinear control system problems. Nonlinear control systems have been a "hot" area of research since the middle of the eighties, since when many valuable nonlinear control theory results have been obtained. In the late eighties and early nineties, *neural networks, which are in fact nonlinear systems with many inputs and many outputs*, emerged as a universal technological tool of the future. However, many questions remain to be answered due to the high level of complexity encountered in the study of nonlinear systems.

In the last section of this chapter, we comment on other important areas of control theory such as algebraic methods in control systems, discrete events systems, intelligent control, fuzzy control, large scale systems, and so on.

10.1 Time-Varying Systems

A time-varying, continuous-time, linear control system in the state space form is represented by

$$\dot{\mathbf{x}}(t) = \mathbf{A}(t)\mathbf{x}(t) + \mathbf{B}(t)\mathbf{u}(t), \quad \mathbf{x}(t_0) = \mathbf{x}_o$$
$$\mathbf{y}(t) = \mathbf{C}(t)\mathbf{x}(t) + \mathbf{D}(t)\mathbf{u}(t)$$

(10.1)

Its coefficient matrices are time functions, which makes these systems much more challenging for analytical studies than the corresponding time-invariant ones.

It can be shown that the solution of (10.1) is given by (Chen, 1984)

$$\mathbf{x}(t) = \Phi(t, t_0)\mathbf{x}(t_0) + \int_{t_0}^{t} \Phi(t, \tau)\mathbf{B}(\tau)\mathbf{u}(\tau)d\tau \qquad (10.2)$$

where $\Phi(t, t_0)$ is the state transition matrix. For an *input-free system*, the transition matrix relates the state of the system at the initial time and the state of the system at any given time, that is

$$\mathbf{x}(t) = \Phi(t, t_0)\mathbf{x}(t_0) \qquad (10.3)$$

It is easy to establish from (10.3) that the state transition matrix has the following properties:

(1) $\Phi(t, t_0)$ satisfies the system differential equation

$$\frac{\partial}{dt}\Phi(t, t_0) = \mathbf{A}(t)\Phi(t, t_0), \qquad \Phi(t_0, t_0) = \mathbf{I} \qquad (10.4)$$

(2) $\Phi(t, t_0)$ is nonsingular, which follows from

$$\Phi^{-1}(t, t_0) = \Phi(t_0, t) \qquad (10.5)$$

(3) $\Phi(t, t_0)$ satisfies

$$\Phi(t_2, t_0) = \Phi(t_2, t_1)\Phi(t_1, t_0) \qquad (10.6)$$

Due to the fact that the system matrix, $\mathbf{A}(t)$, is a function of time, *it is not possible, in general, to find the analytical expression for the system state transition matrix* so that the state response equation (10.2) can be solved only numerically.

Since the coefficient matrices $\mathbf{A}(t), \mathbf{B}(t), \mathbf{C}(t), \mathbf{D}(t)$ are time functions, three essential system concepts presented in Chapters 4 and 5—stability, controllability, and observability—have to be redefined for the case of time-varying systems.

The stability of time-varying systems cannot be defined in terms of the system eigenvalues as for time-invariant systems. Furthermore, several stability definitions have to be introduced for time-varying systems, such as bounded-input bounded-output stability, stability of the system's equilibrium points, global stability, uniform stability, and so on (Chen, 1984). The stability in the sense of Lyapunov, applied to the system equilibrium points, $\mathbf{x}_e(t)$, defined by

$$\mathbf{A}(t)\mathbf{x}_e(t) = 0 \qquad (10.7)$$

indicates that the corresponding equilibrium points are stable if the system state transition matrix is bounded, that is

$$\|\Phi(t, t_0)\| \leq \text{const} < \infty, \qquad \forall t > t_0 \qquad (10.8)$$

Since the system transition matrix has no analytical expression, it is hard, in general, to test the stability condition (10.8).

Similarly, the controllability and observability of time-varying systems are tested differently to the corresponding ones of time-invariant systems. It is necessary to use the notions of controllability and observability Grammians of time-varying systems, respectively, defined by

$$W_c(t_0, t_1) = \int_{t_0}^{t_1} \Phi(t_0, \tau)\mathbf{B}(\tau)\mathbf{B}^T(\tau)\Phi(t_0, \tau)d\tau \qquad (10.9)$$

and

$$W_o(t_0, t_1) = \int_{t_0}^{t_1} \Phi^T(\tau, t_0)\mathbf{C}^T(\tau)\mathbf{C}(\tau)\Phi(\tau, t_0)d\tau \qquad (10.10)$$

The controllability and observability tests are defined in terms of these Grammians and the state transition matrix. Since the system state transition matrix is not known in its analytical form, we conclude that it is very hard to test the controllability and observability of time-varying systems. The reader interested in this topic is referred to Chen (1984), Klamka (1991), and Rugh (1993).

Corresponding results can be presented for discrete-time, time-varying, linear systems defined by

$$\mathbf{x}(k+1) = \mathbf{A}(k)\mathbf{x}(k) + \mathbf{B}(k)\mathbf{u}(k), \qquad \mathbf{x}(k_0) = \mathbf{x}_o$$
$$\mathbf{y}(k) = \mathbf{C}(k)\mathbf{x}(k) + \mathbf{D}(k)\mathbf{u}(k) \qquad (10.11)$$

The state transition matrix for the system (10.11) is given by

$$\Phi(k, k_0) = \mathbf{A}(k)\mathbf{A}(k-1)\cdots\mathbf{A}(k_0+1)\mathbf{A}(k_0) \qquad (10.12)$$

It is interesting to point out that, unlike the continuous-time result, *the discrete-time transition matrix of time-varying systems is in general singular.* It is nonsingular if and only if the matrix $\mathbf{A}(i)$ is nonsingular for $\forall i = k_0, k_0 + 1, ..., k$.

Similarly to the stability study of continuous-time, time-varying systems, in the discrete-time domain one has to consider several stability definitions. The eigenvalues are no longer indicators of system stability. The stability of the system equilibrium points can be tested in terms of the bounds imposed on the system state transition matrix. The system controllability and observability conditions are given in terms of the discrete-time controllability and observability Grammians. The interested reader can find more about discrete-time, time-invariant and time-varying linear systems in Ogata (1987).

10.2 Stochastic Linear Control Systems

Stochastic linear control systems can be defined in several frameworks, such as jump linear systems, Markov chains, systems driven by white noise, to name a few. From the control theory point of view, linear control systems driven by white noise are the most interesting. Such systems are described by the following equations

$$\dot{\mathbf{x}}(t) = \mathbf{A}\mathbf{x}(t) + \mathbf{B}\mathbf{u}(t) + \mathbf{G}\mathbf{w}(t), \quad E\{\mathbf{x}(t_0)\} = \overline{\mathbf{x}}_0$$

$$\mathbf{y}(t) = \mathbf{C}\mathbf{x}(t) + \mathbf{v}(t)$$

(10.13)

where $\mathbf{w}(t)$ and $\mathbf{v}(t)$ are white noise stochastic processes, which represent the system noise (disturbance) and the measurement noise (inaccuracy of sensors or their inability to measure the state variables perfectly). White noise stochastic processes are mathematical fictions that represent real stochastic processes that have a large frequency bandwidth. They are good approximate mathematical models for many real physical processes such as wind, white light, thermal noise, unevenness of roads, and so on. *The spectrum of white noise is constant at all frequencies.* The corresponding constant is called the white noise intensity. Since the spectrum of a signal is the Fourier transform of its covariance, the covariance matrices of the system (plant) white noise and measurement white noise are given by

$$E\{\mathbf{w}(t)\mathbf{w}^T(\tau)\} = \mathbf{W}\delta(t - \tau), \quad E\{\mathbf{v}(t)\mathbf{v}^T(\tau)\} = \mathbf{V}\delta(t - \tau) \qquad (10.14)$$

It is also assumed that the mean values of these white noise processes are equal to zero, that is

$$E\{\mathbf{w}(t)\} = \mathbf{0}, \qquad E\{\mathbf{v}(t)\} = \mathbf{0} \tag{10.15}$$

even though this assumption is not crucial.

The problem of finding the optimal control for (10.13) such that a given performance criterion is optimized will be studied in Section 10.3. Here we consider only an input-free system, $\mathbf{u}(t) = \mathbf{0}$, subjected to (i.e. corrupted by) white noise

$$\dot{\mathbf{x}}(t) = \mathbf{A}\mathbf{x}(t) + \mathbf{G}\mathbf{w}(t), \qquad E\{\mathbf{x}(t_0)\} = \overline{\mathbf{x}}_0 \tag{10.16}$$

and present results for the mean and variance of the state space variables. In order to obtain a valid result, i.e. in order that the mean and variance describe completely the stochastic nature of the system, we have to assume that white noise stochastic processes are Gaussian. It is well known that *Gaussian stochastic processes are completely described by their mean and variance* (Kwakernaak and Sivan, 1972).

Applying the expected value (mean) operator to equation (10.16), we get

$$E\{\dot{\mathbf{x}}(t)\} = \mathbf{A}E\{\mathbf{x}(t)\} + \mathbf{G}E\{\mathbf{w}(t)\}, \qquad E\{\mathbf{x}(t_0)\} = \overline{\mathbf{x}}_0 \tag{10.17}$$

Denoting the expected value $E\{\mathbf{x}(t)\} = \mathbf{m}(t)$, and using the fact that the mean value of white noise is zero, we get

$$\dot{\mathbf{m}}(t) = \mathbf{A}\mathbf{m}(t), \qquad \mathbf{m}(t_0) = E\{\mathbf{x}(t_0)\} = \overline{\mathbf{x}}_0 \tag{10.18}$$

which implies that the mean of the state variables of a linear stochastic system driven by white noise is represented by a pure deterministic input-free system, the solution of which has the known simple form

$$\mathbf{m}(t) = E\{\mathbf{x}(t)\} = e^{\mathbf{A}(t-t_0)}\mathbf{m}(t_0) = e^{\mathbf{A}(t-t_0)}\overline{\mathbf{x}}_0 \tag{10.19}$$

In order to be able to find the expression for the variance of state trajectories for the system defined in (10.16), we also need to know a value for the initial variance of $\mathbf{x}(t_0)$. Let us assume that the initial variance is $\mathbf{Q}(t_0) = \mathbf{Q}_0$. The variance is defined by

$$Var\{\mathbf{x}(t)\} = E\left\{[\mathbf{x}(t) - \mathbf{m}(t)][\mathbf{x}(t) - \mathbf{m}(t)]^T\right\} = \mathbf{Q}(t) \tag{10.20}$$

It is not hard to show (Kwakernaak and Sivan, 1972; Sage and White, 1977) that *the variance of the state variables of a continuous-time linear system driven by white noise satisfies the famous continuous-time matrix Lyapunov equation*

$$\dot{\mathbf{Q}}(t) = \mathbf{Q}(t)\mathbf{A}^T + \mathbf{A}\mathbf{Q}(t) + \mathbf{G}\mathbf{W}\mathbf{G}^T, \quad \mathbf{Q}(t_0) = \mathbf{Q}_0 \qquad (10.21)$$

Note that if the system matrix $\mathbf{A}$ is stable, the system reaches the steady state and the corresponding state variance is given by the algebraic Lyapunov equation of the form

$$0 = \mathbf{Q}\mathbf{A}^T + \mathbf{A}\mathbf{Q} + \mathbf{G}\mathbf{W}\mathbf{G}^T \qquad (10.22)$$

which is, in fact, the steady state counterpart to (10.21).

Example 10.1: MATLAB and its function `lyap` can be used to solve the algebraic Lyapunov equation (10.22). In this example, we find the variance of the state variables of the F-8 aircraft, considered in Section 5.7, under wind disturbances. The matrix $\mathbf{A}$ is given in Section 5.7. Matrices $\mathbf{G}$ and $\mathbf{W}$ are given by Teneketzis and Sandell (1977)

$$\mathbf{G} = [-46.3 \quad 1.214 \quad -1.214 \quad -9.01]^T, \quad \mathbf{W} = 0.000315$$

Note that the wind can be quite accurately modeled as a white noise stochastic process with the intensity matrix $\mathbf{W}$ (Teneketzis and Sandell, 1977). The MATLAB statement

```
Q=lyap(A,G*W*G');
```

produces the unique solution (since the matrix $\mathbf{A}$ is stable) for (10.22) as

$$\mathbf{Q} = \begin{bmatrix} 0.4731 & -0.0050 & 0.0106 & 0.0326 \\ -0.0050 & 0.0001 & -0.0002 & -0.0009 \\ 0.0106 & -0.0002 & 0.0009 & 0.0009 \\ 0.0326 & -0.0009 & 0.0009 & 0.0068 \end{bmatrix}$$

$\diamond$

Similarly, one is able to obtain corresponding results for discrete-time systems, i.e. the state trajectories of a linear stochastic discrete-time system driven by Gaussian white noise

$$\mathbf{x}(k+1) = \mathbf{A}\mathbf{x}(k) + \mathbf{G}\mathbf{w}(k), \quad E\{\mathbf{x}(0)\} = \overline{\mathbf{x}}_0, \quad Var\{\mathbf{x}(0)\} = \mathbf{Q}_0 \qquad (10.23)$$

satisfy the following mean and variance equations

$$m(k+1) = \mathbf{A}m(k), \quad m(0) = m_0 \ \Rightarrow m(k) = \mathbf{A}^k m_0 \qquad (10.24)$$

$$Q(k+1) = \mathbf{A}Q(k)\mathbf{A}^T + \mathbf{GWG}^T, \quad Q(0) = Q_0 \qquad (10.25)$$

Equation (10.25) is known as the *discrete-time matrix difference Lyapunov equation*. If the matrix $\mathbf{A}$ is stable, then one is able to define the steady state system variance in terms of the solution of the algebraic discrete-time Lyapunov equation. This equation is obtained by setting $Q(k+1) = Q(k) = Q$ in (10.25), that is

$$Q = \mathbf{A}Q\mathbf{A}^T + \mathbf{GWG}^T \qquad (10.26)$$

The MATLAB function `dlyap` can be used to solve (10.26). The interested reader can find more about the continuous- and discrete-time Lyapunov matrix equations, and their roles in system stability and control, in Gajić and Qureshi (1995).

10.3 Optimal Linear Control Systems

In Chapters 8 and 9 of this book we have designed dynamic controllers such that the closed-loop systems display the desired transient response and steady state characteristics. The design techniques presented in those chapters have sometimes been limited to trial and error methods while searching for controllers that meet the best given specifications. Furthermore, we have seen that in some cases it has been impossible to satisfy all the desired specifications, due to contradictory requirements, and to find the corresponding controller.

Controller design can also be done through rigorous mathematical optimization techniques. One of these, which originated in the sixties (Kalman, 1960)—called modern optimal control theory in this book—is a time domain technique. During the sixties and seventies, the main contributor to modern optimal control theory was Michael Athans, a professor at the Massachusetts Institute of Technology (Athans and Falb, 1966). Another optimal control theory, known as H_∞, is a trend of the eighties and nineties. H_∞ optimal control theory started with the work of Zames (1981). It combines both the time and frequency domain optimization techniques to give a unified answer, which is optimal from both the time domain and frequency domain points of view (Francis, 1987). Similarly to

H_∞ optimal control theory, the so-called H_2 optimal control theory optimizes systems in both time and frequency domains (Doyle *et al.*, 1989, 1992; Saberi *et al.*, 1995) and is the trend of the nineties. Since H_∞ optimal control theory is mathematically quite involved, in this section we will present results only for the modern optimal linear control theory due to Kalman. It is worth mentioning that very recently a new philosophy has been introduced for system optimization based on linear matrix inequalities (Boyd *et al.*, 1994).

In the context of the modern optimal linear control theory, we present results for the deterministic optimal linear regulator problem, the optimal Kalman filter, and the optimal stochastic linear regulator. Only the main results are given without derivations. This is done for both continuous- and discrete-time domains with emphasis on the infinite time optimization (steady state) and continuous-time problems. In some places, we also present the corresponding finite time optimization results. In addition, several examples are provided to show how to use MATLAB to solve the corresponding optimal linear control theory problems.

10.3.1 Optimal Deterministic Regulator Problem

In modern optimal control theory of linear deterministic dynamic systems, represented in continuous-time by

$$\dot{\mathbf{x}}(t) = \mathbf{A}(t)\mathbf{x}(t) + \mathbf{B}(t)\mathbf{u}(t), \quad \mathbf{x}(t_0) = \mathbf{x}_0 \tag{10.27}$$

we use linear state feedback, that is

$$\mathbf{u}(t) = -\mathbf{F}(t)\mathbf{x}(t) \tag{10.28}$$

and optimize the value for the feedback gain, $\mathbf{F}(t)$, such that the following performance criterion is minimized

$$J = \min_{\mathbf{u}(t)} \left\{ \frac{1}{2} \int_{t_0}^{t_f} \left[\mathbf{x}^T(t)\mathbf{R}_1\mathbf{x}(t) + \mathbf{u}^T(t)\mathbf{R}_2\mathbf{u}(t) \right] dt \right\}, \quad \mathbf{R}_1 \geq 0, \ \mathbf{R}_2 > 0 \tag{10.29}$$

This choice for the performance criterion is quite logical. It requires minimization of the "square" of input, which means, in general, minimization of the input energy required to control a given system, and minimization of the "square" of the state variables. Since the state variables—in the case when a linear system is obtained through linearization of a nonlinear system—represent deviations from the nominal system trajectories, control engineers are interested in minimizing the "square" of this difference, i.e. the "square" of $x(t)$. In the case when the linear mathematical model (10.27) represents the "pure" linear system, the minimization of (10.29) can be interpreted as the goal of bringing the system as close as possible to the origin ($x(t) = 0$) while optimizing the energy. This regulation to zero can easily be modified (by shifting the origin) to regulate state variables to any constant values.

It is shown in Kalman (1960) that the linear feedback law (10.28) produces the global minimum of the performance criterion (10.29). The solution to this optimization problem, obtained by using one of two mathematical techniques for dynamic optimization—dynamic programming (Bellman, 1957) and calculus of variations—is given in terms of the solution of the famous Riccati equation (Bittanti *et al.*, 1991; Lancaster and Rodman, 1995). It can be shown (Kalman, 1960; Kirk, 1970; Sage and White, 1977) that the required optimal solution for the feedback gain is given by

$$\mathbf{F}_{opt}(t) = -\mathbf{R}_2^{-1}\mathbf{B}^T(t)\mathbf{P}(t) \tag{10.30}$$

where $\mathbf{P}(t)$ is the positive semidefinite solution of the matrix differential Riccati equation

$$-\dot{\mathbf{P}}(t) = \mathbf{A}^T(t)\mathbf{P}(t) + \mathbf{P}(t)\mathbf{A}(t) + \mathbf{R}_1 - \mathbf{P}(t)\mathbf{B}(t)\mathbf{R}_2^{-1}\mathbf{B}^T(t)\mathbf{P}(t), \quad \mathbf{P}(t_f) = 0 \tag{10.31}$$

In the case of time invariant systems and for an infinite time optimization period, i.e. for $t_f \to \infty$, the differential Riccati equation becomes an algebraic one

$$0 = \mathbf{A}^T\mathbf{P} + \mathbf{P}\mathbf{A} + \mathbf{R}_1 - \mathbf{P}\mathbf{B}\mathbf{R}_2^{-1}\mathbf{B}^T\mathbf{P} \tag{10.32}$$

If the original system is both controllable and observable (or only stabilizable and detectable) the unique positive definite (semidefinite) solution of (10.32) exists, such that the closed-loop system

$$\dot{\mathbf{x}}(t) = \left(\mathbf{A} - \mathbf{B}\mathbf{R}_2^{-1}\mathbf{B}^T\mathbf{P}\right)\mathbf{x}(t), \quad \mathbf{x}(t_0) = \mathbf{x}_0 \tag{10.33}$$

is asymptotically stable. In addition, the optimal (minimal) value of the performance criterion is given by (Kirk, 1970; Kwakernaak and Sivan, 1972; Sage and White, 1977)

$$J_{opt} = J_{min} = \frac{1}{2}\mathbf{x}_0^T\mathbf{P}\mathbf{x}_0 \tag{10.34}$$

Example 10.2: Consider the linear deterministic regulator for the F-8 aircraft whose matrices $\mathbf{A}$ and $\mathbf{B}$ are given in Section 5.7. The matrices in the performance criterion together with the system initial condition are taken from Teneketzis and Sandell (1977)

$$\mathbf{R}_1 = diag\{0.01 \quad 0 \quad 3260 \quad 3260\}, \quad \mathbf{R}_2 = 3260$$

$$\mathbf{x}_0 = \begin{bmatrix} 100 & 0 & 0.2 & 0 \end{bmatrix}^T$$

The linear deterministic regulator problem is also known as the linear-quadratic optimal control problem since the system is linear and the performance criterion is quadratic. The MATLAB function lqr and the corresponding instruction

        [F,P,ev]=lqr(A,B,R1,R2);

produce values for optimal gain $\mathbf{F}$, solution of the algebraic Riccati equation $\mathbf{P}$, and closed-loop eigenvalues. These quantities are obtained as

$$\mathbf{F} = \begin{bmatrix} -0.004 & 0.5557 & -0.2521 & 0.0590 \end{bmatrix}$$

$$\mathbf{P} = 10^4 \begin{bmatrix} 0.0000 & -0.0016 & -0.0003 & -0.0003 \\ -0.0016 & 1.6934 & 0.1499 & 0.2199 \\ -0.0003 & 0.1499 & 0.8211 & -0.0713 \\ -0.0003 & 0.2199 & -0.0713 & 0.1361 \end{bmatrix}$$

$$\lambda_{1,2} = -0.9631 \pm j3.0061, \quad \lambda_{3,4} = -0.0373 \pm j0.0837$$

The optimal value for the performance criterion can be found from (10.34), which produces $J_{opt} = 743.9707$.

◇

For linear *discrete-time* control systems, a corresponding optimal control theory result can be obtained. Let the discrete-time performance criterion for an infinite time optimization problem of a time-invariant, discrete-time, linear system

$$\mathbf{x}(k+1) = \mathbf{A}\mathbf{x}(k) + \mathbf{B}\mathbf{u}(k), \quad \mathbf{x}(0) = \mathbf{x}_0 \tag{10.35}$$

be defined by

$$J = \frac{1}{2} \sum_{k=0}^{\infty} \left[\mathbf{x}^T(k)\mathbf{R}_1\mathbf{x}(k) + \mathbf{u}^T(k)\mathbf{R}_2\mathbf{u}(k) \right], \quad \mathbf{R}_1 \geq 0, \quad \mathbf{R}_2 > 0 \tag{10.36}$$

then the optimal control is given by

$$u(k) = -\left(R_2 + B^T PB\right)^{-1} B^T PA x(k) = -F_{opt} x(k) \qquad (10.37)$$

where P satisfies the discrete-time algebraic Riccati equation

$$P = A^T PA + R_1 - A^T PB \left(B^T PB + R_2\right)^{-1} B^T PA \qquad (10.38)$$

It can be shown that the standard controllability–observability conditions imply the existence of a unique stabilizing solution of (10.38) such that the closed-loop system

$$\dot{x}(k) = (A - BF_{opt}) x(k) \qquad (10.39)$$

is asymptotically stable. The optimal performance criterion in this case is also given by (10.34) (Kwakernaak and Sivan, 1972; Sage and White, 1977; Ogata, 1987).

10.3.2 Optimal Kalman Filter

Consider a stochastic continuous-time system disturbed by white Gaussian noise with the corresponding measurements also corrupted by white Gaussian noise, that is

$$\dot{x}(t) = Ax(t) + Gw(t), \quad E\{x(t_0)\} = \overline{x}_0, \quad Var\{x(t_0)\} = Q_0 \qquad (10.40)$$
$$y(t) = Cx(t) + v(t)$$

Since the system is disturbed by white noise, the state space variables are also stochastic quantities (processes). Under the assumption that both the system noise and measurement noise are Gaussian stochastic processes, then so too are the state variables. Thus, the state variables are stochastically completely determined by their mean and variance values.

Since the system measurements are corrupted by white noise, exact information about state variables is not available. The Kalman filtering problem can be formulated as follows: find a dynamical system that produces as its output the best estimates, $\hat{x}(t)$, of the state variables $x(t)$. The term "the best estimates" means those estimates for which the variance of the estimation error

$$e(t) = x(t) - \hat{x}(t) \qquad (10.41)$$

is minimized. This problem was originally solved in a paper by Kalman and Bucy (1961). However, in the literature it is mostly known simply as the Kalman filtering problem.

The Kalman filter is a stochastic counterpart to the deterministic observer considered in Section 5.6. It is a dynamical system built by control engineers and driven by the outputs (measurements) of the original system. In addition, it has the same order as the original physical system.

The optimal Kalman filter is given by (Kwakernaak and Sivan, 1972)

$$\dot{\hat{x}}(t) = A\hat{x}(t) + K_{opt}(t)(y(t) - C\hat{x}(t)) \tag{10.42}$$

where the optimal filter gain satisfies

$$K_{opt}(t) = Q(t)C^T V^{-1} \tag{10.43}$$

The matrix $Q(t)$ represents the minimal value for the variance of the estimation error $e(t) = x(t) - \hat{x}(t)$, and is given by the solution to the filter differential Riccati equation

$$\dot{Q}(t) = AQ(t) + Q(t)A^T + GWG^T - Q(t)C^T V^{-1} CQ(t), \quad Q(t_0) = Q_o \tag{10.44}$$

Assuming that the filter reaches steady state, the differential Riccati equation becomes the algebraic one, that is

$$AQ + QA^T + GWG^T - QC^T V^{-1} CQ = 0 \tag{10.45}$$

so that the optimal Kalman filter gain K_{opt} as given by (10.43) and (10.45) is constant at steady state.

Note that in the case when an input is present in the state equation, as in (10.13), the Kalman filter has to be driven by the same input as the original system, that is

$$\dot{\hat{x}}(t) = A\hat{x}(t) + Bu(t) + K_{opt}(t)(y(t) - C\hat{x}(t)) \tag{10.46}$$

The expression for the optimal filter gain stays the same, and is given by (10.43).

Example 10.3: Consider the F-8 aircraft example. Its matrices A, B, and C are given in Section 5.7, and matrices G and W in Example 10.1. From the paper by Teneketzis and Sandell (1977) we have the value for the intensity matrix of the measurement noise

$$\mathbf{V} = \begin{bmatrix} 0.000686 & 0 \\ 0 & 40 \end{bmatrix}$$

The optimal filter gain at steady state can be obtained by using the MATLAB function lqe, which stands for the linear-quadratic estimator (filter) design. This name is justified by the fact that we consider a linear system and intend to minimize the variance ("square") of the estimation error. Thus, by using

[K,Q,ev]=lqe(A,G,C,W,V);

we get *steady state values* for optimal Kalman filter gain **K**, minimal (optimal) error variance **Q**, and closed-loop filter eigenvalues. For the F-8 aircraft, these are given by

$$\mathbf{K} = \begin{bmatrix} 21.6433 & -0.6020 & 0.6021 & 4.5056 \\ 0.0081 & -0.0001 & 0.0001 & 0.0004 \end{bmatrix}^T$$

$$\mathbf{Q} = \begin{bmatrix} 0.3229 & -0.0024 & 0.0053 & 0.0148 \\ -0.0024 & 0.0001 & -0.0001 & -0.0004 \\ 0.0053 & -0.0001 & 0.0004 & 0.0004 \\ 0.0148 & -0.0004 & 0.0004 & 0.0031 \end{bmatrix}$$

$$\lambda_1 = -3.7491, \quad \lambda_2 = -2.6410, \quad \lambda_{3,4} = -0.0104 \pm j0.0760$$

◇

Note that the closed-loop filter in Example 10.3 is asymptotically stable, where the closed-loop structure is obtained by rearranging (10.42) or (10.46) as

$$\dot{\mathbf{x}}(t) = (\mathbf{A} - \mathbf{K}_{opt}\mathbf{C})\hat{\mathbf{x}}(t) + \mathbf{B}\mathbf{u}(t) + \mathbf{K}_{opt}\mathbf{y}(t) \tag{10.47}$$

It can be seen from this structure that the optimal closed-loop Kalman filter is driven by both the system measurements and control. A block diagram for this system-filter configuration is given in Figure 10.1.

Similarly to the continuous-time Kalman filter, it is possible to develop corresponding results for the *discrete-time Kalman filter*. The interested reader can find the corresponding results in several books (e.g. Kwakernaak and Sivan, 1972; Ogata, 1987).

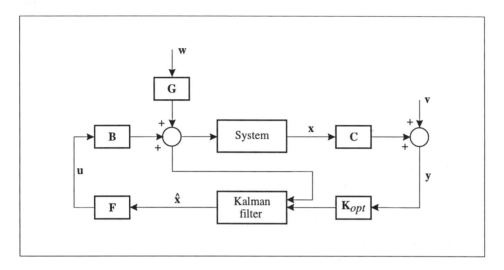

10.1: System-filter configuration

10.3.3 Optimal Stochastic Regulator Problem

In the optimal control problem of linear stochastic systems, represented in continuous time by (10.13), the following stochastic performance criterion is defined

$$J = E\left\{ \lim_{t_f \to 0} \frac{1}{t_f} \int_0^{t_f} \left[\mathbf{x}^T(t)\mathbf{R}_1\mathbf{x}(t) + \mathbf{u}^T(t)\mathbf{R}_2\mathbf{u}(t)\right] dt \right\}, \quad \mathbf{R}_1 \geq 0, \quad \mathbf{R}_2 > 0$$

(10.48)

The goal is to find the optimal value for $\mathbf{u}(t)$ such that the performance measure (10.48) is minimized along the trajectories of dynamical system (10.13).

The solution to the above problem is greatly simplified by the application of the so-called *separation principle. The separation principle says: perform first the optimal estimation, i.e. construct the Kalman filter, and then do the optimal regulation,* i.e. find the optimal regulator as outlined in Subsection 10.3.1 with $\mathbf{x}(t)$ replaced by $\hat{\mathbf{x}}(t)$. For more details about the separation principle and its complete proof see Kwakernaak and Sivan (1972). By the separation principle the optimal control for the above problem is given by

$$\mathbf{u}_{opt}(t) = -\mathbf{F}_{opt}\hat{\mathbf{x}}(t) \tag{10.49}$$

where $\mathbf{F}_{opt}$ is obtained from (10.30) and (10.32), and $\hat{\mathbf{x}}(t)$ is obtained from the Kalman filter (10.46). The optimal performance criterion at steady state can be obtained by using either one of the following two expressions (Kwakernaak and Sivan, 1972)

$$J_{opt} = \mathrm{trace}\{\mathbf{PKVK}^T + \mathbf{QR}_1\} = \mathrm{trace}\{\mathbf{PGWG}^T + \mathbf{QF}^T\mathbf{R}_2\mathbf{F}\} \tag{10.50}$$

Example 10.4: Consider again the F-8 aircraft example. The matrices $\mathbf{F}_{opt}$ and $\mathbf{P}$ are obtained in Example 10.2 and the matrices $\mathbf{K}_{opt}$ and $\mathbf{Q}$ are known from Example 10.3. Using any of the formulas given in (10.50), we obtain the optimal value for the performance criterion as $J_{opt} = 25.0425$. The optimal control is given by (10.49) with the optimal estimates $\hat{\mathbf{x}}(t)$ obtained from the Kalman filter (10.46).

◇

The solution to the *discrete-time* optimal stochastic regulator is also obtained by using the separation principle, i.e. by combining the results of optimal filtering and optimal regulation. For details, the reader is referred to Kwakernaak and Sivan (1972) and Ogata (1987).

10.4 Linear Time-Delay Systems

The dynamics of linear systems containing time-delays is described by delay-differential equations (Driver, 1977). The state space form of a time-delay linear control system is given by

$$\dot{\mathbf{x}}(t) = \mathbf{A}x(t) + \mathbf{A}_D x(t - T) + \mathbf{B}u(t) \tag{10.51}$$

where T represents the time-delay. This form can be generalized to include state variables delayed by $2T, 3T, ...$, time-delay periods. For the purpose of this introduction to linear time-delay systems it is sufficient to consider only models given by (10.51).

Taking the Laplace transform of (10.51) we get

$$s\mathbf{X}(s) - \mathbf{A}\mathbf{X}(s) - \mathbf{A}_D\mathbf{X}(s)e^{-sT} = \mathbf{x}(0^-) + \mathbf{B}\mathbf{U}(s) \qquad (10.52)$$

which produces the *characteristic equation for linear time-delay systems* in the form

$$\det\left(s\mathbf{I} - \mathbf{A} - \mathbf{A}_D e^{-sT}\right) = 0$$

$$= s^n + a_{n-1}\left(e^{-sT}\right)s^{n-1} + \cdots + a_1\left(e^{-sT}\right)s + a_0\left(e^{-Ts}\right) = 0 \qquad (10.53)$$

Note that the coefficients $a_i, i = 0, 1, 2, ..., n - 1$, are functions of e^{-sT}, i.e. of the complex frequency s, and therefore the characteristic equation is not in the polynomial form as in the case of continuous-time, time-invariant linear systems without time-delay. This implies that the transfer function for time-delay linear systems is not a rational function. Note that the rational functions can be represented by a ratio of two polynomials.

An important feature of the characteristic equation (10.53) is that it has in general, infinitely many solutions. Due to this fact the study of time-delay linear systems is much more mathematically involved than the study of linear systems without time-delay.

It is interesting to point out that the stability theory of time-delay systems comes to the conclusion that asymptotic stability is guaranteed if and only if all roots of the characteristic equations (10.53) are strictly in the left half of the complex plane, even though the number of roots may in general be infinite (Mori and Kokame, 1989; Su *et al.*, 1994). In practice, stability of time-delay linear time-invariant systems can be examined by using the Nyquist stability test (Kuo, 1991), as well as by employing Bode diagrams and finding the corresponding phase and gain stability margins.

Studying the controllability of time-delay linear systems is mathematically very complex as demonstrated in the corresponding chapter in the book by Klamka (1991) on system controllability. Analysis, optimization, and applications of time-delayed systems are presented in Malek-Zavarei and Jamshidi (1987). For control problems associated with these systems the reader is referred to Marshall (1977).

Note that in some cases linear time-delay control systems can be related to the sampled data control systems (Ackermann, 1985), which are introduced in Chapter 2, and to the discrete-time linear control systems whose state space form is considered in detail in Chapter 3 and some of their properties are studied throughout this book. Approximations of the time-delay element e^{-Ts} for small values of T are considered in Section 9.7.

10.5 System Identification and Adaptive Control

System identification is an experimental technique used to determine mathematical models of dynamical systems. Adaptive control is applied for systems that change their mathematical models or some parameters with time. Since system identification is included in every adaptive control scheme, in this section we present some essential results for both system identification and adaptive control.

10.5.1 System Identification

The identification procedure is based on data collected by applying known inputs to a system and measuring the corresponding outputs. Using this method, a family of pairs $(\mathbf{y}(t_i), \mathbf{u}(t_i))$, $i = 1, 2, 3, \ldots$ is obtained, where t_i stand for the time instants at which the results are recorded (measured). In this section, we will present only one identification technique, known as the *least-square estimation method*, which is relevant to this book since it can be used to identify the system transfer function of time-invariant linear systems. Many other identification techniques applicable either to deterministic or stochastic systems are presented in several standard books on identification (see for example Ljung, 1987; Soderstrom and Stoica, 1989). For simplicity, in this section we study the transfer function identification problem of single-input single-output systems in the discrete-time domain.

Consider an nth-order time-invariant, discrete-time, linear system described by the corresponding difference equation as presented in Section 3.3.1, that is

$$y(k+n) + a_{n-1}y(k+n-1) + \cdots + a_0 y(k)$$
$$= b_{n-1}u(k+n-1) + b_{n-2}u(k+n-2) + \cdots + b_0 u(k) \tag{10.54}$$

It is assumed that parameters

$$\mathbf{a} = \begin{bmatrix} a_{n-1} & a_{n-2} & \cdots & a_0 \end{bmatrix}, \quad \mathbf{b} = \begin{bmatrix} b_{n-1} & b_{n-2} & \cdots & b_0 \end{bmatrix} \tag{10.55}$$

are not known and ought to be determined using the least-square identification technique.

Equation (10.54) can be rewritten as

$$y(k + n) = -a_{n-1}y(k + n - 1) - \cdots - a_0 y(k)$$
$$+ b_{n-1}u(k + n - 1) + b_{n-2}u(k + n - 2) + \cdots + b_0 u(k)$$
$$(10.56)$$

and put in the vector form

$$y(k + n) = \mathbf{f}(k + n - 1)\begin{bmatrix} \mathbf{a} \\ \mathbf{b} \end{bmatrix} \qquad (10.57)$$

where $\mathbf{a}$ and $\mathbf{b}$ are defined in (10.55), and $\mathbf{f}(k + n - 1)$ is given by

$$\mathbf{f}(k + n - 1) = [-y(k + n - 1) \quad \cdots \quad -y(k) \quad u(k + n - 1) \quad \cdots \quad u(k)]$$
$$(10.58)$$

Assume that for the given input, we perform N measurements and that the actual (measured) system outputs are known, that is

$$\mathbf{Y_a}(k, N) = \begin{bmatrix} y_a(k + n) \\ y_a(k + n - 1) \\ \dots \\ y_a(k + n - N + 1) \\ y_a(k + n - N) \end{bmatrix} \qquad (10.59)$$

The problem now is how to determine 2n parameters in $\mathbf{a}$ *and* $\mathbf{b}$ *such that the actual system outputs* $\mathbf{Y_a}(k, N)$ *are as close as possible to the mathematically computed system outputs that are represented by formula (10.57).*

We can easily generate N equations from (10.57) as

$$\mathbf{Y}(k, N) = \begin{bmatrix} y(k + n) \\ y(k + n - 1) \\ \dots \\ y(k + n - N + 1) \\ y(k + n - N) \end{bmatrix} = \begin{bmatrix} \mathbf{f}(k + n - 1) \\ \mathbf{f}(k + n - 2) \\ \dots \\ \mathbf{f}(k + n - N + 1) \\ \mathbf{f}(k + n - N) \end{bmatrix} \begin{bmatrix} \mathbf{a} \\ \mathbf{b} \end{bmatrix} = \mathbf{\Psi}(k, N)\begin{bmatrix} \mathbf{a} \\ \mathbf{b} \end{bmatrix}$$
$$(10.60)$$

Define the estimation (identification) error as

$$\mathbf{E}(k, N) = \mathbf{Y_a}(k, N) - \mathbf{Y}(k, N) \qquad (10.61)$$

The least-square estimation method requires that the choice of the unknown parameters **a** and **b** minimizes the "square" of the estimation error, that is

$$\min_{\mathbf{a},\mathbf{b}} J = \frac{1}{2} \min_{\mathbf{a},\mathbf{b}} \left\{ \mathbf{E}^T(k,N)\mathbf{E}(k,N) \right\} \qquad (10.62)$$

Using expressions for the vector derivatives (see Appendix C) and (10.60), we can show that

$$\min_{\mathbf{a},\mathbf{b}} \{J\} \Rightarrow \frac{\partial J}{\partial \begin{bmatrix} \mathbf{a} \\ \mathbf{b} \end{bmatrix}} = 0 \Rightarrow \Psi^T(k,N)\Psi(k,N)\begin{bmatrix} \mathbf{a} \\ \mathbf{b} \end{bmatrix} - \Psi^T(k,N)\mathbf{Y}_a(k,N) = 0$$

$$(10.63)$$

which produces the least-square optimal estimates for the unknown parameters as

$$\begin{bmatrix} \mathbf{a} \\ \mathbf{b} \end{bmatrix} = \left\{ \Psi^T(k,N)\Psi(k,N) \right\}^{-1} \Psi^T(k,N)\mathbf{Y}_a(k,N) \qquad (10.64)$$

Note that the input signal has to be chosen such that the matrix inversion defined in (10.64) exists.

Sometimes it is sufficient to estimate (identify) only some parameters in a system or in a problem under consideration in order to obtain a complete insight into its dynamical behavior. Very often the identification (estimation) process is combined with known physical laws which describe some, but not all, of the system variables and parameters. It is interesting to point out that MATLAB contains a special toolbox for system identification.

10.5.2 Adaptive Control

Adaptive control schemes in closed-loop configurations represent nonlinear control systems even in those cases when the systems under consideration are linear. Due to this fact, it is not easy to study adaptive control systems analytically. However, due to their practical importance, adaptive controllers are widely used nowadays in industry since they produce satisfactory results despite the fact that many theoretical questions remain unsolved.

Two major configurations in adaptive control theory and practice are *self-tuning regulators* and *model-reference adaptive schemes*. These configurations are represented in Figures 10.2 and 10.3. For self-tuning regulators, it is assumed that the system parameters are constant, but unknown. On the other hand, for the model-reference adaptive scheme, it is assumed that the system parameters change over time.

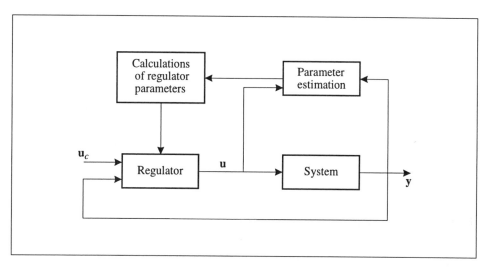

10.2: Self-tuning regulator

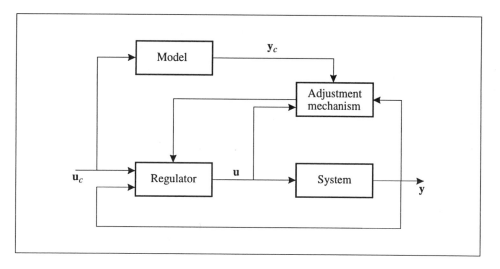

10.3: Model-reference adaptive control scheme

It can be seen from Figure 10.2 that for self-tuning regulators the "separation principle" is used, i.e. the problem is divided into independent estimation and regulation tasks. In the regulation problem, the estimates are used as the true values of the unknown parameters. The command signal u_c must be chosen such that unknown system parameters can be estimated. The stability of the closed-loop systems and the convergence of the proposed schemes for self-tuning regulators are very challenging and interesting research areas (Wellstead and Zarrop, 1991).

In the model-reference adaptive scheme, a desired response is specified by using the corresponding mathematical model (see Figure 10.3). The error signal generated as a difference between desired and actual outputs is used to adjust system parameters that change over time. It is assumed that system parameters change much slower than system state variables.

The adjusted parameters are used to design a controller, a feedback regulator. There are several ways to adjust parameters; one commonly used method is known as the MIT rule (Astrom and Wittermark, 1989). As in the case of self-tuning regulators, model-reference adaptive schemes still have many theoretically unresolved stability and convergence questions, even though they do perform very well in practice.

For detailed study of self-tuning regulators, model-reference adaptive systems, and other adaptive control schemes and techniques applicable to both deterministic and stochastic systems, the reader is referred to Astrom and Wittenmark (1989), Wellstead and Zarrop (1991), Isermann et al. (1992), and Krstić et al. (1995).

10.6 Nonlinear Control Systems

Nonlinear control systems are introduced in this book in Section 1.6, where we have presented a method for their linearization. Mathematical models of time-invariant nonlinear control systems are given by

$$\dot{\mathbf{x}}(t) = \mathcal{F}(\mathbf{x}(t), \mathbf{u}(t)), \quad \mathbf{x}(t_0) = \mathbf{x}_0$$
$$\mathbf{y}(t) = \mathcal{G}(\mathbf{x}(t), \mathbf{u}(t))$$
(10.65)

where $\mathbf{x}(t)$ is a state vector, $\mathbf{u}(t)$ is an input vector, $\mathbf{y}(t)$ is an output vector, and $\mathcal{F}$ and $\mathcal{G}$ are nonlinear matrix functions. These control systems are in general very difficult to study analytically. Most of the analytical results come from the mathematical theory of classic nonlinear differential equations, the theory of which has been developing for more than a hundred years (Coddington and Levinson, 1955). In contrast to classic differential equations, due to the presence of the input vector in (10.65) one is faced with the even more challenging so-called controlled differential equations. The mathematical and engineering theories of controlled nonlinear differential equations are the modern trend of the eighties and nineties (Sontag, 1990).

Many interesting phenomena not encountered in linear systems appear in nonlinear systems, e.g. hysteresis, limit cycles, subharmonic oscillations, finite escape time, self-excitation, multiple isolated equilibria, and chaos. For more details about these nonlinear phenomena see Siljak (1969) and Khalil (1992).

It is very hard to give a brief presentation of any result and/or any concept of nonlinear control theory since almost all of them take quite complex forms. Familiar notions such as system stability and controllability for nonlinear systems have to be described by using several definitions (Klamka, 1991; Khalil, 1992).

One of the most interesting results of nonlinear theory is the so-called *stability concept in the sense of Lyapunov*. This concept deals with the stability of system equilibrium points. The equilibrium points of nonlinear systems are defined by

$$\mathbf{0} = \mathcal{F}(\mathbf{x}_e(t), \mathbf{u}(t)) \implies \mathbf{x}_e(t) \tag{10.66}$$

Roughly speaking, an equilibrium point is stable in the sense of Lyapunov if a small perturbation in the system initial condition does not cause the system trajectory to leave a bounded neighborhood of the system equilibrium point. The Lyapunov stability can be formulated for time invariant nonlinear systems (10.65) as follows (Slotine and Li, 1991; Khalil, 1992; Vidyasagar, 1993).

Theorem 10.1 *The equilibrium point $\mathbf{x}_e = 0$ of a time invariant nonlinear system is stable in the sense of Lyapunov if there exists a continuously differentiable scalar function $V(\mathbf{x})$ such that along the system trajectories the following is satisfied*

$$V(\mathbf{x}) > 0, \quad V(0) = 0$$

$$\dot{V}(\mathbf{x}) = \frac{dV}{dt} = \frac{\partial V}{\partial \mathbf{x}} \frac{d\mathbf{x}}{dt} \leq 0 \tag{10.67}$$

Thus, the problem of examining system stability in the sense of Lyapunov requires finding a scalar function known as the *Lyapunov function $V(\mathbf{x})$*. Again, this is a very hard task in general, and one is able to find the Lyapunov function $V(\mathbf{x})$ for only a few real physical nonlinear control systems.

In this book we have presented in Section 1.6 the procedure for linearization of nonlinear control systems. Another classic method, known as the describing function method, has proved very popular for analyzing nonlinear control systems (Slotine and Li, 1991; Khalil, 1992; Vidyasagar, 1993).

10.7 Comments

In addition to the classes of control systems extensively studied in this book, and those introduced in Chapter 10, many other theoretical and practical control areas have emerged during the last thirty years. For example, decentralized control (Siljak, 1991), learning systems (Narendra, 1986), algebraic methods for multivariable control systems (Callier and Desoer, 1982; Maciejowski, 1989), robust control (Morari and Zafiriou, 1989; Chiang and Safonov, 1992; Grimble, 1994; Green and Limebeer, 1995), control of robots (Vukobratović and Stokić, 1982; Spong and Vidyasagar, 1989; Spong et al., 1993), differential games (Isaacs, 1965; Basar and Olsder, 1982; Basar and Bernhard, 1991), neural network control (Gupta and Rao, 1994), variable structure control (Itkis, 1976; Utkin, 1992), hierarchical and multilevel systems (Mesarović et al., 1970), control of systems with slow and fast modes (singular perturbations) (Kokotović and Khalil, 1986; Kokotović et al., 1986; Gajić and Shen, 1993), predictive control (Soeterboek, 1992), distributed parameter control, large-scale systems (Siljak, 1978; Gajić and Shen, 1993), fuzzy control systems (Kandel and Langholz, 1994; Yen et al., 1995), discrete event systems (Ho, 1991; Ho and Cao, 1991), intelligent vehicles and highway control systems, intelligent control systems (Gupta and Sinha, 1995; de Silva, 1995), control in manufacturing (Zhou and DiCesare, 1993), control of flexible structures, power systems control (Anderson and Fouad, 1984), control of aircraft (McLean, 1991), linear algebra and numerical analysis control algorithms (Laub, 1985; Bittanti et al., 1991; Petkov et al., 1991; Bingulac and Vanlandingham, 1993; Patel et al., 1994), and computer-controlled systems (Astrom and Wittenmark, 1990).

Finally, it should be emphasized that control theory and its applications are studied within all engineering disciplines, and as well as in applied mathematics (Kalman et al., 1969; Sontag, 1990) and computer science.

10.8 References

Ackermann, J., *Sampled-Data Control Systems*, Springer-Verlag, London, 1985.

Anderson, P. and A. Fouad, *Power System Control and Stability*, Iowa State University Press, Ames, Iowa, 1984.

Astrom, K. and B. Wittenmark, *Adaptive Control*, Wiley, New York, 1989.

Astrom, K. and B. Wittenmark, *Computer-Controlled Systems: Theory and Design*, Prentice Hall, Englewood Cliffs, New Jersey, 1990.

Athans, M. and P. Falb, *Optimal Control: An Introduction to the Theory and its Applications*, McGraw-Hill, New York, 1966.

Basar, T. and P. Bernhard, H^∞—*Optimal Control and Related Minimax Design Problems*, Birkhauser, Boston, Massachusetts, 1991.

Basar, T. and G. Olsder, *Dynamic Non-Cooperative Game Theory*, Academic Press, New York, 1982.

Bellman, R., *Dynamic Programming*, Princeton University Press, Princeton, New Jersey, 1957.

Bingulac, S. and H. Vanlandingham, *Algorithms for Computer-Aided Design of Multivariable Control Systems*, Marcel Dekker, New York, 1993.

Bittanti, S., A. Laub, and J. Willems, *The Riccati Equation*, Springer-Verlag, Berlin, 1991.

Boyd, S., L. El Ghaoui, E. Feron, and V. Balakrishnan, *Linear Matrix Inequalities in System and Control Theory*, SIAM, Philadelphia, 1994.

Callier, F. and C. Desoer, *Multivariable Feedback Systems*, Springer-Verlag, Stroudsburg, Pennsylvania, 1982.

Chen, C., *Linear System Theory and Design*, Holt, Rinehart and Winston, New York, 1984.

Chiang, R. and M. Safonov, *Robust Control Tool Box User's Guide*, The Math Works, Inc., Natick, Massachusetts, 1992.

Coddington, E. and N. Levinson, *Theory of Ordinary Differential Equations*, McGraw-Hill, New York, 1955.

de Silva, C., *Intelligent Control: Fuzzy Logic Applications*, CRC Press, Boca Raton, Florida, 1995.

Doyle, J., B. Francis, and A. Tannenbaum, *Feedback Control Theory*, Macmillan, New York, 1992.

Doyle, J., K. Glover, P. Khargonekar, and B. Francis, "State space solutions to standard H_2 and H_∞ control problems," *IEEE Transactions on Automatic Control*, vol. AC-34, 831–847, 1989.

Driver, R., *Ordinary and Delay Differential Equations*, Springer-Verlag, New York, 1977.

Francis, B., *A Course in H_∞ Control Theory*, Springer-Verlag, Berlin, 1987.

Gajić, Z. and X. Shen, *Parallel Algorithms for Optimal Control of Large Scale Linear Systems*, Springer-Verlag, London, 1993.

Gajić, Z. and M. Qureshi, *Lyapunov Matrix Equation in System Stability and Control*, Academic Press, San Diego, California, 1995.

Green, M. and D. Limebeer, *Linear Robust Control*, Prentice Hall, Englewood Cliffs, New Jersey, 1995.

Grimble, M., *Robust Industrial Control*, Prentice Hall International, Hemel Hempstead, 1994.

Gupta, M. and D. Rao, (eds.), *Neuro-Control Systems*, IEEE Press, New York, 1994.

Gupta, M. and N. Sinha, (eds.), *Intelligent Control Systems: Concepts and Algorithms*, IEEE Press, New York, 1995.

Ho, Y., (ed.), *Discrete Event Dynamic Systems*, IEEE Press, New York, 1991.

Ho, Y. and X. Cao, *Perturbation Analysis of Discrete Event Dynamic Systems*, Kluwer, Dordrecht, 1991.

Isaacs, R., *Differential Games*, Wiley, New York, 1965.

Isermann, R., K. Lachmann, and D. Matko, *Adaptive Control Systems*, Prentice Hall International, Hemel Hempstead, 1992.

Itkis, U., *Control Systems of Variable Structure*, Wiley, New York, 1976.

Kailath, T., *Linear Systems*, Prentice Hall, Englewood Cliffs, New Jersey, 1980.

Kalman, R., "Contribution to the theory of optimal control," *Boletin Sociedad Matematica Mexicana*, vol. 5, 102–119, 1960.

Kalman, R. and R. Bucy, "New results in linear filtering and prediction theory," *Journal of Basic Engineering, Transactions of ASME, Ser. D.*, vol. 83, 95–108, 1961.

Kalman, R., P. Falb, and M. Arbib, *Topics in Mathematical System Theory*, McGraw-Hill, New York, 1969.

Kandel A. and G. Langholz (eds.), *Fuzzy Control Systems*, CRC Press, Boca Raton, Florida, 1994.

Kirk, D., *Optimal Control Theory*, Prentice Hall, Englewood Cliffs, New Jersey, 1970.

Khalil, H., *Nonlinear Systems*, Macmillan, New York, 1992.

Klamka, J., *Controllability of Dynamical Systems*, Kluwer, Warszawa, 1991.

Kokotović P. and H. Khalil, *Singular Perturbations in Systems and Control*, IEEE Press, New York, 1986.

Kokotović, P., H. Khalil, and J. O'Reilly, *Singular Perturbation Methods in Control: Analysis and Design*, Academic Press, Orlando, Florida, 1986.

Krstić, M., I. Kanellakopoulos, and P. Kokotović, *Nonlinear and Adaptive Control Design*, Wiley, New York, 1995.

Kuo, B., *Automatic Control Systems*, Prentice Hall, Englewood Cliffs, New Jersey, 1991.

Kwakernaak, H. and R. Sivan, *Linear Optimal Control Systems*, Wiley, New York, 1972.

Lancaster, P. and L. Rodman, *Algebraic Riccati Equation*, Clarendon Press, Oxford, 1995.

Laub, A., "Numerical linear algebra aspects of control design computations," *IEEE Transactions on Automatic Control*, vol. AC-30, 97–108, 1985.

Lewis, F., *Applied Optimal Control and Estimation*, Prentice Hall, Englewood Cliffs, New Jersey, 1992.

Ljung, L., *System Identification: Theory for the User*, Prentice Hall, Englewood Cliffs, New Jersey, 1987.

Maciejowski, J., *Multivariable Feedback Design*, Addison-Wesley, Wokingham, 1989.

Malek-Zavarei, M. and M. Jamshidi, *Time-Delay Systems: Analysis, Optimization and Applications*, North-Holland, Amsterdam, 1987.

Marsall, J., *Control of Time-Delay Systems*, IEE Peter Peregrinus, New York, 1977.

McLean, D., *Automatic Flight Control Systems*, Prentice Hall International, Hemel Hempstead, 1991.

Mesarović, M., D. Macko, and Y. Takahara, *Theory of Hierarchical, Multilevel, Systems*, Academic Press, New York, 1970.

Morari, M. and E. Zafiriou, *Robust Process Control*, Prentice Hall, Englewood Cliffs, New Jersey, 1989.

Mori, T. and H. Kokame, "Stability of $\dot{x}(t) = Ax(t) + Bx(t - \tau)$," *IEEE Transactions on Automatic Control*, vol. AC-34, 460–462, 1989.

Narendra, K. (ed.), *Adaptive and Learning Systems—Theory and Applications*, Plenum Press, New York, 1986.

Ogata, K., *Discrete-Time Control Systems*, Prentice Hall, Englewood Cliffs, New Jersey, 1987.

Patel, R., A. Laub, and P. Van Dooren, *Numerical Linear Algebra Techniques for Systems and Control*, IEEE Press, New York, 1994.

Petkov, P., N. Christov, and M. Konstantinov, *Computational Methods for Linear Control Systems*, Prentice Hall International, Hemel Hempstead, 1991.

Rugh, W., *Linear System Theory*, Prentice Hall, Englewood Cliffs, New Jersey, 1993.

Saberi, A., P. Sannuti, and B. Chen, H_2—*Optimal Control*, Prentice Hall International, Hemel Hempstead, 1995.

Sage, A. and C. White, *Optimum Systems Control*, Prentice Hall, Englewood Cliffs, New Jersey, 1977.

Siljak, D., *Nonlinear Systems*, Wiley, New York, 1969.

Siljak, D., *Large Scale Dynamic Systems: Stability and Structure*, North-Holland, New York, 1978.

Siljak, D., *Decentralized Control of Complex Systems*, Academic Press, San Diego, California, 1991.

Slotine, J. and W. Li, *Applied Nonlinear Control*, Prentice Hall, Englewood Cliffs, New Jersey, 1991.

Soderstrom, T. and P. Stoica, *System Identification*, Prentice Hall International, Hemel Hempstead, 1989.

Soeterboek, R., *Predictive Control—A Unified Approach*, Prentice Hall, Englewood Cliffs, New Jersey, 1992.

Sontag, E., *Mathematical Control Theory*, Springer-Verlag, New York, 1990.

Spong, M. and M. Vidyasagar, *Robot Dynamics and Control*, Wiley, New York, 1989.

Spong, M., F. Lewis, and C. Abdallah (eds.), *Robot Control: Dynamics, Motion Planing, and Analysis*, IEEE Press, New York, 1993.

Su, J., I. Fong, and C. Tseng, "Stability analysis of linear systems with time delay," *IEEE Transactions on Automatic Control*, vol. AC-39, 1341–1344, 1994.

Teneketzis, D. and N. Sandell, "Linear regulator design for stochastic systems by multiple time-scale method," *IEEE Transactions on Automatic Control*, vol. AC-22, 615–621, 1977.

Utkin, V., *Sliding Modes in Control Optimization*, Springer-Verlag, Berlin, 1992.

Vukobratović, M. and D. Stokić, *Control of Manipulation Robots: Theory and Applications*, Springer-Verlag, Berlin, 1982.

Vidyasagar, M., *Nonlinear Systems Analysis*, Prentice Hall, Englewood Cliffs, New Jersey, 1993.

Wellstead, P. and M. Zarrop, *Self-Tuning Systems—Control and Signal Processing*, Wiley, Chichester, 1991.

Yen, J., R. Langari, and L. Zadeh, (eds.), *Industrial Applications of Fuzzy Control and Intelligent Systems*, IEEE Press, New York, 1995.

Zames, G., "Feedback and optimal sensitivity: Model reference transformations, multiplicative seminorms, and approximate inverses," *IEEE Transactions on Automatic Control*, vol. AC-26, 301–320, 1981.

Zhou, M. and F. DiCesare, *Petri Net Synthesis for Discrete Event Control of Manufacturing Systems*, Kluwer, Boston, Massachusetts, 1993.

Appendix

In these appendices we present several standard results that the reader might find useful while reading this book and solving the corresponding examples and problems.

In Appendix A the properties (Table A.1) and common transform pairs (Table A.2) of the one-sided Laplace transform are given. The one-sided Laplace transform is used extensively in this book, and represents the main tool for frequency domain analysis and the design of continuous-time, time-invariant, linear control systems.

Appendix B contains the properties of the one-sided $\mathcal{Z}$-transform (Table B.1) and the common $\mathcal{Z}$-transform pairs (Table B.2). The one-sided $\mathcal{Z}$-transform is used throughout this book for the frequency domain study of discrete-time, time-invariant, linear control systems.

Some results from linear algebra encountered in this book in the context of the state space approach to continuous- and discrete-time, time-invariant, linear systems, are summarized in Appendix C. Several definitions, justifications, and some short proofs of important results are given. Furthermore, it has been indicated that linear algebra is a very powerful tool for studying numerous linear control system problems.

Appendix D contains a short introduction to MATLAB in the form of a laboratory manual for linear control systems. The laboratory manual is self-explanatory. The MATLAB programs used in this book are provided on a computer disk, which is distributed through the Math Works, Inc. The programs may also be obtained via anonymous FTP from the Internet site `ece.rutgers.edu` in the directory `/pub/gajic` or by pointing a Web browser to the book's WWW homepage on `http://www.ece.rutgers.edu/~gajic/control.html`.

A. Laplace Transform

In this appendix we present the definition of the Laplace transformation and give its derivative property, which helps in transforming differential equations that describe system dynamics into complex coefficient algebraic equations. The algebraic equations obtained are frequency domain representations of the considered dynamical systems. At the end of the appendix, we present tables with the properties of the Laplace transform and the Laplace transform common pairs.

A.1 Definition of the Laplace Transform and its Main Property

For a continuous-time function $f(t)$, the Laplace transform is defined by

$$\mathcal{L}\{f(t)\} = F(s) = \int_{0^-}^{\infty} f(t)e^{-st}dt \qquad (a.1)$$

This is the so-called *one-sided* or *unilateral Laplace transform* of $f(t)$, and depends only on the values of $f(t)$ for $t \geq 0$, i.e. it is applicable to *causal* signals for which $f(t) = 0$, $t \leq 0$. In general, the *two-sided* or *bilateral* Laplace transform, in which the lower limit in the integral is $-\infty$, can be defined. For the purpose of this course, it is sufficient to define and use only the one-sided Laplace transform. Note that the Laplace transform $F(s)$ depends on a complex variable $s = \sigma + j\omega$, called the complex frequency. The existence condition of the Laplace transform requires the existence (convergence) of the following integral

$$\int_{0^-}^{\infty} |f(t)|e^{-\sigma t}dt < \infty \qquad (a.2)$$

where σ is any nonnegative real number.

The *inverse Laplace transform* is obtained according to the formula

$$f(t) = \mathcal{L}^{-1}\{F(s)\} = \frac{1}{2\pi} \int_{\gamma-j\omega}^{\gamma+j\omega} F(s)e^{st}ds \qquad (a.3)$$

where γ is a real value chosen to the right of all singularities of the function $F(s)$. It recovers the original function $f(t)$ for $t \geq 0$ and gives zero for $t < 0$. The inverse Laplace transform, as given by (a.3), is not easily calculated, and so in order to find the Laplace inverse one uses the familiar procedure of partial fraction expansion and the table of common Laplace transform pairs (see Table A.2).

A very important property of the Laplace transform, which is extremely useful in system theory, is the derivative property. It can be easily shown, *assuming that all initial conditions are equal to zero*, that the Laplace transforms of the derivatives are given by the following simple formulas

$$\mathcal{L}\left\{\frac{df(t)}{dt}\right\} = sF(s)$$

$$\mathcal{L}\left\{\frac{d^2 f(t)}{dt^2}\right\} = s^2 F(s)$$

$$\cdots$$

$$\mathcal{L}\left\{\frac{d^n f(t)}{dt^n}\right\} = s^n F(s)$$

(a.4)

In other words, the operation of taking the nth derivative in time, $n = 1, 2, 3, ...$, corresponds in the complex domain to a multiplication by s^n. In the case when the initial conditions are different from zero, the derivative property is slightly modified. In that case it reads as

$$\mathcal{L}\left\{\frac{d^n f(t)}{dt^n}\right\} = s^n F(s) - f\left(0^-\right)s^{n-1} - f^{(1)}\left(0^-\right)s^{n-2} - \cdots - f^{(n-1)}\left(0^-\right) \quad \text{(a.5)}$$

In the next example, we demonstrate the use of the derivative property (a.5) in the analysis of an electrical system.

Example A.1: For the RLC circuit considered in Section 1.4 the differential input/output equation is given by (1.24), that is

$$\frac{d^2 e_0(t)}{dt^2} + a_1 \frac{de_0(t)}{dt} + a_0 e_0(t) = b_0 e_i(t)$$

$$a_1 = \frac{L + R_1 R_2 C}{R_2 LC}, \quad a_0 = \frac{R_1 + R_2}{R_2 LC}, \quad b_0 = \frac{1}{LC}$$

The Laplace transform of this equation is given by

$$\left(s^2 E_0(s) - se_0\left(0^-\right) - \frac{de_0(0^-)}{dt}\right) + a_1 \left(sE_0(s) - e_0\left(0^-\right)\right) + a_0 E_0(s) = b_0 E_i(s)$$

or, after grouping the terms with $E_0(s)$

$$\left(s^2 + a_1 s + a_0\right) E_0(s) - \left(s e_0\left(0^-\right) + \frac{d e_0(0^-)}{dt} + a_1 e_0\left(0^-\right)\right) = b_0 E_i(s)$$

The output voltage $E_0(s)$ in the complex form is obtained as

$$E_0(s) = \frac{b_0}{s^2 + a_1 s + a_0} E_i(s) + \frac{e_0(0^-)s + \frac{d e_0(0^-)}{dt} + a_1 e_0(0^-)}{s^2 + a_1 s + a_0}$$

Time function $e_0(t)$ can be obtained by applying the inverse Laplace transform to the last equation. It can be seen that the system output response has two components, one coming from the forcing function $E_i(s)$ and the second contributed by the system initial conditions.

◇

A.2 Tables of Properties and Common Pairs

In the following we present two tables: the Laplace transform properties (Table A.1), and the table of the common Laplace transform pairs (Table A.2).

The last property given in Table A.1, known as the *final value theorem*, is very often used in linear control system analysis and design, e.g. in order to find the steady state errors. We would like to point out that the final value theorem is applicable only to time functions for which the limit at infinity exists. For example, $\sin t$ has no limit at infinity so that in this case the final value theorem is not applicable. Note that from Table A.2 the Laplace transform of $\sin t$ has a pair of complex conjugate poles on the imaginary axis. An easy check in the complex domain as to whether or not a time function $f(t)$ has a limit at infinity is to examine the poles of its complex image $F(s)$. The final value theorem of the Laplace transform is applicable if the function $sF(s)$ has no poles on the imaginary axis and in the right half of the complex plane. Thus, if a function $F(s)$ has all asymptotically stable poles and a *simple pole* at the origin the final value theorem may be applied.

$\mathcal{L}\{\alpha_1 f_1(t) + \alpha_2 f_2(t)\}$	$\alpha_1 F_1(s) + \alpha_2 F_2(s)$
$\mathcal{L}\{f(t - t_0)u(t - t_0)\}$	$e^{st_0} F(s)$
$\mathcal{L}\{f(at)\}$	$\frac{1}{|a|} F\left(\frac{s}{a}\right)$
$\mathcal{L}\{t^n f(t)\}$	$(-1)^n \frac{d^n}{ds^n} F(s)$
$e^{at} f(t)$	$F(s - a)$
$f(t) \cos \omega t$	$\frac{1}{2}[F(s + j\omega) + F(s - j\omega)]$
$\mathcal{L}\{\frac{d}{dt} f(t)\}$	$sF(s) - f(0^-)$
$\mathcal{L}\{\frac{d^2}{dt^2} f(t)\}$	$s^2 F(s) - sf(0^-) - \dot{f}(0^-)$
$\mathcal{L}\{\frac{d^n}{dt^n} f(t)\}$	$s^n F(s) - s^{n-1} f(0^-) - \cdots - f^{(n-1)}(0^-)$
$\mathcal{L}\{f_1(t) * f_2(t)\}$	$F_1(s) F_2(s)$
$\mathcal{L}\{\int_{-\infty}^{t} f(\tau)d\tau\}$	$\frac{1}{s} F(s)$
$\lim_{t \to 0} \{f(t)\}$	$\lim_{s \to \infty} \{sF(s)\}$
$\lim_{t \to \infty} \{f(t)\}$	$\lim_{s \to 0} \{sF(s)\}$

Table A.1: Properties of the Laplace Transform

$\delta(t)$	1
$u(t)$	$\dfrac{1}{s}$
$e^{-\alpha t}u(t)$	$\dfrac{1}{s+1}$
$t^n u(t)$	$\dfrac{n!}{s^{n+1}}$
$t^n e^{-\alpha t}u(t)$	$\dfrac{n!}{(s+\alpha)^{n+1}}$
$u(t)\cos\omega t$	$\dfrac{s}{s^2+\omega^2}$
$u(t)\sin\omega t$	$\dfrac{\omega}{s^2+\omega^2}$
$e^{-\alpha t}u(t)\cos\omega t$	$\dfrac{s}{(s+\alpha)^2+\omega^2}$
$e^{-\alpha t}u(t)\sin\omega t$	$\dfrac{\omega}{(s+\alpha)^2+\omega^2}$
$tu(t)\cos\omega t$	$\dfrac{s^2-\omega^2}{(s^2+\omega^2)^2}$
$tu(t)\sin\omega t$	$\dfrac{2\omega s}{(s^2+\omega^2)^2}$
$te^{-\alpha t}u(t)\cos\omega t$	$\dfrac{(s+\alpha)^2-\omega^2}{((s+\alpha)^2+\omega^2)^2}$
$te^{-\alpha t}u(t)\sin\omega t$	$\dfrac{2\omega(s+\alpha)}{((s+\alpha)^2+\omega^2)^2}$

Table A.2: Common Laplace transform pairs

B. The $\mathcal{Z}$-Transform

B.1 Definition of the $\mathcal{Z}$-Transform and its Main Property

The $\mathcal{Z}$-transform is the discrete-time counterpart of the Laplace transform. It operates on a discrete-time signal $f(kT)$, where T is a sampling period and kT stands for discrete-time—unlike the Laplace transform, which acts on a continuous-time signal $f(t)$. The discrete-time signal $f(kT)$ arises in practice by sampling a continuous-time signal $f(t)$ every T seconds. In another case, when the system is inherently discrete, we denote such signals by $f(k)$. For simplicity, we omit in this book the parameter T so that $f(k)$ *also stands for* $f(kT)$. Many analogies can be drawn between the Laplace and $\mathcal{Z}$-transforms, and for linear time-invariant systems the results of these two transforms can be presented in parallel. However, there are some minor differences between the transform theory for discrete-time signals and the corresponding theory for continuous-time signals. Here, we present only its definition and the derivative property. Table B.1 contains more details about the $\mathcal{Z}$-transform and its basic properties.

Consider the discrete time causal (equal to zero for negative times) signal $f(k)$, where k, representing discrete time, belongs to a set of integers. The *one-sided $\mathcal{Z}$-transform* of this signal is defined by

$$F(z) = \mathcal{Z}\{f(k)\} = \sum_{k=0}^{\infty} f(k)z^{-k} \tag{b.1}$$

From a mathematical standpoint we may consider this transformation only if the power series defined in (b.1) absolutely converges. This will be the case if

$$\lim_{n \to \infty} \sum_{k=0}^{n} |f(k)|\rho^{-k} = M < \infty \tag{b.2}$$

where ρ belongs to a set of positive real numbers, and M is a positive constant, which may depend on ρ (Kamen, 1990). The above condition means that $f(k)$ has the one-sided $\mathcal{Z}$-transform if

$$|z| > \rho_{min} \tag{b.3}$$

where ρ_{min} denotes the minimal element in the set of real positive numbers such that the convergence condition is satisfied. The set of complex numbers

z satisfying inequality (b.3) defines the *region of absolute convergence* of the $\mathcal{Z}$-transform $F(z)$, which means that $F(z)$ is not defined outside of this region.

The definition formula for finding an inverse of the $\mathcal{Z}$-transform is given, in general, by a complex contour integral

$$f(k) = \frac{1}{2\pi j} \oint_\Gamma F(z)z^{k-1}dz \qquad (b.4)$$

where Γ is a circle of radius greater than ρ_{min} that encircles all singularities (poles of $F(s)z^{k-1}$) in the region of convergence (Kamen, 1990). The evaluation of this integral can be done by using Cauchy's residue theorem. The result of this theorem is that a contour integral of a function of z that is analytic inside contour Γ, except at a finite number of isolated singularities z_i, is given by

$$f(k) = \frac{1}{2\pi j} \oint_\Gamma F(z)z^{k-1}dz = \sum \text{residues of } F(z)z^{k-1} \qquad (b.5)$$

The general method for finding the inverse transformation by using the definition integral is very often computationally involved. For rational functions (represented by ratios of two polynomials), two additional methods can be employed to obtain an inverse of the $\mathcal{Z}$-transform, namely *expansion by long division* and *partial-fraction expansion*. Note that the inverse $\mathcal{Z}$-transform is not unique.

Expansion by Long Division

The idea here is to calculate $f(k)$ by expanding function $F(z)$ into a series of powers of z^{-1}, that is, as

$$F(z) = f(0) + f(1)z^{-1} + f(2)z^{-2} + \cdots + f(k)z^{-k} + \cdots \qquad (b.6)$$

The coefficients of this series are the values of $f(t)$ at sampling instants kT, $k = 0, 1, 2, \dots$.

Partial Fraction Expansion

Given the $\mathcal{Z}$-transform $F(z)$, we first form

$$F_1(z) = \frac{F(z)}{z} \qquad (b.7)$$

The function $F_1(z)$ is then expanded into partial fractions. For example, for distinct poles, we get

$$F_1(z) = \frac{c_1}{z + p_1} + \frac{c_2}{z + p_2} + \cdots + \frac{c_n}{z + p_n}$$

Now each element of the expanded version of $F(z)$

$$F(z) = \frac{c_1 z}{z + p_1} + \frac{c_2 z}{z + p_2} + \cdots + \frac{c_n z}{z + p_n}$$

has its inverse given by discrete-time exponential functions (see Table B.2), that is

$$f(k) = c_1 (p_1)^k + c_2 (p_2)^k + \cdots + c_n (p_n)^k$$

The derivative property for the $\mathcal{Z}$-transform can be obtained analogously to the corresponding property of the Laplace transform. Namely, the nth derivative in discrete-time, which is represented by a left shift in time for n discrete-time instants, in the frequency domain corresponds to a multiplication by z^n, i.e. *assuming that all initial conditions are zero*, we have

$$\mathcal{Z}\{f(k+n)\} = z^n F(z) \tag{b.8}$$

If the initial conditions are different from zero, then the derivative property is expanded into

$$\mathcal{Z}\{f(k+n)\} = z^n F(z) - f(0)z^n - f(1)z^{n-1} - \cdots - f(n-1)z \tag{b.9}$$

B.2 Relation between the $\mathcal{Z}$ and Laplace Transforms

A relation between the $\mathcal{Z}$ and Laplace transforms can be obtained by using the following reasoning. A sequence of samples $f(k)$ is associated with a train of impulses, which is obtained by sampling $f(t)$ by an ideal sampler (Kuo, 1992)

$$f^*(t) = f(0)\delta(t) + f(1)\delta(t-T) + f(2)\delta(t-2T) + \cdots = \sum_{k=0}^{\infty} f(k)\delta(t-kT)$$
$$\tag{b.10}$$

where $\delta(t - kT)$ represents the impulse delta function. The Laplace transform of this sequence is easily found as

$$F^*(s) = \sum_{k=0}^{\infty} f(k)e^{-kTs} \tag{b.11}$$

A star is used to indicate that the Laplace transform operator is applied to the discrete-time function $f^*(t)$. If we replace e^{Ts} by z, the previous equation becomes

$$F^*(s) = \sum_{k=0}^{\infty} f(k)z^{-k} = \mathcal{Z}\{f(k)\} = F(z), \quad s = \frac{1}{T}\ln z \tag{b.12}$$

which shows that the $\mathcal{Z}$-transform of $f(t)$ is equal to its starred Laplace transform $F^*(s)$.

Note that $F^*(s)$ is a periodic function with respect to the sampling frequency $\omega_s = 2\pi/T$, that is

$$F^*(s) = F^*(s + jl\omega_s), \quad l = \text{any integer} \tag{b.13}$$

Another expression for $F^*(s)$ can be derived either by using contour integration (Kuo, 1992) or by the following reasoning (Kuo, 1995, pp. 109–110). Let

$$f^*(t) = f(t)\delta_T(t)$$

where

$$\delta_T(t) = \sum_{n=-\infty}^{n=+\infty} \delta(t - nT) = \frac{1}{T} \sum_{n=-\infty}^{n=+\infty} e^{-j\frac{2\pi nt}{T}}$$

Then

$$f^*(t) = \frac{1}{T} \sum_{n=-\infty}^{n=+\infty} f(t)e^{-jn\omega_s t}$$

which implies

$$F^*(s) = \frac{1}{T} \sum_{n=-\infty}^{n=+\infty} F(s + jn\omega_s) \tag{b.14}$$

This expression is valid assuming that $f(t)$ has no a jump discontinuity at $t = 0$. If there is such a discontinuity, the quantity $f(0^+)/2$ has to be added to the right-hand side of (b.14).

Using formulas (b.13) and (b.14) it follows that

$$C(s) = G(s)F^*(s) \tag{b.15}$$

which implies

$$C^*(s) = \frac{1}{T} \sum_{n=-\infty}^{n=+\infty} C(s + jn\omega_s) = \frac{1}{T} \sum_{n=-\infty}^{n=+\infty} G(s + jn\omega_s)F^*(s + jn\omega_s)$$

$$\tag{b.16}$$

$$\frac{1}{T} \left\{ \frac{1}{T} \sum_{n=-\infty}^{n=+\infty} G(s + jn\omega_s) \right\} F^*(s) = G^*(s)F^*(s)$$

The last expression is particularly important for finding transfer functions of sampled data control systems (Section 2.5).

B.3 Tables of Properties and Common Pairs

$\mathcal{Z}\{a_1 f_1(k) + a_2 f_2(k)\}$	$a_1 F_1(z) + a_2 F_2(z)$
$\mathcal{Z}\{f(k - k_0)u(k - k_0)\}$	$\frac{1}{z^{k_0}} F(z)$
$\mathcal{Z}\{f(k + 1)\}$	$z F(z) - z f(0)$
$\mathcal{Z}\{f(k + 2)\}$	$z^2 F(z) - z^2 f(0) - z f(1)$
$\mathcal{Z}\{f(k + k_0)\}$	$z^{k_0} F(z) - z^{k_0} f(0) - z^{k_0 - 1} f(1) - \cdots - z f(k_0 - 1)$
$\mathcal{Z}\{k f(k)\}$	$-z \frac{d}{dz} F(z)$
$\mathcal{Z}\{k^2 f(k)\}$	$z \frac{d}{dz} F(z) + z^2 \frac{d^2}{dz^2} F(z)$
$\mathcal{Z}\{a^k f(k)\}$	$F\left(\frac{z}{a}\right)$
$\mathcal{Z}\{f(k) \cos \omega kT\}$	$\frac{1}{2}\left[F\left(z e^{j\omega T}\right) + F\left(z e^{-j\omega T}\right)\right]$
$\mathcal{Z}\{f(k) \sin \omega kT\}$	$\frac{1}{2}\left[F\left(z e^{j\omega T}\right) - F\left(z e^{-j\omega T}\right)\right]$
$\mathcal{Z}\{f_1(k) * f_2(k)\}$	$F_1(z) F_2(z)$
$\lim_{k \to 0} f(k)$	$\lim_{z \to \infty} \{F(z)\}$
$\lim_{k \to \infty} f(k)$	$\lim_{z \to 1} \left\{\frac{z-1}{z} F(z)\right\}$

Table B.1: Properties of the $\mathcal{Z}$-transform

$\delta(k)$	1
$u(k)$	$\dfrac{z}{z-1}$
$a^k u(k)$	$\dfrac{z}{z-a}$
$ku(k)$	$\dfrac{z}{(z-1)^2}$
$k^2 u(k)$	$\dfrac{z(z+1)}{(z-a)^2}$
$ka^k u(k)$	$\dfrac{az}{(z-a)^2}$
$k^2 a^k u(k)$	$\dfrac{az(z+a)}{(z-a)^3}$
$u(k)\cos\omega kT$	$\dfrac{z^2 - z\cos\omega T}{z^2 - 2z\cos\omega T + 1}$
$u(k)\sin\omega kT$	$\dfrac{z\sin\omega T}{z^2 - 2z\cos\omega T + 1}$
$a^k u(k)\cos\omega kT$	$\dfrac{z^2 - az\cos\omega T}{z^2 - 2az\cos\omega T + a^2}$
$a^k u(k)\sin\omega kT$	$\dfrac{az\sin\omega T}{z^2 - 2az\cos\omega T + a^2}$

Table B.2: Common $\mathcal{Z}$-transform pairs

References

Kamen, E., *Introduction to Signals and Systems*, Macmillan, New York, 1990.

Kuo, B., *Digital Control Systems*, Saunders College Publishing, New York, 1992.

Kuo, B., *Automatic Control Systems*, Prentice Hall, Englewood Cliffs, New Jersey, 1995.

C. Some Results from Linear Algebra

Linear algebra plays a very important role in linear system control theory and applications (Laub, 1985; Skelton and Iwasaki, 1995). Here we review some standard and important linear algebra results.

Definite Matrices

Definition C.1: A square matrix $\mathbf{M}$ is *positive definite* if all of its eigenvalues have positive real parts, $Re\{\lambda_i(\mathbf{M})\} > 0$. It is *positive semidefinite* if $Re\{\lambda_i(\mathbf{M})\} \geq 0, \forall i$. In addition, *negative definite* matrices are defined by $Re\{\lambda_i(\mathbf{M})\} < 0, \forall i$ and *negative semidefinite* by $Re\{\lambda_i(\mathbf{M})\} \leq 0, \forall i$.

Null Space

Definition C.2: The null space of a matrix $\mathbf{M}$ of dimensions $m \times n$ is the space spanned by vectors $\mathbf{v}$ that satisfy $\mathbf{Av} = \mathbf{0}$.

Systems of Linear Algebraic Equations

Theorem C.1 *Consider a consistent (solvable) system of linear algebraic equations in n unknowns*

$$\mathbf{Mx} = \mathbf{b} \tag{c.1}$$

with $\dim\{\mathbf{M}\} = m \times n$. *Equation (c.1) has a solution if and only if (consistency condition)*

$$\text{rank}\{[\mathbf{M}\ \ \mathbf{b}]\} = \text{rank}\{\mathbf{M}\} \tag{c.2}$$

In addition, if $\text{rank}\{\mathbf{M}\} = m$, *then (c.1) always has a solution. For* $n = m$ *and* $\text{rank}\{\mathbf{M}\} = m$ *the solution obtained is unique.*

Determinant of a Matrix Product

The following results hold for the determinant of a matrix product

$$\det\{\mathbf{M}_1\mathbf{M}_2\} = \det\{\mathbf{M}_1\}\det\{\mathbf{M}_2\} \tag{c.3}$$

For the proof of the above statement the reader is referred to Stewart (1973). This result can be generalized to the product of a finite number of matrices.

Determinant of Matrix Inversion

By using the rule for the determinant of a product we are able to establish the following formula

$$\det\{\mathbf{M}^{-1}\} = \frac{1}{\det\{\mathbf{M}\}} \tag{c.4}$$

This can be proved as follows

$$\det\{MM^{-1}\} = \det\{I\} = 1 = \det\{M\}\det\{M^{-1}\} \Rightarrow \det\{M^{-1}\} = \frac{1}{\det\{M\}}$$

Inversion of a Matrix Product

Consider the problem of finding the matrix inversion to $M_1 M_2$. The inversion is a matrix whose product with $M_1 M_2$ produces an identity matrix, that is

$$M_1 M_2 [\text{Inverse}] = I \tag{c.5}$$

It can be checked that the inverse of the form $M_2^{-1} M_1^{-1}$ satisfies (c.5). This inversion of a product formula can be easily generalized to the product of a finite number of matrices.

Spectral Theorem

Theorem C.2 *If* M *is a symmetric matrix, then its eigenvalues are real and* M *is diagonalizable, i.e. there exists a similarity transformation* P *such that* $P^{-1}MP$ *is diagonal. Furthermore, the transformation is unitary, i.e.* $P^{-1} = P^T$.

Proof of this theorem can be found in many standard books on linear algebra and matrices (see for example Lancaster and Tismenetsky, 1985).

Integral of a Matrix Exponent

The following matrix integral formula is useful in some applications

$$\int_0^T e^{Mt} dt = \left(e^{MT} - I\right)M^{-1} \tag{c.6}$$

provided that the matrix M is nonsingular. In addition, if all eigenvalues of matrix M are asymptotically stable, then

$$\int_0^\infty e^{Mt} dt = -M^{-1}$$

Cayley–Hamilton Theorem

One of the most powerful theorems of linear algebra and matrix analysis is the Cayley–Hamilton theorem. It simply says the following.

Theorem C.3 *Every square matrix satisfies its characteristic polynomial.*

This means that if the characteristic polynomial of $\mathbf{M}$ is

$$\Delta(\lambda) = \lambda^n + a_{n-1}\lambda^{n-1} + \cdots + a_1\lambda + a_0 = 0$$

then, in addition to the above single algebraic equation, the following $n \times n$ algebraic equations are satisfied

$$\Delta(\mathbf{M}) = \mathbf{M}^n + a_{n-1}\mathbf{M}^{n-1} + \cdots + a_1\mathbf{M} + a_0\mathbf{I} = \mathbf{0}^{n \times n} \qquad \text{(c.7)}$$

The Cayley–Hamilton theorem can be used to find efficiently the transition matrices for both continuous- and discrete-time systems. From (c.7) one is able to express $\mathbf{M}^n$ as a function of lower powers of $\mathbf{M}$. The same can be done for $\mathbf{M}^{n+m}$ where $m = 1, 2, 3, ...$, that is

$$\begin{aligned} \mathbf{M}^{n+m} &= \mathcal{F}\left(\mathbf{M}^{n-1}, \mathbf{M}^{n-2}, \ldots, \mathbf{M}, \mathbf{I}\right) \\ &= \alpha_0(t)\mathbf{I} + \alpha_1(t)\mathbf{M} + \cdots + \alpha_{n-1}(t)\mathbf{M}^{n-1}, \quad m = 1, 2, 3, \ldots \end{aligned} \qquad \text{(c.8)}$$

By using (c.8), the expression for the transition matrix of the discrete-time systems defined in (3.74) can be found as

$$\Phi_d(k) = \mathbf{M}_d^k = \alpha_0(k)\mathbf{I} + \alpha_1(k)\mathbf{M}_d + \cdots + \alpha_{n-1}(k)\mathbf{M}_d^{n-1} \qquad \text{(c.9)}$$

Similarly, all powers of n and higher in the infinite series expansion for the continuous-time transition matrix given in (3.37) can be replaced using (c.8), which implies

$$e^{\mathbf{M}t} = \alpha_0(t)\mathbf{I} + \alpha_1(t)\mathbf{M} + \cdots + \alpha_{n-1}(t)\mathbf{M}^{n-1} \qquad \text{(c.10)}$$

The unknown coefficients in (c.9) can be obtained from

$$\lambda^k = \alpha_0(k) + \alpha_1(k)\lambda + \alpha_2(k)\lambda^2 + \cdots + \alpha_{n-1}(k)\lambda^{n-1}, \quad \lambda = \lambda_1, \lambda_2, ..., \lambda_n \qquad \text{(c.11)}$$

Assuming that the eigenvalues are distinct we get from (c.11) n linearly independent equations for n unknown coefficients $\alpha_i(k), i = 0, 1, 2, ..., n-1$. Note that in the case of multiple eigenvalues a set of independent equations can be generated by taking the derivatives of (c.11) *with respect to* λ.

The coefficients $\alpha_i(t), i = 0, 1, 2, ..., n-1$, can be obtained from

$$e^{\lambda t} = \alpha_0(t) + \alpha_1(t)\lambda + \alpha_2(t)\lambda^2 + \cdots + \alpha_{n-1}(t)\lambda^{n-1} \qquad \text{(c.12)}$$

Using different values of $\lambda_i, i = 1, 2, ..., n$, we are able to generate a system of n linear algebraic equations that determine uniquely the required coefficients $\alpha_i(t), i = 0, 1, 2, ..., n-1$. In the case of multiple eigenvalues, the corresponding derivatives of (c.12) have to be taken *with respect to* λ.

Vector Derivatives

The following formulas for vector derivatives are used in Chapter 10.

$$\frac{\partial}{\partial \mathbf{y}}(\mathbf{My}) = \mathbf{M}$$

$$\frac{\partial}{\partial \mathbf{y}}(\mathbf{y}^T \mathbf{M}^T) = \mathbf{M}^T$$

$$\frac{\partial}{\partial \mathbf{y}}(\mathbf{y}^T \mathbf{My}) = \mathbf{My} + \mathbf{M}^T \mathbf{y}$$

Viete's Formulas

Given a polynomial equation

$$a_0 x^n + a_1 x^{n-1} + \cdots + a_{n-1} x + a_n = a_0(x - x_1)(x - x_2) \cdots (x - x_n) = 0$$

then its solutions satisfy

$$x_1 + x_2 + \cdots + x_n = -\frac{a_1}{a_0}$$

$$x_1 x_2 + x_1 x_3 + \cdots + x_{n-1} x_n = \frac{a_2}{a_0}$$

$$x_1 x_2 x_3 + x_1 x_2 x_4 + \cdots + x_{n-2} x_{n-1} x_n = -\frac{a_3}{a_0} \qquad \text{(c.13)}$$

$$\cdots$$

$$x_1 x_2 \cdots x_n = (-1)^n \frac{a_n}{a_0}$$

References

Laub, A., "Numerical linear algebra aspects of control design computations," *IEEE Transactions on Automatic Control*, vol. AC-30, 97–108, 1985.

Lancaster, P. and M. Tismenetsky, *The Theory of Matrices*, Academic Press, Orlando, Florida, 1985.

Skelton, R. and T. Iwasaki, "Increased roles of linear algebra in control education," *IEEE Control Systems Magazine*, vol. 15, 76–90, 1995.

Stewart, G., *Introduction to Matrix Computations*, Academic Press, New York, 1973.

D. MATLAB Manual

MATLAB Tutorial—Linear Systems and Control

The main linear algebra MATLAB functions are presented in Table A.5.

B=A'	Matrix transpose
C=A+B	Matrix addition (subtraction)
C=A*B	Matrix multiplication
expm(A)	Matrix exponent of $\mathbf{A}$, i.e. $e^{\mathbf{A}}$
inv(A)	Matrix inversion of $\mathbf{A}$
eig(A)	Eigenvalues of $\mathbf{A}$
[X,D]=eig(A)	Eigenvectors and eigenvalues of $\mathbf{A}$
rank(A)	Calculates the rank of $\mathbf{A}$
p=poly(A)	Characteristic polynomial of $\mathbf{A}$
r=roots(p)	Root of the polynomial equation
det(A)	Determinant of $\mathbf{A}$

Table A.5: Linear algebra functions

The system/control MATLAB functions of particular interest for undergraduate students are listed below.

(1) [NUM,den]=ss2tf(A,B,C,D,in)
    Finds the system transfer function from the in-th input, that is
    $$\mathbf{H}(s) = \mathbf{C}(s\mathbf{I} - \mathbf{A})^{-1}\mathbf{B} + \mathbf{D} = \mathbf{NUM}(s)/den(s)$$
    where $\mathbf{NUM}(s)$ contains r (number of outputs) rows.

(2) [A,B,C,D]=tf2ss(num,den)
    State space form of the transfer function.

(3) [z,p,k]=ss2zp(A,B,C,D,in)
    This function gives the factored expression for the transfer function

$$H(s) = \frac{k(s + z_1)(s + z_2) \cdots (s + z_m)}{(s + p_1)(s + p_2) \cdots (s + p_n)}$$

(4) `[A,B,C,D]=zp2ss(z,p,k)`

(5) `[z,p,k]=tf2zp(num,den)`

(6) `[num,den]=zp2tf(z,p,k)`
   Functions (4)–(6) are self-explanatory.

(7) `co=ctrb(A,B)`
   Calculates the controllability matrix defined in (5.20).

(8) `ob=obsv(A,C)`
   Calculates the observability matrix defined in (5.6).

(9) `[y,x]=impulse(A,B,C,D,in,t)`
   The impulse response at time t from the in-th input. Also,
   `[y,x]=impulse(num,den,t)`

(10) `[y,x]=step(A,B,C,D,in,t)`
   The step response at time t from the in-th input. Also,
   `[y,x]=step(num,den,t)`

(11) `[y,x]=lsim(A,B,C,D,U,t)`
   The system response at time $t = k \times \Delta t$ due to arbitrary inputs whose values at the discrete time instants k are defined in the matrix $\mathbf{U} \in R^{k \times m}$. Also,
   `[y,x]=lsim(A,B,C,D,U,t,X0)`
   where x_0 stands for the system initial condition.

(12) `[Ad,Bd]=c2d(A,B,T)`
   From the continuous-time to the discrete-time linear system, where T stands for the sampling period.

(13) `r=rlocus(A,B,C,D,k)`
   The root locus of the control system given in the state space form with $u = -\mathbf{K}y$. Also,
   `r=rlocus(num,den,k)`

(14) `plot(r,'-')`
   Plotting function.

(15) `[mag,phase]=bode(A,B,C,D,in,w)`
   Bode diagram, where `mag,phase,` and w represent magnitude, phase, and frequency, respectively. Also,
   `[mag,phase]=bode(num,den,w)`

(16) `[re,im]=nyquist(A,B,C,D,iu,w)`

Nyquist plot. Also,

`[re,im]=nyquist(num,den,w)`

(17) `[Gm,Ph,wcp,wcg]=margin(num,den)`

The phase and gain margins and the corresponding crossover frequencies.

(18) `gram(A,B)`

Calculates the controllability Grammian defined in (5.31). The same function is used to calculate the observability Grammian given in (5.63) with $\mathbf{A}$ replaced by $\mathbf{A}^T$ and $\mathbf{B}$ replaced by $\mathbf{C}^T$.

Note that the functions `impulse`, `step`, and `lsim` applied to discrete-time domain systems have a prefix `d`. The controllability/observability functions have the same form in both continuous-time and discrete-time domains.

Basic MATLAB Functions[1]

Here we present basic, general, MATLAB functions related to the basic mathematical operations and procedures of inputting data, starting and quitting MATLAB.

Entering and Quitting MATLAB

To enter MATLAB type `matlab`. To quit MATLAB type `quit` or `exit`.

The HELP Facility

MATLAB has an extensive built-in help system. If help is required on any function, simply type `help` followed by the function in question. MATLAB displays a brief text indicating the function use and examples of its usage. A HELP facility is available, providing on-line information on most MATLAB topics. To get a list of HELP topics type `help`. To get HELP on a specific topic, type `help topic`. For example, `help eig` provides information on the use of the eigenvalue function.

MATLAB was written so that it can be used much in the same way as you would if you were writing on paper. For example, to find the solution to 21/3, you would type

```
>>21/3
ans =
      7
```

Note that the double right caret, >>, indicates that MATLAB is ready for your

---

[1]   This part is mostly written by T. McCrimmon, a former Rutgers University undergraduate student.

input. Also note that any comment used in MATLAB starts with a percent sign. It is important to point out that no variable is either defined or used in the above operation. To assign the above value to the variable x, just use the assignment operator "=". Example,

```
>>x=21/3
x=
      7
```

Now the variable x has a value assigned to it.

There are two things to remember here. One is that x will hold that value until it is changed, until it is cleared, or until the MATLAB session is terminated. Typing the command who causes MATLAB to show all of the variables that have been declared thus far. The command whos lists all variables that have been declared plus additional information such as the type of variable, size of the variable, and the amount of RAM being consumed due to all of the variables that have been assigned. Issuing the clear command purges all of the declared variables from memory. The second thing to remember is that MATLAB defaults to being case sensitive, that is, x does not equal to X.

To see the present value of a variable, simply type the variable name and hit return.

```
>>x
x=
      7
```

The basic mathematics operators are
+ addition
− subtraction
* multiplication
/ division
^ power

Brackets are reserved for the identification of vectors and matrices. One of the special operators is the semicolon. The prior examples have not included the semicolon and the result was immediately displayed. Issuing the semicolon at the end of the line stops the result from being displayed. The advantages of this will become obvious when writing and debugging scripts.

```
>>x^2;
>>% does not display the result
```

```
>>x
x=
     49
```

The ellipses (three or more periods) is the concatenation function which is for continuing equations on the next line that will not fit on one line.

```
>>x=1+2+3+4...
+5+6  returns
x=
     21
```

Numbers and Arithmetic Expressions

Conventional decimal notation, with optional decimal point and leading minus sign, is used for numbers. A power-of-ten scale factor can be included as a suffix. For example

```
3              -99            0.0001
9.64595   1.606E-20  6.066e23
```

MATLAB includes several predefined variables such as i, j, pi, inf, NaN. The i and j variables are the square root of -1, pi is π, inf is ∞, and NaN is not a number, e.g. divide by zero. MATLAB will return an NaN error message if a divide by zero occurs.

Vectors and Matrices

Entering a vector or matrix is a painless operation. Simply use the brackets to delimit the elements of the vector, separated by at least one space. Do not use a comma to separate the elements. For example

```
>>x=[1 2 3 4]
```

results in the output

```
x =
      1   2   3   4
```

Entering matrices is basically the same as entering vectors except that each row is delimited by a semicolon.

```
>>y=[1 2;3 4]
y =
      1   2
      3   4
```

A matrix may also be entered this way (carriage returns replace the semicolons)

```
>>y=[1 2
     3 4]
y =
        1    2
        3    4
```

```
>>y(2,2)
```

results in

```
ans=
        4
```

Using the parenthesis in this fashion allows the user to access any element in the matrix, in this case, the element in the second row and second column.

Matrix Operations

Transpose

The special character ' (apostrophe) denotes the transpose of a matrix. The statements

```
>>A=[1 2 3;4 5 6;7 8 0]
>>B=A'
```

result in

```
A =
        1    2    3
        4    5    6
        7    8    0
B =
        1    4    7
        2    5    8
        3    6    0
```

Addition and Subtraction

Addition and subtraction of matrices are denoted by + and −. For example, with the above matrices, the statement

```
>>C=A+B
```

results in

C =

2	6	10
6	10	14
10	14	0

Matrix Multiplication

Multiplication is denoted by *. For example, the statement

`>>C=A*B`

multiplies matrices **A** and **B** and stores the obtained result in matrix **C**.

Matrix Powers

The expression A^p raises **A** to the p-th power and is defined if **A** is a square matrix and p is a scalar.

Eigenvalues and Eigenvectors

If **A** is an $n \times n$ matrix, the n scalars λ that satisfy $\mathbf{Ax} = \lambda\mathbf{x}$ are the eigenvalues of **A**. They are found by using the function eig. For example

`>>A=[2 1;0 3]`
`>>eig(A)`

produces

ans =

2

3

Eigenvectors are obtained with the statement

`>>[X,D]=eig(A)`

in which case the diagonal elements of **D** are the eigenvalues and the columns of **X** are the corresponding eigenvectors.

Characteristic Polynomial

The coefficients of the characteristic polynomial of the matrix **A** are obtained by using the function poly. The characteristic equation, a polynomial equation, is solved by using the function roots. For example, the statement

`>>poly(A)`

for the matrix **A** given by

`>>A=[1 2 3;4 5 6;7 8 0]`

produces

```
p =
    1 -6 -72 -27
```
The characteristic polynomial is given by $s^3 - 6s^2 - 72s - 27$. The roots of the characteristic equations are obtained using the function `roots` as

```
>>r=roots(p)
```

producing
```
r =
    12.1229
    -5.7345
    -0.3884
```

Polynomials

Entering polynomials is as simple as entering a vector. The following polynomial $h(s) = 5s^4 + 10s^2 + 18s + 23$ would be entered starting with the highest order first, as follows:

```
>>h=[5 0 10 18 23];
```

We have already seen that the function `roots(h)` finds all solutions of the polynomial equation $h(s) = 0$. The other useful MATLAB functions dealing with polynomials are:

1. `polyval(h,10)`
   evaluates the polynomial $h(s)$ at $s = 10$;
2. `[r,p,k]=residue(a,b)`
   performs partial fraction expansion where a = numerator, b = denominator, r = residues, p = poles, k = direct (constant) term;
3. `c=conv(a,b)`
   multiplication of polynomials $a(s)$ and $b(s)$. Note that the vector c contains coefficients of the polynomial $c(s)$ in descending order;
4. `[q,r]=deconv(c,a)`
   divides polynomial $c(s)$ by $a(s)$ with the quotient given by $q(s)$ and remainder by $r(s)$.

Examples:

```
>>p=[1 5 6];
>>polyval(p,1)
```

produces

```
ans=
      12
```

evaluates the polynomial $s^2 + 5s + 6$ at $s = 1$.

```
>>r=roots(p)

r=
      -3
      -2
```

finds roots of $p(s) = 0$.

```
>>a=[1 2 3];
>>b=[4 5 6];
>>c=conv(a,b)
c=
      4 13 28 27 18
```

Plots

The plot command creates linear *x*–*y* plots. If **y** is a vector, plot(y) produces a linear plot of the elements of **y**. Notice that the data are autoscaled and that *x*–*y* axes are drawn.

Loops

MATLAB provides several methods of looping. The for, while, and elseif are the most useful. Each one of these requires an end statement at the end of the loop. They are used much in the same way as one would in any program language with the exception of incrementing. This is done using the colon. The following is a quick set of examples using these methods of looping.

Example 1:

```
>>for k=1:2:10
for m=1:5
if k==m
A(k,m)=2;
elseif abs(k-m)==1
A(k,m)=-1;
else
A(k,m)=0;
end
end
end
```

Example 2:
```
>>count=0
>>n=10
>>while (n-1)>=2
count=count+1;
n=n-1;
end
```

Script Files

Script files, heretofore known as m-files, are simply text files containing all of the code necessary to perform some function. As stated before, a text editor is required for the creation and editing of m-files. The name m-files comes from the fact that the extension for all scripts must be m, e.g. script.m is an acceptable filename. The exclamation mark tells MATLAB that the following line is to be executed by the operating system. Not only is this useful in running the text editor from within MATLAB, but operating system commands, such as accessing an external port from an m-file, may also be run from within MATLAB. M-files can also become a function, where arguments are passed from MATLAB to the m-file and processed. Also m-files maybe written to execute interactively, prompting the user for information or data. MATLAB was written in C and uses many of the I/O functions found in C, increasing its flexibility even more.

To access an m-file simply execute the following:
```
>>!editor script_name.m
```
! causes MATLAB to move the command to the operating system.

editor is the name of your editor.

script_name.m is the filename; make sure that .m is the extension.

Normally, when writing an m-file, several lines of comments are placed at the top of the file indicating the purpose and use of the file. The help function can access these lines and display them as it would any of the other functions in the system. So, later on, long after you have forgotten how to use the file, simply type help filename (no extension is required) and those comments at the top of the file are displayed.

Example: The created m-file finds the poles, zeros, and evaluates $H(s)$ at $s = -10$, where $H(s)$ is given by

$$H(s) = \frac{s^3 + 6.4s^2 + 11.29s + 6.76}{s^4 + 14s^3 + 46s^2 + 64s + 40}$$

The filename is sample1.m

```
num=[1 6.4 11.29 6.76];
den=[1 14 46 64 40];
pole=roots(den);
zero=roots(num);
value=(polyval(num,-10))/(polyval(den,-10));
```

To execute this m-file type the filename without the extension

```
>>sample1
```

and the file will run by itself.

Some MATLAB Functions for Advanced Courses in Controls

Many problems in control theory require solutions of the algebraic Lyapunov and Riccati equations. These equations can be solved by using MATLAB functions are, lyap, dlyap, lqr, dlqr.

The function are solves the *algebraic Riccati equation* defined in (10.32) with S=B*inv(R2)*B'. Executing

```
>>P=are(A,S,R1)
```

produces the positive semidefinite solution of the algebraic Riccati equation.

The algebraic Riccati equation can be also solved by using the function lqr (linear-quadratic regulator defined in (10.27)–(10.34)). In addition to solving the algebraic Riccati equation (10.32), this function also calculates the optimal feedback gain **F**. The required solution is obtained by

```
>>[F,P]=lqr(A,B,R1,R2)
```

One can also use the function lqr2, which is identical to lqr, but produces better accuracy. Note that the discrete-time linear-quadratic regulator problem can be solved similarly by using operator dlqr.

The *algebraic Lyapunov equation* $\mathbf{AX} + \mathbf{XA}^T = -\mathbf{Q}$ can be solved by using the MATLAB function lyap, for example

```
>>X=lyap(A,Q)
```

Note that the matrix **A** ought to be stable in order to get the unique solution. Also note that the above equation is the variance (or filter) type Lyapunov equation, on the contrary to the regulator type Lyapunov equation given by $\mathbf{A}^T\mathbf{X} + \mathbf{XA} = -\mathbf{Q}$. Solution of the last equation is obtained by

```
>>X=lyap(A',Q)
```

that is, one must transpose the matrix **A** in this case.

The function `lyap` can be used to solve a more general "Lyapunov type" equation frequently known as Sylvester's equation $\mathbf{AX} + \mathbf{XB} = -\mathbf{C}$. This equation can be solved as

```
>>X=lyap(A,B,C)
```

Discrete-time Lyapunov algebraic equation can be solved by using the MATLAB function `dlyap`.

BUSINESS REPLY MAIL

FIRST CLASS MAIL PERMIT NO. 82 NATICK, MA

POSTAGE WILL BE PAID BY ADDRESSEE

The MathWorks, Inc.
24 Prime Park Way
Natick, MA USA 01760-9889

Index